NATO ASI Series

Advanced Science Institutes Series

A series presenting the results of activities sponsored by the NATO Science Committee, which aims at the dissemination of advanced scientific and technological knowledge, with a view to strengthening links between scientific communities.

The Series is published by an international board of publishers in conjunction with the NATO Scientific Affairs Division

A Life Sciences B Physics	Plenum Publishing Corporation London and New York
C Mathematical and Physical Sciences D Behavioural and Social Sciences E Applied Sciences	Kluwer Academic Publishers Dordrecht, Boston and London
F Computer and Systems Sciences G Ecological Sciences H Cell Biology I Global Environmental Change	Springer-Verlag Berlin Heidelberg New York London Paris Tokyo Hong Kong Barcelona

NATO-PCO DATABASE

The electronic index to the NATO ASI Series provides full bibliographical references (with keywords and/or abstracts) to more than 30000 contributions from international scientists published in all sections of the NATO ASI Series. Access to the NATO-PCO DATABASE is possible in two ways:

– via online FILE 128 (NATO-PCO DATABASE) hosted by ESRIN, Via Galileo Galilei, I-00044 Frascati, Italy.

– via CD-ROM "NATO-PCO DATABASE" with user-friendly retrieval software in English, French and German (© WTV GmbH and DATAWARE Technologies Inc. 1989).

The CD-ROM can be ordered through any member of the Board of Publishers or through NATO-PCO, Overijse, Belgium.

The ASI Series Books Published as a Result of Activities of the Special Programme on SENSORY SYSTEMS FOR ROBOTIC CONTROL

This book contains the proceedings of a NATO Advanced Research Workshop held within the activities of the NATO Special Programme on Sensory Systems for Robotic Control, running from 1983 to 1988 under the auspices of the NATO Science Committee.

The books published so far as a result of the activities of the Special Programme are:

Vol. F25: Pyramidal Systems for Computer Vision. Edited by V. Cantoni and S. Levialdi. 1986.

Vol. F29: Languages for Sensor-Based Control in Robotics. Edited by U. Rembold and K. Hörmann. 1987.

Vol. F33: Machine Intelligence and Knowledge Engineering for Robotic Applications. Edited by A. K. C. Wong and A. Pugh. 1987.

Vol. F42: Real-Time Object Measurement and Classification. Edited by A. K. Jain. 1988.

Vol. F43: Sensors and Sensory Systems for Advanced Robots. Edited by P. Dario. 1988.

Vol. F44: Signal Processing and Pattern Recognition in Nondestructive Evaluation of Materials. Edited by C. H. Chen. 1988.

Vol. F45: Syntactic and Structural Pattern Recognition. Edited by G. Ferraté, T. Pavlidis, A. Sanfeliu and H. Bunke. 1988.

Vol. F50: CAD Based Programming for Sensory Robots. Edited by B. Ravani. 1988.

Vol. F52: Sensor Devices and Systems for Robotics. Edited by A. Casals. 1989.

Vol. F57: Kinematic and Dynamic Issues in Sensor Based Control. Edited by G. E. Taylor. 1990.

Vol. F58: Highly Redundant Sensing in Robotic Systems. Edited by J. T. Tou and J. G. Balchen. 1990.

Vol. F63: Traditional and Non-Traditional Robotic Sensors. Edited by T. C. Henderson. 1990.

Vol. F64: Sensory Robotics for the Handling of Limp Materials. Edited by P. M. Taylor. 1990.

Vol. F65: Mapping and Spatial Modelling for Navigation. Edited by L. F. Pau. 1990.

Vol. F66: Sensor-Based Robots: Algorithms and Architectures. Edited by C. S. G. Lee. 1991.

Series F: Computer and Systems Sciences, Vol. 66

Sensor-Based Robots: Algorithms and Architectures

Edited by

C. S. George Lee

School of Electrical Engineering
Purdue University
West Lafayette, Indiana 47907-0501, USA

Springer-Verlag Berlin Heidelberg New York
London Paris Tokyo Hong Kong Barcelona
Published in cooperation with NATO Scientific Affairs Division

Proceedings of the NATO Advanced Research Workshop on Sensor-Based Robots: Algorithms and Architectures, held in Château de Bonas, France, October 12–14, 1988.

ISBN-13: 978-3-642-75532-3 e-ISBN-13: 978-3-642-75530-9
DOI: 10.1007/978-3-642-75530-9

Softcover reprint of the hardcover 1st edition 1991

45/3140-543210 – Printed on acid-free-paper

Preface

Most industrial robots today have little or no sensory capability. Feedback is limited to information about joint positions, combined with a few interlock and timing signals. These robots can function only in an environment where the objects to be manipulated are precisely located in the proper position for the robot to grasp (i.e., in a structured environment). For many present industrial applications, this level of performance has been adequate. With the increasing demand for high performance sensor-based robot manipulators in assembly tasks, meeting this demand and challenge can only be achieved through the consideration of: 1) efficient acquisition and processing of internal/external sensory information, 2) utilization and integration of sensory information from various sensors (tactile, force, and vision) to acquire knowledge in a changing environment, 3) exploitation of inherent robotic parallel algorithms and efficient VLSI architectures for robotic computations, and finally 4) system integration into a working and functioning robotic system. This is the intent of the Workshop on *Sensor-Based Robots: Algorithms and Architectures* — to study the fundamental research issues and problems associated with sensor-based robot manipulators and to propose approaches and solutions from various viewpoints in improving present day robot manipulators in the areas of sensor fusion and integration, sensory information processing, and parallel algorithms and architectures for robotic computations.

This Workshop was held on October 12-14, 1988, at Chateau de Bonas, Bonas, France, and was held in conjunction with another NATO Advanced Research Workshop on *Knowledge-Based Robot Control*, organized by Professor George N. Saridis of Rensselaer Polytechnic Institute and Professor Harry E. Stephanou of George Mason University, which was held at the same location on October 10-12, 1988. Both Workshops addressed a common theme on October 12 — Sensor Fusion. The purpose of holding these two Workshops back to back was to reinforce each Workshop's findings and to integrate the results since they are closely interrelated.

A total of 30 participants attended the Workshop with 14 speakers, 12 participants, and 4 committee members. Each day of the Workshop was devoted primarily to a brief presentation of research results followed by a discussion in each of the three major areas in sensor-based robots: *sensor fusion and integration, vision algorithms and architectures,* and *neural networks, parallel algorithms and control architectures*. This book includes all the twelve papers that were presented at the Workshop.

A total of five papers were presented at the Workshop addressing problems in sensor fusion and integration, such as sensing with uncertainty, sensor modeling, description, representation, and integration of sensory information in multisensor environment. Only three papers were included in this book and the other two papers were included in the NATO ARW book, *Knowledge-Based Robot Control*, edited by Professors G. N. Saridis and H. E. Stephanou. The first paper, "An Integrated Sensor System for Robots," by Rembold and Levi, describes an experimental autonomous mobile system with sensors, called KAMRO (KArlsruhe Mobile RObot), for manufacturing applications. The paper details the architecture and functions of the sensor system of KAMRO. The second paper, "Robot Tactile Perception," by Buttazzo, Bicchi, and Dario, describes an active or exploratory sensing strategy for a tactile sensor in a 4 DOF "hand." The paper describes an approach for decomposing complex tactile operations into elementary sensory-motor actions, each of which extracts a specific feature from the explored object. The third paper, "Uncertainty in Robot Sensing," by Grant, describes approaches and possible solutions for dealing with the inherent uncertainty that is associated with the modeling, planning and motion of manipulators and workpieces.

For the vision algorithms and architectures session, algorithms and architectures of model-based and/or knowledge-based vision systems were addressed to add intelligence to robotic systems. A total of four papers were presented in this area. The paper, "Robotic Vision Knowledge System," by Wong, describes the use of local features and geometric constraints for constructing knowledge-based vision system for object recognition. The paper, "Algorithm for Visible Surface Pattern Generation — a Tool for 3D Object Recognition," by Majumdar, Rembold, and Levi, describes the use of a CAD model for modeling and the manipulation of 3D objects which can be transformed and used for vision recognition. The paper, "Knowledge-Based Robot Workstation: Supervisor Design," by Kelley, describes a knowledge-based system for planning and scheduling tasks to be executed on various robotic workstations. The paper, "Robot/Vision System Calibrations in Automated Assembly," by King, Puskorius, Yuan, Meier, Jeyabalan, and Feldkamp, describes a fully-implemented vision-guided robotic system. The robot (Merlin robot) is equipped with a pair of CCD cameras for automated assembly tasks.

For the neural networks, parallel algorithms and control architectures session, a total of five papers were presented. The paper, "A Unified Modeling of Neural Networks Architectures," by Kung and Hwang, proposes a unified modeling formulation for a variety of artificial neural networks (ANNs), which leads to a basic structure for a universal simulation tool and neurocomputer architecture. The paper, "Practical Neural Computing for Robots: Prospects for Real-Time Operation," by Aleksander, describes the use of a neural machine called WISARD for pattern classification and its extension to experiential knowledge-based tasks. The paper, "Self-Organizing Neuromorphic Architecture for Manipulator Inverse Kinematics," by Barhen and Gulati, proposes a novel neural learning formalism, based on "terminal attractors" for solving a large class of inverse problems, including the inverse kinematics of redundant robots. The paper, "Robotics Vector Processor Architecture for Real-Time Control," by Orin, Sadayappan, Ling, and Olson, describes a restructurable VLSI robotic vector processor (RVP) architecture, which exploits the parallelism in the low-level matrix-vector operations in robot arm kinematics and dynamics computation. Interconnection of multiple RVPs can be used to match the computational requirements of specific robot control strategies. The paper, "On the Parallel Algorithms for Robotic Computations," by Lee, describes the inherent parallelism in robotic computation which was exploited to develop efficient parallel algorithms to be computed on SIMD machines for controlling robots. Finally, a report on the group discussion entitled "Neural Networks in Robotics" was written by Torras.

The presentations and discussions at this Workshop only present a small sample of solutions for an important research area of algorithms and architectures for sensor-based robots. I expect the research in this area to continue to grow, and more NATO Advanced Research Workshops about this area may be appropriately scheduled in the near future.

Finally, I would like to take this opportunity to thank Dr. Norm Caplan of the National Science Foundation (USA) for his continued encouragement throughout the process of organizing and realizing this Workshop. I also would like to thank the Organizing Committee, Professor R. L. Kashyap of Purdue University, USA, Professor F. Nicolo of University of Rome, Italy, Professor U. Rembold of Universität Karlsruhe, FRG, and Professor H. E. Stephanou of George Mason University, USA, for their hard work for putting the program together. Special thanks are also due to Ms. Dee Dee Dexter for her clerical work associated with the Workshop and for putting all the manuscripts together. Last but not the least is Professor G. N. Saridis who deserves special thanks for his advice on organizing the Workshop, without whose continued push for perfection, the Workshop would not have been a success.

C. S. George Lee

CONTENTS

Part I

Sensor Fusion and Integration

indicate that many basic problems still have to be solved until a real autonomous mobile vehicle can be created; e.g. the development of an integrated sensor system for the robot is a very complex effort. To recognize stationary and moving objects from a driving vehicle is several orders of magnitude more complex than the identification of workpieces by a stationary camera system. In most cases the autonomous system needs various sensors. For processing of multi-sensor signals, science has not found no good solution to date. An additional problem imposes the presentation and processing of the knowledge needed for operating the sensor system. Unexpected obstacles have to be recognized by the sensor and interpreted. If necessary, an alternate coarse of action has to be planned.

Seldom, an autonomous system is used for driving missions only. In general, it has retrieve parts from a storage, to bring them to a work table and to assemble them to a product, Fig. 2. All work has to be done autonomously, according to a defined manufacturing plan which is given to the system. In this article, the sensor module for an autonomous mobile system is being described, whereby the functions are explained with the help of the Karlsruhe Autonomous Mobile Assembly Robot (KAMRO).

2 *Autonomous Mobile Systems for Manufacturing*

There are various applications for autonomous systems in manufacturing. Most of the early projects concerned with this subject, involved the conception and implementation of vehicles for the movement of materials and workpieces. Hitherto, the efforts only succeeded in developing semi-automatic vehicles which can follow a path laid out by a guide system, such as an induction loop or a painted stripe on the floor. This type of guidance needs a simple sensoric and control strategy to steer the vehicle. The developments allowed to significantly increase the flexibility of manufacturing systems, whereby various manufacturing orders may be processed by a different combination of machine tools. Thus, it is possible to conceive simple programmable manufacturing facilities. However, the motion of the vehicle is confined by the guide system.

With autonomous mobile robots it is possible to develop manufacturing plants of great flexibility. Any combination of machine tools may be selected according to a virtual manufacturing concept. E.g. an autonomous assembly system equipped with robot arms is capable of working at various assembly stations. For welding or riveting tasks, the robot can move along a large object, such as the hull of a ship and perform the desired operations. An increase in flexibility can only be obtained by the use of knowledge based planning, execution and supervision modules which are sensor supported. In addition, omnidirectional drive systems have to be conceived, capable of giving the vehicle a three-dimensional flexibility, including turning on a spot.

3 *Components of an Autonomous Mobile System*

An autonomous system must be capable of planning and executing a task according to a given assignment. When a plan is available, its execution can start. A complex sensor system must be activated which leads and supervises the travel of the vehicle. Furthermore, it is necessary to recognize and solve conflicts with the help of a knowledge processing module. The basic components of an autonomous intelligent robot are shown in Fig. 3. To conceive and build these components, expertise of many disciplines such as physics, electronics, computer science, mechanical engineering, etc. is required. It is very important to design good interfaces between the functional components of the system.

The most difficult task is building the software. This is a universal problem with automation efforts involving computers. Designing software for autonomous vehicles is, however, complicated by the fact that very little is known about their basic concepts. An autonomous vehicle must have the following capabilities:

- autonomous planning and preparation of actions according to a given task
- independent execution and supervision of the actions
- understanding of the environment and interpretation of the results from sensor information
- independent reaction to unforeseen events
- passive and active learning capabilities

Figure 4 shows the planning and control system of the autonomous vehicle. It consists of several hardware and software modules which are interconnected to a functional unit.

The planner obtains information to assemble a product. In order to execute the assignment knowledge about the product is obtained from a CAD database. Furthermore, the robot has to know its environment, operating parameters and sensor hypotheses. This knowledge is obtained from a world model. The information about its work scenario must be current and dynamically updated by the sensor system. The planning is a very difficult and time consuming process and is done off-line with a powerful scientific computer. Since the planner needs live sensor data a link to the computers executing the plan must be provided.

The execution of the plan is done by a distributed vehicle computer. There are several CPUs operating in parallel to expedite the processing of the work assignment. The vehicle computers interprete stepwise the instructions and execute them. In addition, expert knowledge is given to the vehicle computer to process sensor information and to solve conflicts which may arise during the navigation, docking or assembly. Since the size of the vehicle computer is restricted, it only can solve simple problems. In serious situations the main computer will be notified and it in turn tries to find a solution. It will also prepare and issue a situation report for the operator.

The supervisor observes the operation of the vehicle and reports any problems. There are two types of disturbance which may occur, they are of parametric and structured nature. Parametric problems stem from wrong sensor parameters. If properly recognized they can be corrected locally. Structured problems stem from unforeseen changes in the robot world where the location of parts may have changed. In this case the operation has to be replanned off-line by the planner. To perform its task, the supervisor constantly reads and evaluates sensor data. Since conclusions may have to be drawn from measurements of various sensors the evaluation of the sensor data may be very involved.

The control module operates the feedback loops of the robot system, it compares the set points with the controlled variables and tries to correct deviation. Any problems are reported back to the executive and planner and are used for corrective actions if necessary.

In the further discussion of this paper only the sensor system will considered.

4 *The Sensor System*

A sensor system of the Karlsruhe autonomous mobile robot consists of various sensors which are interconnected by a hierarchical control concept. The sensors furnish the planning and supervision modules with information about the status of the robot world. For each of the three major tasks of the vehicle, the navigation, docking and assembly an own sensor system is provided.

The sensoric has the following assignments:

- locating workpieces in storage
- supervising the vehicle navigation
- controlling the docking maneuver
- identifying the workpieces and their location and orientation on the assembly table
- supervising the assembly

- inspecting the completed assembly

For the navigation of the autonomous vehicle, a multisensor system is necessary. A distinction is made between vehicle based internal and external sensors and world based sensors. Internal sensors are incremental decoders in the drive wheels, a compass, inclinometer, etc. External sensors are TV cameras, range finders, approach and contact sensors, etc. World based sensor systems use sonic, infrared, laser or radiotelemetris principles. For the navigation various approches may be used:

- deadreckoning
- navigation under the direction of a compass
- the use of world based sensor systems
- driving under the guidance of floor markers and vehicle based external sensors
- navigation by vehicle based external sensors, such as a camera or a laser range finder
- the use of a combination of navigation principles

A vehicle driving in an obstacle free environment may use any of the first four principles. In case obstacles are entering or leaving the vehicle´s path or when it is possible that the robot may veer off the course, vehicle based external sensors must be used. For example, the vehicle must constantly monitor the path with a camera system. Most advanced autonomous vehicles use a combination of several approaches systems to react to unforeseeable events. Recognition is done by extracting specific features from the picture of the scenario and comparing these with a sensor hypotheses obtained from a world model. For scenes with many and complex objects the support of an expert system is needed for the sensor evaluation.

The docking maneuver will be supported by optical, magnetical or mechanical proximity sensors. Thereby, for coarse positioning a vision system may be used and for fine positioning mechanical feelers.

Recognition of parts for assembly will be done with a 2D vision system which also determines the position and orientation of the object. For the supervision of the assembly process a 3D vision system is required. Operations such as parts mating, fastening and aligning require force/torque, approach and touch sensors. The most important sensor is the vision system; it is connected with the other sensors to a multisensor module to supervise a complex operation such as an assembly.

A multisensor system may be designed according to the following concepts:

- a combination of various types of sensors
- the use of the same type of sensors at various locations
- the use of one sensor for the acquisition of various parameters
- the use of one sensor for interpreting moving scenes

The first two sensor principles are task dependent and must be carefully designed for the specific application. The last two sensor principles are difficult to implement. In all cases the capacity of the sensor channel and the picture evaluation algorithms must be carefully designed. For example, if a sonic sensor is used in conjunction with a laser scanner, the signals of both sensors have to be combined to one channel parameters.

5 *Sensor Data Processing*

In a complex work environment it usually is necessary to employ several sensors to understand an event. The KAMRO system will employ the hierarchical sensor system architecture shown in Fig. 5 (2). There are 3 data processing levels in this schema. They are:

Basic sensor data processing level
Sensor data evaluation level
Strategic sensor planning level

On the sensor data processing level the raw sensor data is evaluated. The algorithms are basically of procedural nature and the knowledge is contained in a well structure form. As an example a vision system will have algorithms to determine the identity, location and orientation of a workpiece. The interaction with the robot on this level is in realtime .
On the sensor data evaluation level for each sensor a matching of the preprocessed data with a sensor hypothesis is done. The hypotheses for the individual sensors are obtained from the world model where they are stored as logical sensor information. It may only be necessary to supply to the matching unit a subset of the available sensor information. On the basis of the hypothesis and the measured data the matching is done. For this operation tolerances be provided. If information is missing or the tolerances are too great the system will reject the results and it may try a new matching operation. In case this attempt still does not rendor any result, the problem will be sent to the next higher level to reach at a desicion.

On the strategic sensor planning level the fusion of sensor data from the lower levels takes place. A comprehensive evaluation of all sensor data is done with the aid of knowledge stored about each sensor. Thereafter, action instructions are given to the robot or a new sensor processing strategy is planned. A blackboard is employed as an information exchange mechanism between the knowledge bases of the individual sensors. It acts as a short term memory and uses the preprocessed sensor data of the matching units as input. This data is evaluated by independent knowledge processing units. The sequence of the knowledge processing is determined by a problem specific control strategy.

The structure of the hierarchical sensor evaluation system is so complex that it must be supervised by an independent control and communication module. It is based on an object oriented control schema. The data exchange between the objects is done via messages. The various modules of the sensor evaluation system can be tied together for solving a specific problem. The data manipulation within the modules is done locally. Thus, changes in one component do not effect other modules.

6 *Description of the Sensor Systems*

KAMRO has three independent sensor systems, one for the navigation, one for docking and one for the assembly. The control of the sensor operation is done via the blackboard architecture described in the previous section. In general, the sensor systems operate independently. However, in some cases communication between the individual sensor systems is necessary. In the following sections the sensor systems are described.

6.1 The Sensors for Navigation

The robot obtains from the planner a route map in which the path to be travelled is defined. In principle the navigation can be done with the help of resolvers located in the omnidirectional wheels. However, with this method travel can only be done in a deadreckoning mode and the vehicle would have to reorient itself constantly with the help of external markers to correct deviation from the defined path.

KAMRO will be equipped with various sensors to help guiding the vehicle along its path. There will be an interface to the world model to constantly update the position of the vehicle. The camera system used tries to understand the environment from sequences of images. In addition a laser triangulation sensor is employed to measure the exact distance of an object. For observing the close proximity of walls and other obstacles an array of sonic sensors is located in a belt fashion about the vehicle. Every wheel of the drive system is equipped with a resolver to control travel strategies of the vehicle. In the following the sensors are explained in more detail.

1 The camera system

The camera system is being designed to interprete the path of the robot from sequences of images taken during the travel. There will be a total of 4 cameras installed, they will be working together as two sets. Each set is capable of processing stereoscopic images of the robot world. An attempt will be made to determine the position of the robot in reference to known objects. For this reason, it is necessary to communicate with the world model.

For the teach-in phase the robot will be sent through the various routes it can take, thereby, features of objetcs marking the path will be extracted. The features are entered into the world map. During navigation the system is determining its location in reference to the tought features of the objects. It will be possible to cross-check the travelled path with information from the resolvers in the wheels.

2 The triangulation laser sensor

The laser sensor aids the camera system in case the exact distance of an object has to be known. The deflection of the laser beam is programmable. Thus, it is possible to direct the beam on a specific object and to scan its surface to obtain the topology. The laser scanner will also be used to supervise the assembly. Its principle is described in section 6.3.

3 Sonic distance sensors

To avoid collision at close distances a belt of commercially available sonic sensors is installed along the outside walls of the vehicle. It is the task of these sensors to detect the proximity of walls, doorways and obstacles. The sensor information may have to be processed in conjunction with the pictures obtained from the cameras and laser system, entailing a sensor fusion. In case obstacles are encountered it is necessary to identify them and to plan a collision avoidance maneuver. For this reason there will be a communication link to a collision avoidance and error recovery module.

4 The resolver for the drive wheels

Since the vehicle has an omnidirectional drive system it must be possible to monitor the rotation of every wheel and to control their rotational speed and direction. The wheels are driven by brushless servomotors having three phase stator windings and an armature equipped with permanent magnets. A resolver is fastened to the motor to decode the rotation. The construction of the resolver is similar to that of a motor, Fig. 6. It has a rotor winding and two stator windings. The latter are located at an angular displacement of

90^0 to each other and pick up the rotating magnetic field of the armature. The sine and cosine functions generated by the two armature windings are used for determining the rotational angle. With the help of a special electronic module foreward and reverse signals are produced and sent with a frequency of 512 pulses per revolution to the drive motors. The control strategy for the 4 drive motors is determined by a special software module which communicates with the world model.

6.2 The Sensors for Docking

Various types of sensor principles may be employed to calibrate the docking position. E.g. a tactile sensor in the gripper of both robot arms can be used to detect with the help of a pin the known position of two holes in the worktable which are adequately (located by a camera) spaced apart. The angular position of the joint angles of the arms, together with the arm dimensions, are used to calculate the location and orientation of the mobile robot in reference to the work table. Beside this contact measuring principle there exist three non-contact techniques. Inductive sensors may be used to search and to register the edge of a thin metal sheet, or a capacitive sensor measures capacity changes if there is a transition between materials with different dielectric parameters. Finally, optical sensors are capable of measuring the exact position and orientation of the mobile robot relative to a work area (e.g. assembly desk). Both, active and passive techniques are used. The last kind of docking sensor is installed in the KAMRO.

The active approach integrates several laser diodes (LEDs) which are arranged in an orthogonal fashion in reference to the assembly table. These diodes form an orthogonal coordinate system. A PSD - sensor is integrated into the gripper. In a first step this sensor is brought into an parallel position relative to the x, y - plane indicated by the three diodes. This can be performed by vertical gripper movements. In case the z-coordinate of the working surface is not known a fourth diod at a kown relative distance from the x y -plan can be employed to obtain this dimension. The z position of the tables can also be obtained with the wrist sensor of the effector. By this method the effector is moved slowly towards the table and upon though the z coordinate is calculated from the joint angles and dimensiones of the arm.

The passive approach uses the triangulation principle. A pulsed laser diode is integrated into the gripper and is used to scan several discrete points of the working area. An CCD-camera which is installed in the wrist (see next section) receives the position of the reflected laser beam. Thereafter, the distance of these scanned points is calculated. By this calculation the position and the orientation of the working area can be determined. This measuring methods can be improved if the scanned region is marked by a dedicated feature (e.g. a cross). In this case, the scanning region can be found very quickly.

6.3 The Sensors for Assembly

Various sensors are used for supervising the assembly. They are a multifunction laser scanner, two wrist cameras, a force/torque sensor, a touch sensor, a sensor for the gripping forces and an approach sensor.

1 Multi-function laser scanner

The measurement of exact distances is done by triangulation. Figure 7 shows the block diagram of the system. This system is entirely computer-driven (two M68020 processors). The two galvanometer mirrors can be rotated to generate any desired (e.g. tracking) field of view and scanning frequency (tracking). This capability of random scanning is important, e.g. for the detection of operational features of a workpiece (e.g. curvature, hole position).

The position of the reflected beam is measured (difference of currents) by a two dimensional position sensitive diode (PSD). It is used to calculate the distance. A special feature of this sensor is that it also can measure the intensity of the reflected light of any point scanned (normalized addition of currents). This features will be used to merge information on distance and intensity without a correspondence problem.

An additional CCD-camera controls the global field of view of the laser scanner by locating regions of interest. When such a region has been pinpointed the laser scanner can be automatically turned into this direction by the robot to take measurements. The pictures taken with this camera are used to separate background objects and shadows.

This combined laser scanner needs about 10sec for processing 256^2 scanning points (distance and intensity). However, the raw data have to be corrected mainly for background lights, PSD non linearity pincushion and parallel distortion. The last two types of distortions are generated by the transformation of the equidistant spherical scanning lines into non-equidistant differences in cartesian coordinates.

The essential objective of this sensor system is the automatic fusion of distance and intensity data. The fusion is done for various operators obtained from the feature extraction. For example, edges are easily detected from intensity images; whereas from distance measurements they are hardly detectable because the reflected laser beam is split by geometric edges.

The final classification of an object is done from distance, PSD-intensity and CCD camera intensity data. Fig. 8 shows the steps to be taken to obtain corrected measurement data and to identify an image. In the following the steps are described in more details.

step 1: Simultaneous generation of the distance and laser intensity image by the PSD. Thereafter, extraction of the background shadows from the laser intensity image.

step 2: Extraction of the object contours from the laser intensity image. These contours stem from geometric object features (physical edges) or changes of the reflectivity of the object.

step 3: Extraction of the geometric contoures; resulting in a segmentation of the distance image.

step 4: Evaluation of surface features (e.g. curvatures) from the areas which were determined in step 3 (geometrical segmentation).

step 5: Separation of the volumetric and reflective contours in the laser intensity image.

step 6: Tracking of the shadow regions by the use of the laser beam and exposure of these marked regions to the CCD-camera. This camera has a field of view which is different from the scanning field of the laser scanner. Therefore, it is possible to separate objects which are not visible in the two laser images.

step 7: Recognition of objects in the background which are concealed by front objects. The latter are only visible to the laser scanner.

2 Two wrist cameras

The global two dimensional view of the assembly is obtained by the camera which is fixed above the workstation. A three dimensional view of the workpieces to be mated and the details of three dimensional operational features of the parts (before and during the assembly) cannot be generated by the "overhead" camera. These tasks are performed by two small CCD-cameras which are mounted to the wrists of the two KAMRO grippers. The operating range of these two cameras is about 10 cm to 50 cm without zooming. Within this range they can obtain accurate distance information and no laser scanner is needed.

The tasks of these cameras are as following

a) Recognition of the operational features of an object and determination of the objects position and orientation. The features extraction is performed with the help of 3-D CAD models. The geometric models are referenced automatically in several steps to evaluate the image obtained by the camera.

   The final evaluation of the object features is done by rules. These rules control the motion of one wrist camera (e.g. top view, side view or front view) in order to find the exact feature parameters. All rules are installed in a so called visibility tree [3].

b) The designation of the hand to be used for the manipulation (left or right gripper) and the control of the work is done by a rule based module.

For the control of the manipulation a measurement tree is used in analogy to the visibility tree mentioned above [4]. The tree defines the kind of measurement to be performed by a camera for every elementary assembly operation (e.g. insert, turn). If one hand is selected to perform a joining operation then the camera of the other hand is used for the manipulation control.

3 Force/torque sensor

This type of sensor is needed to control handling of the object, to supervise assembly, and to protect the robot from overload. It is installed in the wrist of the effector and should be able to render accurate and repeatable results throughout the entire load range of the robot. The sensor itself has to be protected against possible overload, shock, and vibration. The basic principle of our sensor is shown in Fig. 9. It is divided into two parts: a lower ring with four rigid supports and a upper ring with four fastenings points for the gripper. Careful attention must be paid to a good design so that the measurement signals are highly linear with the applied force. The analog strain gauge signals are converted to force and torque information via matrix multiplication.

4 Touch sensor

The dedicated type of a touch sensor to be used is shown in Fig. 10, [5] This sensor is integrated into the gripper finger. The sensor pad (30 mm in diameter) is formed from a silicone rubber sheet that is tensioned across a flat plexiglas surface by means of a retaining ring. Light that is directed onto the plaxiglas from small bulbs is totally reflected at the inside of the flat plexiglas surfaces. When the rubber is pressed into intimate contact with the plaxiglas, by the force exerted on the workpiece, total internal reflection is prevented and the light then leaving the plexiglas is projected by a lens onto a CCD photodetector array (256 x 256). A video image of the tactile image is then available for processing and analysis using procedures of image analysis.

If necessary the pad can be rotated via a drive mechanism to reposition the object held in the hand.

5 Sensor for gripping forces

This sensor monitors the gripping forces the fingers apply to the object to be handled. A conventional strain gauge measurening arrangement is used.

6 Ultrasonic sensor

This sensor is needed to monitor the approach of the gripper to the object. For economical reason it is necessary that the hand moves toward the object very quickly, and as soon as it is close to the object the motion has to be slowed down in order to avoid collision and to assure a save gripping position. Conventional ultrasonic sensors are used. To be able to measure close distances the control circuit provides a fast switching time between emission and reception of the signal.

Acknowledgement

This work was carried out at the Institute of Realtime Computer Systems and Robotics (Prof.-Dr.-Ing. U. Rembold and Prof. Dr.-Ing. R. Dillmann) of the University of Karlsruhe, and at the research group "Technical Expert Systems and Robotics" (Prof. Dr.-Ing. U. Rembold and Prof. Dr. P. Levi) of the Forschungszentrum Informatik, Karlsruhe. The work was supported by the Deutsche Forschungsgemeinschaft.

References

1. Rembold, U.: "The Karlsruhe Autonomous Assembly Robot", IEEE International Conference on Robotics and Automation, April 24-29, 1988, Philadelphia

2. Raczkowsky, J. and Rembold, U.: "The Multisensory System of the KAMRO Robot", Nato Workshop on Highly Redundant Sensors in Robotics, May 16-20, 1988, IL Chiocco, Italy

3. Majumdar, J.; Levi, P.; Rembold, U.: 3-D Model Based Robot Vision by Matching Scene Description with Object Model from a CAD-Modeller, Proc. of the 3rd ICAR, Versailles, October 1987

4. Majumdar, J.; Levi, P.; Rembold, U.: Use of Control Knowledge Sources for Elementary Operations of a Two Arm Assembly Robot, proc. of the IEEE International Workshop on Intelligent Robots and Systems, Tokyo, Oct. 31 - Nov. 02, 1988

5. Technical Report No. TR6, ESPRIT Project 278 "Tactile Sensing for Integrated Sensor-Based Robot Systems", December 1987

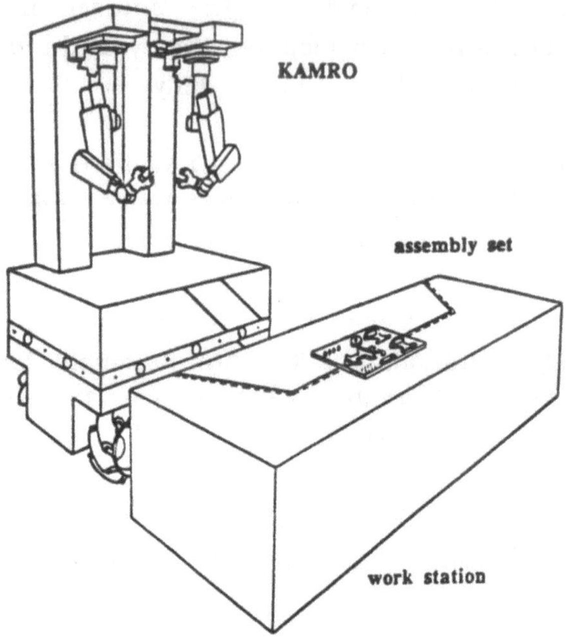

Fig. 1: The Karlsruhe autonomous mobile assembly robot KAMRO

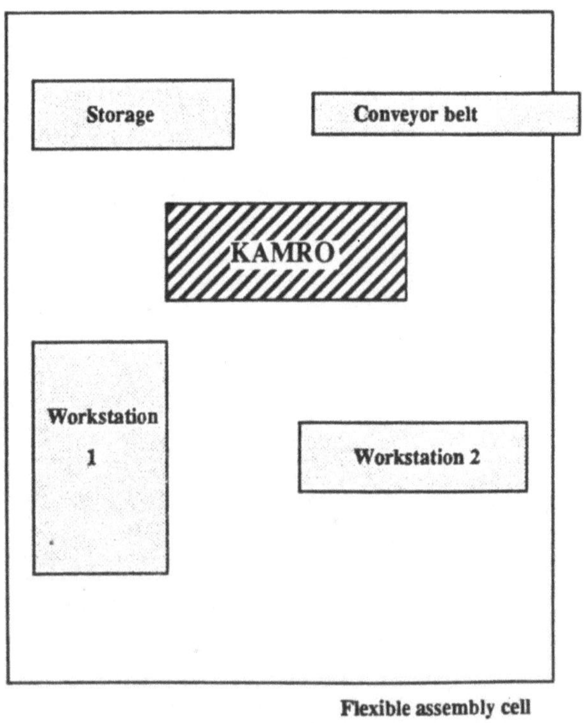

Fig. 2: Typical application of KAMRO

Components of an autonomous robot

1. **Mechanics and drive system**
2. **Sensor system**
   - internal sensors
   - external sensors
3. **Planner and navigator**
   - planner
   - navigator
   - expert system
   - knowledge base
   - meta knowledge
4. **World model**
   - static component
   - dynamic component
5. **Knowledge acquisition and world modeling**
6. **The computer system**

Fig. 3: Components of an autonomous robot

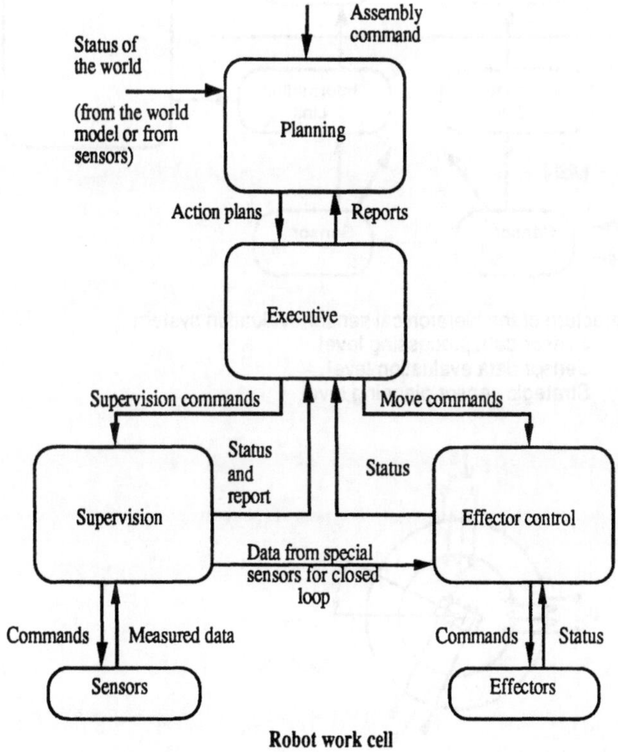

Fig. 4: The overview of the structure of the third generation robot system KAMRO

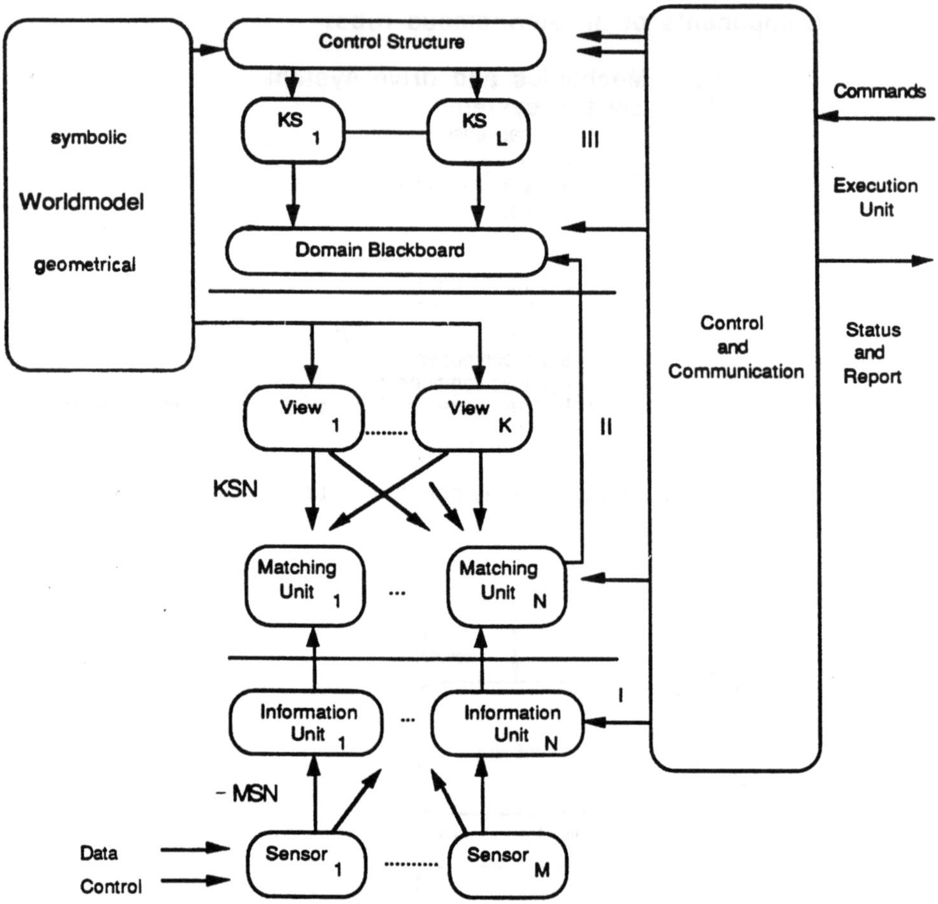

Fig. 5: The Basic structure of the hierarchical sensor evaluation system
I. Sensor data processing level
II. Sensor data evaluation level
III. Strategic sensor planning level

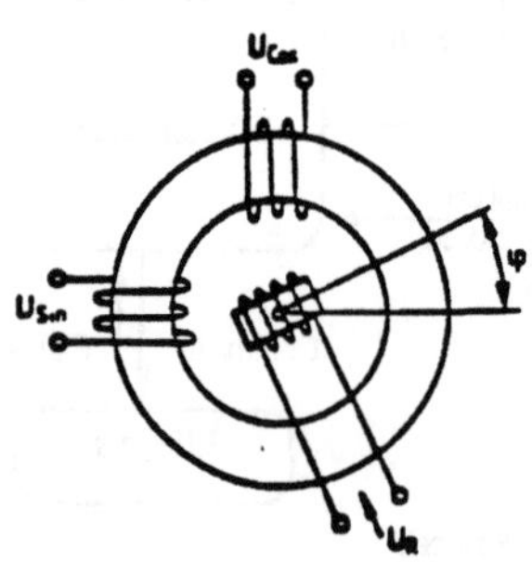

Fig. 6: Arrangement of the stator and rotor windings of the resolver

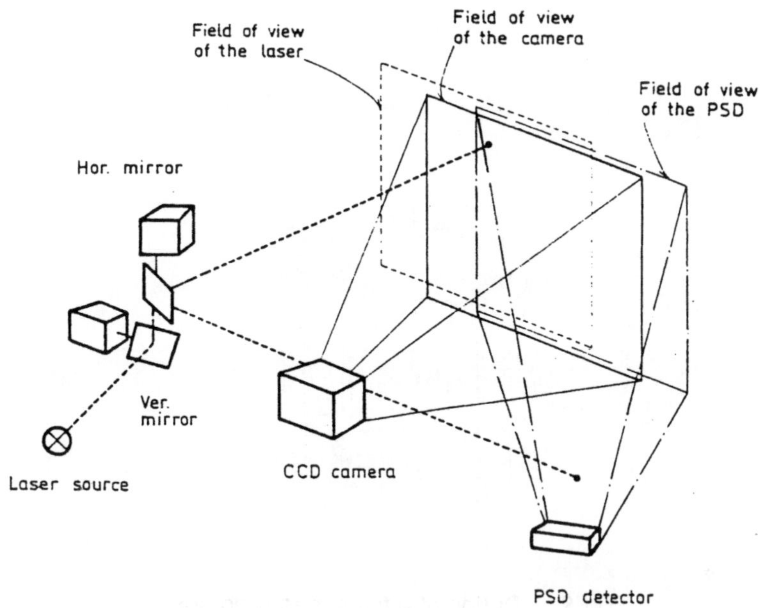

Fig. 7: Schematical structure of the triangulation scanner

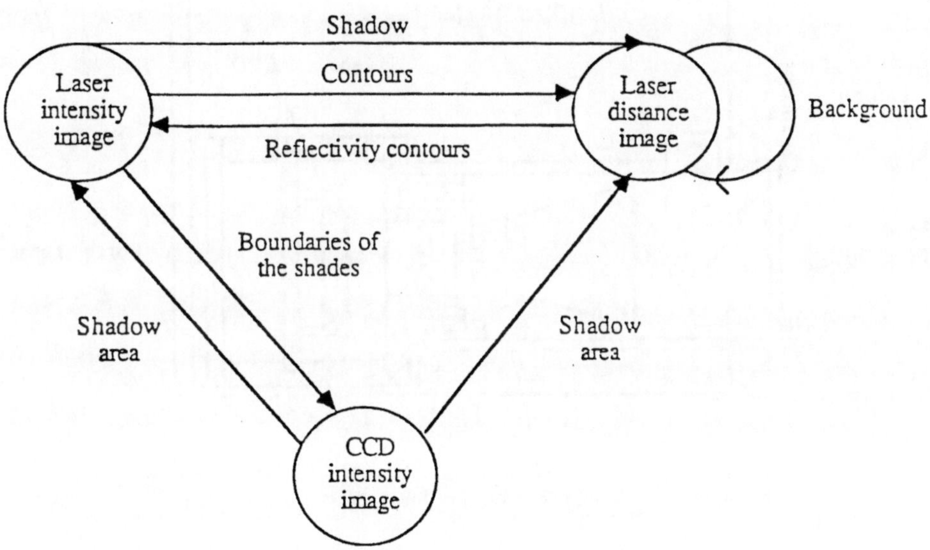

Fig. 8: Measuring principle of the multifunction laser scanner

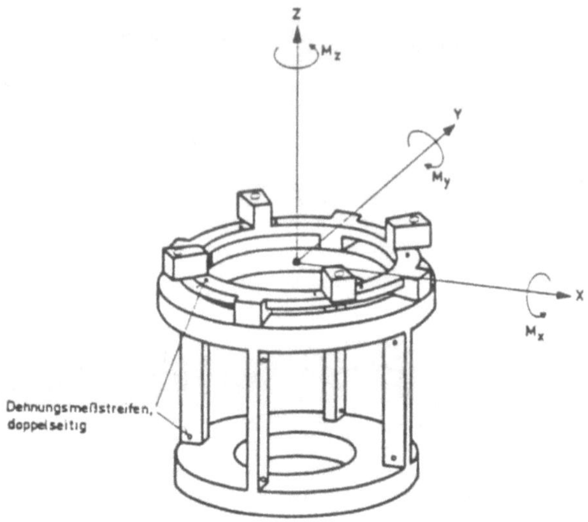

Fig. 9: Design of a force measuring box

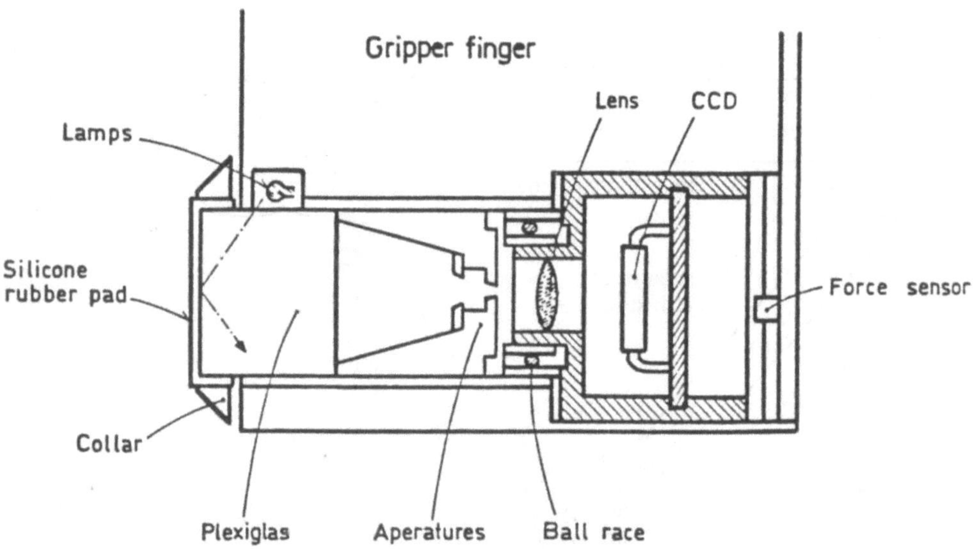

Fig. 10: Tactile sensor design

ROBOT TACTILE PERCEPTION

G. Buttazzo[†], A. Bicchi[‡], P. Dario[†]

Centro "E. Piaggio",
Faculty of Engineering, University of Pisa,
Via Diotisalvi, 2   56100 Pisa
Italy

Abstract

In this paper we discuss some fundamental issues related to the development of an artificial tactile sensing system intended for investigating robotic active touch. The analysis of some psychological and psychophysical aspects of human tactile perception, and a system design approach aimed at effectively integrating the motor and sensory functions of the robot system, suggested to conceptually organize tactile exploratory tasks into a hierarchical structure of sensory-motor acts. Our approach is to decompose complex tactile operations into elementary sensory-motor acts, that we call "TACTILE SUBROUTINES", each aimed at the extraction of a specific feature from the explored object. This approach simplifies robot control and allows a modular implementation of the system architecture: each function can be developed independently and new capabilities can easily be added to the system. All tactile exploratory procedures are selected and coordinated by a high-level controller, which also operates the integration of tactile data coming from sensors and from lower levels of the hierarchy.

Some experimental results will be presented demonstrating the feasibility and usefulness of tactile sensing in exploratory operations. A recently developed sensor will be briefly presented, which exploits force/torque information measured directly at the tip of the robot end-effector. This sensor is able to detect, besides the position of the contact point, the normal and tangential components of the contact force. Methods for characterizing the surface of manipulated objects, according to their hardness, texture and friction properties will also be discussed.

† *Scuola Superiore S. Anna, Pisa, Italy*
‡ *Dept. of Mech. Eng. - DIEM, University of Bologna, Italy*

NATO ASI Series, Vol. F 66
Sensor-Based Robots: Algorithms and Architectures
Edited by C. S. George Lee

1. Introduction

Tactile perception is a fundamental capability for a robot that has to execute manipulative and explorative tasks. The interactive behavior of touch allows humans to extract several features from the external world, that cannot be detected by vision or other senses. Examples of such features are: hardness, elasticity, roughness, texture, temperature, thermal conductivity and local geometrical characteristics, such as holes, edges, cavities, sharp regions, etc.

It is important to point out that extracting such features from an object is not a capability of a specific sensor, but it is rather a capability of the whole system. Performing explorative tasks involves the execution of sensory-motor procedures, in which tactile information is used to sense and drive the movements of the fingers. Touch is intrinsically active and involves dynamic sensing, where movements are utilized for augmenting and driving sensory information. The coordination of sensory activity and motor activity is not just a summation of capabilities, but it considerably improves robot performance, by increasing the perceptual skill of the system and extending the set of characteristics that can be extracted from the external world.

In passive perception (mostly followed in vision), sensors and actuators are physically separated: sensors are fixed devices which statically observe the world and send information to a central controller at a very low sampling rate; once sensory data are analyzed, a motor action for the manipulator is planned. In this approach, data processing and motion planning are two distinct processes, which do not overlap in time. Motion planning is based on sensory information, but once the trajectory of the arm has been calculated it cannot be changed.

In active perception, especially in tactile perception, sensing and control are tied together. Sensors are often mounted on actuators and are used by the system to probe the environment and precisely control the movements. Trajectories are computed in real time using sensor-based control techniques.

Assigning perceptual capabilities to exploratory acts rather than to static sensors is a novel concept in robotics, that has not received much attention among scientists so far, unless at the level of speculation. Recently, however, this issue has been considered more seriously, and some implementation has been attempted [1][2][3].

Our goal is to build an autonomous tactile robot system capable to perform active exploration and fine manipulation of real objects, for their recognition.

Based on the analysis of some aspects of human tactile perception, our approach is to decompose complex tactile exploratory procedures into a sequence of elementary sensory-motor acts, that we call "TACTILE SUBROUTINES", each aimed at the extraction of a specific object feature [4].

In humans, it is possible to identify a number of typical tactile procedures that are performed with the fingers every time we want to detect some particular feature from an object. For example, if we are interested to know the hardness of a material, we repeatedly press our fingertip against the object surface, paying attention to the force we exert and to the object deformation. If we are interested in object texture, we gently slide the fingertip along the object surface and we pay attention to the tactile sensation coming from our epidermal sensors. As another example, if we want to reconstruct the shape of an object, we follow the object contour, keeping in mind the trajectories of the contact points achieved by the fingertip.

In this context, we define a "TACTILE SUBROUTINE" as a motor action executed on a sensor, guided by the tactile information coming from the sensor itself, according to a control strategy which depends on the sensor and on the feature that has to be extracted.

This approach considerably simplifies robot control and allows a modular implementation of the system architecture: we can develop one tactile subroutine at a time and freeze it in the system as "innate behavior". To add capabilities to the robot we simply insert new subroutines in the system. All tactile subroutines are selected and coordinated by a high-level controller, which also operates data integration and directs the global exploratory strategy.

2. System description

The tactile system we developed for investigating tactile perception consists of the following components:

- a PUMA 560 robot arm, controlled by its dedicated microprocessor (UNIMATE) and programmed in VAL II;
- a miniaturized force/torque (F/T) fingertip sensor, working as a sensitive probe for tactile exploratory tasks;
- a piezoelectric polymer (PVF2) sensor, implementing a sort of artificial finger nail, intended to rub rough surfaces for texture detection.

Other components of the system are two PC's, utilized for sensor preprocessing, and a DEC micro VAX II, used as a system supervisor for tactile data integration and high level control.

The complete architecture of the system is depicted in Figure 1. The fingertip F/T sensor is mounted on the PUMA wrist and the nail-sensor is attached to the fingertip. Each sensor is connected to a PC. PC1 is intended to process the information coming from the F/T sensor and to control the execution of tactile exploratory procedures; PC2 is dedicated to the nail

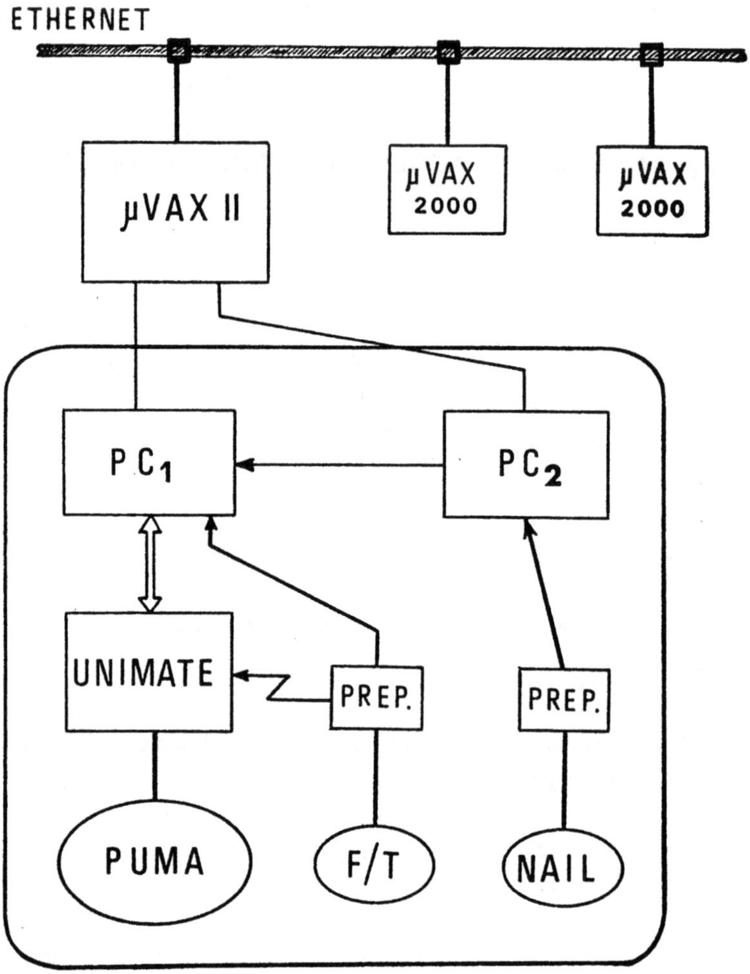

Fig 1. The system architecture.

sensor and works as a slave in the communication with PC1. It continuously read the signal produced by the nail during the sliding movements and computes a number of parameters useful to characterize roughness. The two PC's communicate via serial line. PC1 is also connected to the PUMA processor through a 16 bit input/output parallel port for managing sensor-based movements.

An additional parallel interconnection exists between fingertip sensor and PUMA processor, which implements a sort of reflex pulse for stopping the PUMA in case of dangerous situations (overloads on the sensors) that could damage the system.

2.1 The F/T sensor

This sensor has been designed to be easily incorporated as a sensitive fingertip in an articulated robot hand, but in the system presented in this paper it is used as a tactile probe for exploratory tasks and it is mounted on a single rigid "finger". This finger is connected to the PUMA wrist through a compliant adaptor: in fact, a certain amount of flexibility is mandatory for controlling interaction forces between robot and environment. A schematic description of the sensor is shown in Figure 2.

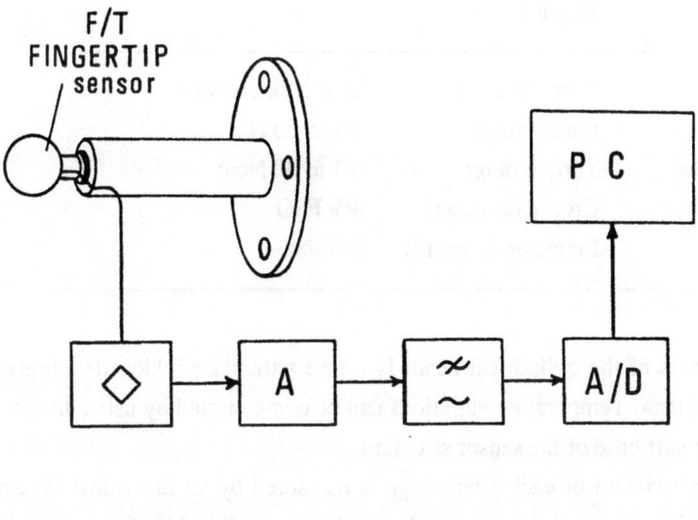

Fig. 2. The fingertip force/torque sensor with its conditioning units.

The device has the purpose of measuring the three orthogonal components of the resultant force and the three orthogonal components of the resultant torque applied to its mechanical structure. The measurement principle is the mechano-electric transduction of the elastic strain of a monolithic cylinder beam to which the load is applied. The transduction is carried out by 6 strain-gages only.

The top of the cylindrical structure is threaded so that different types of fingertips can easily be adapted to the sensor. When an external force is exerted on the fingertip, the mechanical structure of the sensor deflects, causing the strain-gage response.

The electric resistance variation of each strain gage, due to the strains imposed to the cylinder by the load, is separately measured. This information can be processed in the form of six orthogonal components of the applied force/moment by solving the set of linear equations which model the elastic compliance of the structure; the equations can be obtained by using beam theory or by calibrating the cell experimentally. Conventional algorithms for linear system solution, e.g. Gaussian elimination, are adequate for this purpose. However, the peculiar arrangement of the strain gages on the cylindrical surface of the sensor allows a more time-efficient algorithm, almost decoupling the cell readings [5].

The small size of the sensor, the low cost, along with its simple structure, make it attracting for being integrated in the mechanical structure of robot hands or robot end-effectors for fine manipulation.

Some performance figures experimentally obtained from a prototype sensor, using a non-engineered technology, are listed in table 1.

Table 1

Active cell size:	10 x 10 x 16 mm^3
Force range:	0.1 to 30 N
Torque range:	0.1 to 30 Ncm
Crosstalk (max):	4% FSO
Precision (repeat.):	2% FSO

The thickness of the cylindrical beam is a free parameter which determines the loading range of the sensor. Temperature variations can be compensated by using an extra strain-gage, bonded to the stiff base of the sensor structure.

Resistance variation of each strain-gage is measured by an individual Wheatstone bridge (module ◊ in Fig. 2); the 6 output signals are then amplified (A), filtered out by a low pass filter (~), multiplexed and finally converted into digital form (A/D). The F/T sensor is connected to a PC through a Data Acquisition Card, which performs multiplexer addressing and analog to digital conversion.

2.2 The PVF2 sensor

A piezoelectric sensor, made by PVF2 polymer, is utilized as a dynamic sensor for implementing a sort of artificial fingertip nail, intended to rub rough surfaces for texture detection [2]. The nail structure consists of a properly shaped plastic sheet, adapted to the upper surface of the fingertip, from which it protrudes for about 5 mm (Figure 3).

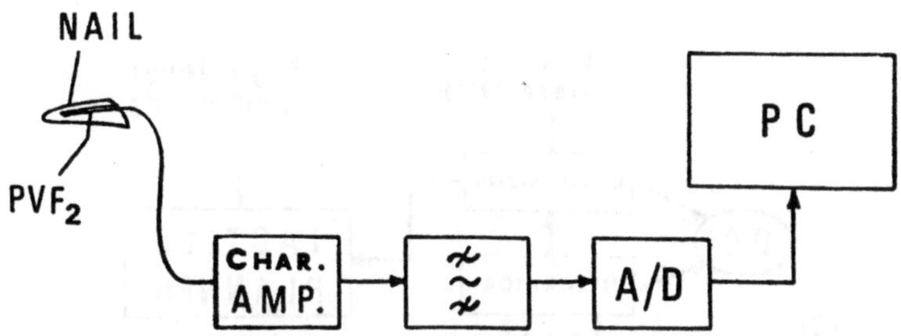

Fig. 3. The PVF2 nail sensor with its conditioning units.

This arrangement allows to add compliance to the sensor and to increase sensor sensitivity to mechanical vibrations. The PVF2 film (25 micron thick), used in a bilaminate configuration, is located between nail and fingertip, bonded to the inner side of the nail.

When the nail is slid along a rough surface, the nail structure vibrates, producing strain in the PVF2 sensor, which generates an amount of charge proportional to the strain. This charge is amplified by a charge amplifier and the output voltage signal is digitized by an A/D converter and processed by another PC.

The upper frequency limit of the digitized signal, established by the sampling rate of the system, is almost 5 KHz, and it proved to be sufficient for all practical surface explorations.

As for all piezoelectric sensors, the lower frequency limit of the nail signal is not zero, but a few hundreds mHz. This is due to the finite time constant of the piezoelectric sensor, that derives from its finite internal resistance. In this particular case, such intrinsic limitation turns out to be a positive feature of the sensor: in fact, as a consequence of a non-zero lower frequency limit, the nail cannot respond to very slow mechanical deflections. Therefore the high frequency components of the signal due to the roughness of the explored surface are detected, while the low frequency "noise" caused by the variation of the contact force during the sliding movement is filtered out.

An approach involving dynamic tactile sensing for texture detection using a PVF2 sensor has also been reported by Cutkosky [6].

3. Functional architecture

Based on the functions that the robot system is intended to implement, the software architecture has been organized in three control levels, as illustrated in Figure 4.

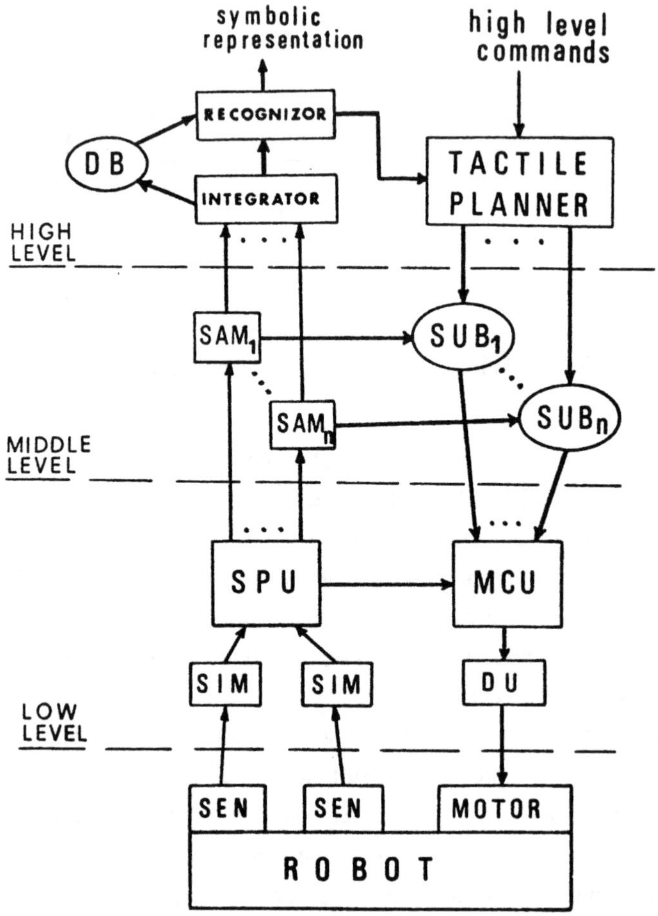

Fig. 4. Hierarchical functional architecture of the system

Level 1

The lowest level of this hierarchy includes all VAL II programming and all assembler routines for sensor acquisition, processor communication and actuator driving. This level is designed to execute simple commands sent by the middle-level controller. Such commands may include position commands in joint space or in cartesian space, or force/torque commands.

The Sensor Interface Module (SIM) realizes the interface between sensor and computer, providing analog to digital conversion and data acquisition. The Sensor Processing Unit (SPU) performs a first stage of processing and provides the Motor Control Unit (MCU) with feedback signals for motor control. According to the middle-level commands and to the feedback signals, the MCU computes the proper output data, which are converted in analog voltages and then sent to the Driver Unit for driving the motors. VAL II programs are included in this module.

Level 2

The middle level is the level in which elementary sensory-motor operations (tactile subroutines) are frozen in separated modules (SUB_i) as behavior of the system. Each subroutine has the role of managing the execution of an exploratory procedure aimed at the extraction of an object feature. The exploratory strategy depends on the feature that has to be extracted and on the sensor used in the exploration.

A dedicated Signal Analyzer Module (SAM), one for each subroutine, performs a compression of sensory data coming from the lower level, by computing some significant and synthetic parameters utilized as feedback signals for the middle-level controller. The same parameters are also combined in a next stage for computing a quantity representative for the feature extracted by the tactile subroutine. All outputs produced at this level are sent to the high level for further processing.

Level 3

The purpose of the high level in this architecture is to plan an exploratory strategy according to the input task and to attempt a recognition or a classification of the objects explored by the robot system. All data and parameters computed by the middle level converge in a module, called Integrator, whose task is to merge all sparse sensory data into few synthetic quantities compatible with the information stored in the Data Base.

The real recognition process is performed by the Recognizor module, which compares the parameters extracted by the Integrator with sample parameters stored in the Data Base. This sample parameters are extracted from a number of sample objects in a previous learning phase, carried out by using the same procedure.

In this way, the system learns how to build its own model of the world, since only internally processed information is utilized to construct and update the Data Base. In this way, systematic errors and imperfect calibration do not affect the system performace significantly, and the recognition process comes out more robust.

The Tactile Planner selects the next tactile subroutine for optimizing the recognition process, according to the input task and to the local features recognized during tactile exploration (given as feedback information in the high-level controller).

4. Experimental results

Three tactile subroutines have been implemented on this system, HARDNESS, TEXTURE and FRICTION, aimed at the extraction of hardness, texture and friction coefficients respectively.

In all experiments the objects were fixed on the table, located in a position known *a priori* by the robot, since no vision system was used to identify absolute positions in the robot work-space. All tactile subroutines were coordinated by PC1.

4.1 Hardness procedure

Starting from an initial configuration, the arm moves slowly toward the object, in order to press the object surface with the sensor tip. When the contact force detected by the F/T sensor exceeds a given threshold, say F1, the PUMA stops its motion and sends the coordinates of its wrist to PC1. After the transmition is completed, the PUMA slowly increases the contact force on the object (as allowed by the compliance of the wrist adaptor) and when the force on the F/T sensor reaches a second threshold F2, the PUMA stops again and sends the new wrist location to PC1.

Based on the information received from the PUMA and on the elastic properties of the compliant wrist adaptor, PC1 determines the position displacement D of the F/T sensor during the pushing procedure and computes the following ratio:

$$H = \frac{F2 - F1}{D}$$

In the case of soft materials, the displacemet D caused by object deformation will be relatively large, while for hard objects D will result much smaller. Thus, the parameter H represents a rough estimate of object hardness.

Figure 5 shows the results obtained by executing the procedure on several sample objects, having the same shape (parallelepipedal), the same thickness (10 mm), and modulus of elasticity comprised between 10^5 and 10^9 Pa.

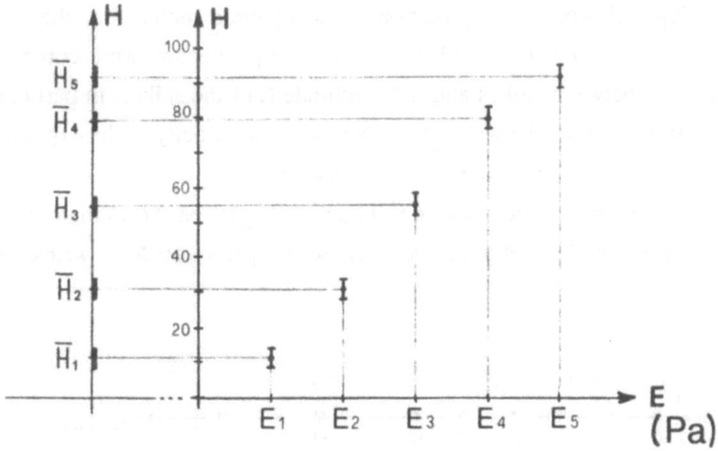

Fig. 5. Statistical evaluation of the H parameter, executing the procedure on five sample objects.

Repeatibility was also tested by running the procedure several times on each object: the standard deviation computed over 20 tests on the same object did not exceed the value of 4% F.S.O.

4.2 Texture procedure

This tactile subroutine was executed by rubbing the PVF2 nail sensor (located at the tip of the F/T sensor) on the object surface with a predetermined force. Since fine texture details are better perceived by exploring planar surfaces, we used flat objects only. Moreover, in order to symplify signal processing and easily describe roughness by few sinthetic parameters, we decided to test the system by using "wrinkly patterns", prepared by disposing in parallel thin wires on a smooth board. We used spacing between wires and diameter of the wires as parameters for characterizing roughness. Wrinkly patterns have also been used by psychologists to test the human tactile system [7][8]; therefore they also represent a good method for comparing the human perceptual system with an artificial one.

The procedure has been caried out by rubbing the nail sensor on the wrinkly patterns along a straight line, for a length of 35 cm, at the speed of 125 mm/s. The force exerted on the surface was set at 3 N and controlled by PC1, while the nail signal was sampled by PC2 at the frequency of 3.2 KHz. The aim of the experiments was to test the ability of the system in discriminating spacing and thickness of the wrinkles.

Signal processing following the exploration of each pattern included a filtering phase, a thresholding phase and an evaluation phase, where two parameters were computed on the signal: the distance **d** between spikes and the amplitude **A** of the spikes. In particular, if **n** is the number of samples acquired between two spikes, **v** is the velocity of the exploration and **f** the sampling frequency, spacing is given by: **d = nv/f**.

Results of these experiments are reported in Figure 6: Figure 6a shows the parameter **d** vs. the real spacing of the wrinkles, while Figure 6b shows the parameter **A** vs. wrinkle thickness.

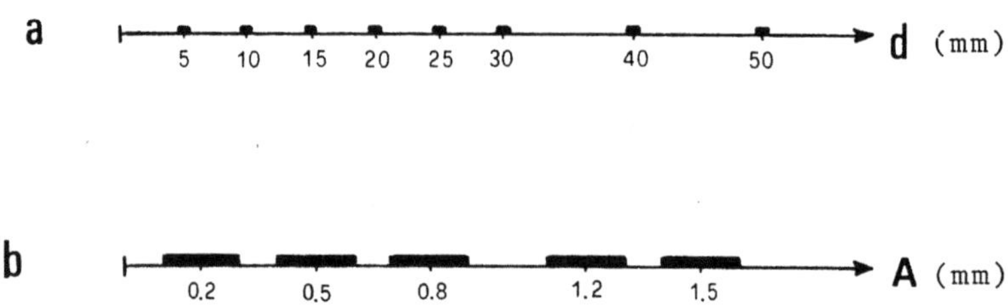

Fig. 6. Statistical evaluation the Texture Procedure.
Fig. 6a: parameter **A** related to wrinkle amplitude
Fig. 6b: parameter **d** realted to wrinkle spacing.

The system exhibited a precision of about 0.2 mm in perceiving distancies, but its tactile acuity (i.e. the smallest distance at which the system is able to discriminate two wrinkles as distinct) resulted of 0.5 mm. This greater value can be explained by considering that a spike produced by a wrinkle fades out in about 3-4 ms, depending on the elasticity of the nail, and at the speed of 125 mm/s the nail advances of about 0.5 mm.

The standard deviation calculated for the parameter **A** was much greater than the standard deviation calculated for **d**. The main factor affecting the value of **A** is the mechanical vibration of the robot arm during the exploratory procedure. However, the system was able to discriminate five wrinkles (0.2, 0.5, 0.8, 1.2 and 1.5 mm thick) with an error smaller than 15%, and four wrinkles (0.2, 0.6, 1, 1.5 mm thick) with an error smaller than 2%.

4.3 Friction procedure

This procedure involves automatically testing the friction properties of an object, in order to estimate its static and dynamic friction coefficients. This information is very useful for programming operations like grasping or manipulation of objects, which often rely on the forces the friction is able to withstand; beyond that, it can be used in order to characterize different objects, contributing to their recognition.

The way the friction coefficient is estimated is inspired by the observation of human behavior: we usually proceed by touching the object with a finger, pressing on it moderately and then exerting on the finger a force tending to slide it over the object surface; this force is increased until the fingertip actually slips, after which the operation is over.

To replicate such an operation, an automatic system needs the capability of sensing both the normal and tangential forces exerted at the contact point. This feature, which is not possessed by most conventional tactile sensors, is realized by the so-called Intrinsic Tactile (IT) sensor, as described by Bicchi and Dario [5]. An IT sensor consists basically of a force/torque sensor integrated within the fingertip surface, so that all the components of the force system generated by contact pressures are measured. If the geometrical description of the fingertip surface is known, it is possible to apply simple algorithms (as the original one proposed by Salisbury [9], or a more precise one described in Bicchi [10]) so as to obtain the following information:

a) the location of the contact point on the fingertip surface;

b) the intensity and direction of the contact force, and hence

c) the values of the normal and tangential (friction) components of the contact force.

Using the miniaturized F/T sensor mounted on the Puma arm and a spherical fingertip of radius 10 mm, fixed in turn to the F/T sensor, we performed several experiments aimed at automatically measuring the coefficients of static (μ_s) and dynamic (μ_d) friction of different objects in contact with the fingertip. The fingertip was initially brought to touch the object surface with a normal force of about 0.5 Kg; then the robot arm started to force it to move in the tangential direction, increasing this force linearly with time. The values of normal and tangential components of contact force, detected by the IT sensor during this phase and the following slippage, were stored in a buffer memory. Once arm motion is stopped, data are elaborated and presented in graphic form as shown in fig.7.

The diagram showed in fig.7 refers to an experiment with a rubber object (with relatively high friction), and presents the plot of friction ratio R_f (i.e. the ratio between the tangential and the normal component of contact force) vs. time. Each small square in the plot corresponds to a value of R_f measured at a sampling rate of 10 Hz. In the diagram of fig.7 two parts can be easily recognized: in the first part R_f increases almost linearly, until a maximum is reached, after which the friction ratio drops to a lower value; in the second part R_f is approximately constant. The interpretation of such plots is straightforward: the friction force increases until R_f reaches the static friction limit μ_s, then motion (slip) starts, and, according to the Coulomb model of friction, the friction ratio drops to μ_d.

Due to the fact that accidental perturbations of the mechanical system and of the sensor measurements superimpose random oscillations to the experimental curves, their

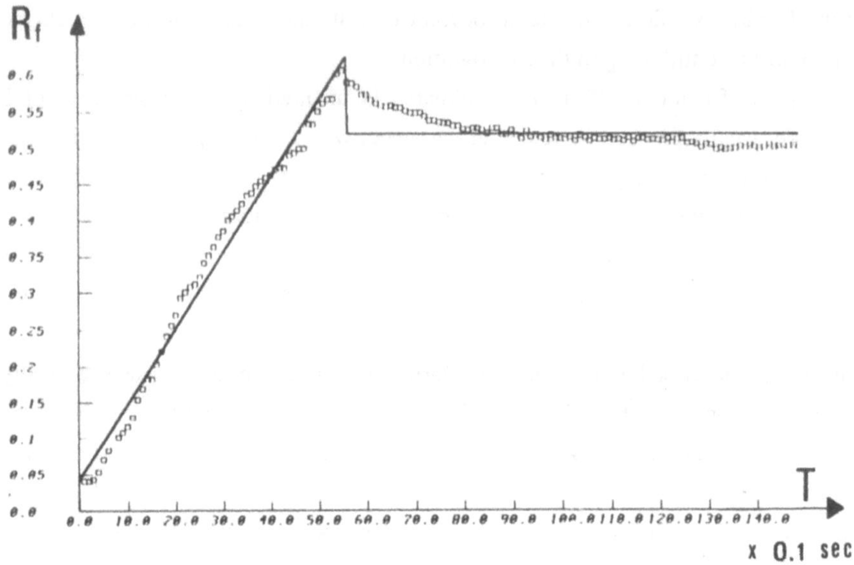

Fig. 7. Friction ratio during a sliding movement

interpretation in terms of quantitative estimates of ms and md is not obvious. Repeated experiments on the same objects resulted in data having a common pattern, but several local discrepancies. An algorithm for interpreting such data that resulted in fairly repeatable estimates is the following: the set of measurements is splitted in two parts corresponding to a tentative slippage instant T_S; the first subset of data is fitted with the best line in least-squares sense, and the second subset is aproximated with a constant value equal to its average value. The sum of the averaged squared errors in each data subset is assumed as a measure of approximation; at varying T_S, the slippage instant is found as the one minimizing the approximation error.

The resulting linear approximation is presented in Fig.7 with a superimposed solid line. The maximum value of friction ratio reached before slippage is assumed as the static friction coefficient; the average value of the following phase is the exstimated dynamic friction coefficient.

Based on the above technique, an automatic sorting of objects, having different friction characteristics, has been attempted. Objects belonging to three classes, with low, intermediate and high friction, were examined in random order by the system and the friction of their surfaces measured with the above described methods. The objects were then recognized as belonging to one out of the three classes: the incidence of errors in these tests was virtually null.

5. Conclusions

A sensorized robot system able to perform specific exploratory procedures (tactile subroutines) on objects in order to extract information useful for their description, has been described.

The approach we have proposed is an attempt of replicating in an artificial system some of the sensory-motor paradigms used by humans in exploratory tasks. Obviously, many simplifications were introduced to reduce the complexity of control and the amount of computation on the sensor signals.

In spite of the limitations of the present work and the rather simple structure of the system, results show the validity of this approach. Studying one finger exploratory strategies based on the decomposition of complex human tactile perceptual activities in a sequence of elementary sensory-motor acts, seems to be promising and to encourage further investigation in the field.

Acknowledgements

The authors wish to thank R. Di Leonardo and F. Vivaldi for their help in building the sensors. This work has been supported by the Scuola Superiore S. Anna of Pisa and the University of Bologna.

References

[1] Bajcsy, R., "What Can We Learn From One Finger Experiments?", in Robotics Research, Brady and Paul, editors, MIT Press, 1984, pp. 509-527.

[2] Buttazzo, G., P. Dario and R. Bajcsy, "Finger Based Explorations", in David Casasent, editor, Intelligent Robots and Computer Vision: Fifth in a Series, pp. 338-345, Proceedings of SPIE Vol. 726, Cambridge, MA, 1986.

[3] Stansfield, S.A., "Primitives, features, and exploratory procedures: Building a robot tactile perception system". Proc. of IEEE Int. Conf. on Robotics and Automation, S. Francisco, CA, 1986, pp. 1274-1279.

[4] Dario, P. and G. Buttazzo, "An Anthropomorphic Robot Finger for Investigating Artificial Tactile Perception", The Int. Journal of Robotics Research, Vol. 6, no. 3, Fall 1987, MIT Press.

[5] Bicchi, A. and P.Dario, "Intrinsic Tactile Sensing for Artificial Hands"; Robotics Research, R.Bolles and B.Roth Editors, MIT Press, 1987.

[6] Cutkosky, M.R., "Dynamic tactile sensing", Proceeding of RO.MAN.SY, Udine, September 1988.

[7] Loomis, J.M., "Tactile pattern perception", Perception, 10, 5-27, 1981.

[8] Lederman, S.J., "The perception of texture by touch", in Tactual Perception, W. Schiff and E. Foulke, eds., Cambridge University Press, 1982, pp. 130-167.

[9] Salisbury, J.K., "Interpretation of Contact Geometries from Force Measurements"; Robotics Research, M.Brady and R.Paul Editors, MIT Press, 1984.

[10] Bicchi, A., "Methods and Devices for Dextrous Manipulation Sensory Control", Doctoral Dissertation, University of Bologna (in preparation), 1988.

UNCERTAINTY IN ROBOT SENSING

E. Grant
Acting Director of Research

The Turing Institute
George House
36 North Hanover Street
Glasgow G1 2AD
Scotland
UK

ABSTRACT

This paper deals with sensing uncertainty in a robot world. Sensors typically provide signals that are both incomplete and ambiguous. Three pieces of research are described which attempt effective solutions to this common problem but using three different approaches. The first piece of work uses vision to demonstrate the construction and integration of a dynamic world model for mobile robot navigation. The second, provides an adaptive rule-based controller for an inverted pendulum and cart problem and the third, sensory integration of vision and taction for the purposes object recognition.

The theme for the first piece of work is that the most effective solutions are obtained when maximising the amount of representational data available. The theme of the second is that broad qualitative partitioning of a state-space can avoid problems of ambiguity and noise without performance decrement. Indeed, the use of broad qualitative partitions is shown to lead to the development of heuristic adaptive controllers for complex dynamic systems that offer far greater than flexibility than those based on classical methodologies. The final theme is that machine learning can play a powerful role in the generation of sensor-based models.

NATO ASI Series, Vol. F 66
Sensor-Based Robots: Algorithms and Architectures
Edited by C. S. George Lee

INTRODUCTION

Uncertainty exists in numerous forms in present robot systems because robots must operate in the real world. In order to operate in this world, robots must cope with the inherent uncertainty that is associated with the modelling, planning and motion of manipulators and parts. A review of the literature shows that two schools of thought exist regarding the solution to reducing uncertainty. Although both are knowledge-based, it is difficult to imagine how the first could reduce the effects of uncertainty since it requires structuring the robot world more than at present, and enough rigidity is already imposed on current robotic systems. This method also advocates the building of stiff, precise robots [3,4], further structuring. In highly structured systems it is the accumulative effect of small errors caused by, kinematic, kinetics and sensor data that makes such systems fail continuously.

The second method proposes robots incorporating learning and reasoning through the use of sensory integration [1]. It is the route to reducing uncertainty that is reported on here. Through applying knowledge-based methods to data acquired from various sensors, robust sensory integration techniques were developed that improve robot system flexibility, generality and reliability [9]. Throughout the paper, numerous examples are given of the types of uncertainties commonly associated with robot sensors. Generally, these either take the form of uncertainty factors associated with data collection, e.g. noise, or, through poor data interpretation by humans.

Brady [2] defined robotics as *the intelligent connection of perception to action.* As such, the control process requires three fundamental elements:

- perception
- action
- intelligence

Perception, is defined here as integrated sensing. All sensors provide data which is incomplete or ambiguous. To solve this fundamental problem we have two options. First, to try to make perfect sensors. Second, to make effective use of several sensors so allowing individual sensors to complement each other in a graceful and integrated fashion. The usual vehicle for such integration is referred to as a world model. Action is simply referred to as any mechanical operation in the world. Intelligence, in the context of this paper, is viewed as a process involving the learning of the relationship between perception and action.

A brief overview is given of three pieces of work related to the theme of sensing for intelligent control. The first shows how sensory data collected from different geometries can be merged into a simple world model in the domain of mobile robotics. The second, how integrated qualitative sensing provides control in a dynamic domain. Initially, the classic control problem of the inverted pendulum was chosen to demonstrate the principle and conduct experiments on a purpose built test-rig. Later, the theorems developed were applied to a satellite control problem to test their robustness. In each instance sensory data was used to induce control rules. The third shows how rule induction is useful for integrating different sensing modalities, and for classifying objects.

WORLD MODELLING FOR MOBILE ROBOTS

Because they operate in technically complex environments mobile robots are considered to be a perfect platform for developing knowledge-based methods for sensory integration and researching into uncertainty. These robots are now entering a new phase of development that will see them become free-roving, be able to avoid obstacles and dock. No longer will they be constrained to their present wire-guided, pre-defined routes [6]. Although we are only in the initial stages of mobile robot development, it is recognised that the complexity of their technology must equal the complexity of the environment in which they operate.

Sensory integration alone is not well understood, therefore it is a barrier to progress. Integration requires consistent and repeatable algorithms that deal with uncertainty, technically these are difficult to develop. The AI community considers this as a major research area of interest because it has prior experience of uncertainty generation and representation. The data that requires interpretation is obtained from sensing sources that are becoming increasingly more complex leading to computational overheads. Although vision remains the major sensing medium at present, free-roving robots might require data to be integrated into a world model from numerous sensors, e.g. taction, inertial navigation systems, laser ring gyros or acoustic rangefinders. Presently these sensors suffer from a combination of uncertainty factors such as noise, poor resoluton, reflection problems and enormous computational needs. However, continued advances in

Figure 1 A black and white photograph of a teapot

Figure 2 Nine line drawing replies

all the areas highlighted should see successfully planned, collision-free navigation for mobiles, and robot systems in general.

Sensory integration must address the problems of incorporating data into a dynamic world model. This means handling data obtained from different sensors, multiple data from a single sensor or, in the case of compliant tactile sensors, time-varying data from a single sensor [5]. In this paper an attempt is made to reduce sensor uncertainty through using a common data structure as a repository for sensory data. Finally, knowledge-based methods are used to control the data integration, and provide reliable and consistent information.

Two major paradigms predominate the approaches taken to early and intermediate computer vision. In early computer vision, the major paradigm consists of describing the world in terms of explicit edge tokens, these descriptions were then used as a basis for computing useful intrinsic information. For intermediate computer vision, the major paradigm is based around the idea of the 2.5D sketch which describes local surface orientations and discontinuities relative to the viewer using a rich symbolic language. This then represents a crude world model. Unfortunately, these approaches have been fraught with problems. First, the current generation of edge detectors still provide hesitant tokens which suffer from distorsion, omission and false labeling whilst empirical results suggest that edge detection is an ill-posed problem *per se*.

As an indication of the uncertainty surrounding edge detection consider the results obtained from a recently conducted experiment [8]. The experiment consisted of providing twenty emminent vision practitioners with a photograph of a teapot, Figure 1, an acetate sheet and a pen, they were then asked to identify the edges. Figure 2, shows the line drawings of the nine replies received. Obviously, there is a considerable degree of uncertainty with regard to what constitutes edges. Second, and as a consequence, intrinsic image computation based on edges has provided only frail solutions using natural image data. These essentially negative conclusions have been bolstered by the discovery that useful intrinsic image representations may be generated without prior recourse to edge tokens.

TOspace, see Figure 3, is a single 3D iconic data-structure into which may be merged data from surfaces rather than edges. This data may be either visual or tactile in origin. The data-structure has been used to provide navigation information for a mobile

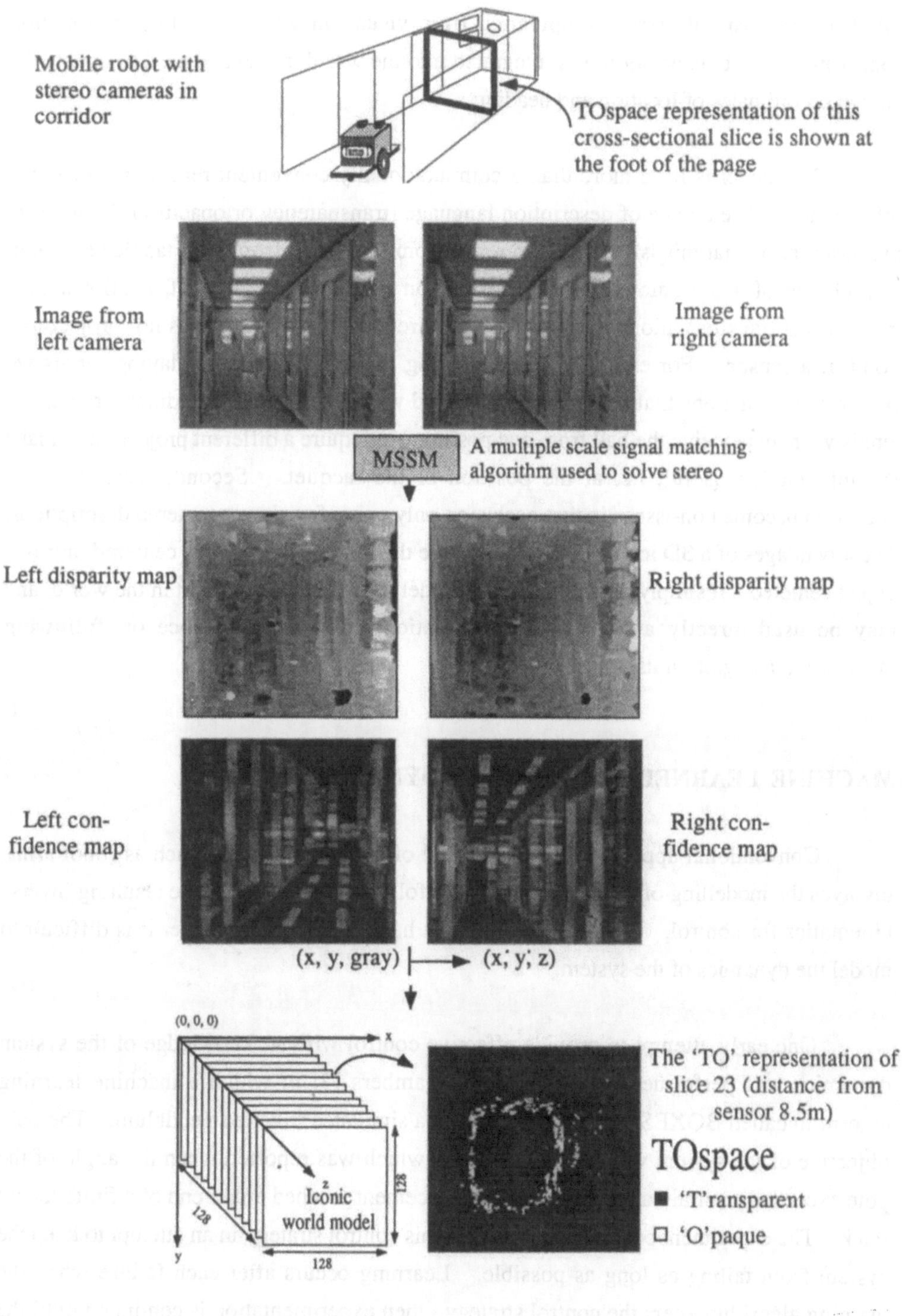

Figure 3 The TOspace world model

robot on the basis of stereo computation from visual images. Absolute information concerning the robots position was integrated into the world model and used to update the known co-ordinates of location and heading.

The model is little more than a computationally convenient repository for range information. The choice of description language (transparency or opacity) follows from the observation that this is what should result from any shape-from-X or tactile sensation. The choice of a 3D data-structure is useful on two accounts. First, it allows easy geometrical transformations whenever we require a solution which does not project to a point at a sensor. For example, when running through a forest or playing tennis we perform many actions that need to be interpreted within a dynamic coordinate frame. In tennis we may perceive the ball from one position but require a different projection to make the information useful, i.e. at the position of the racquet. Second, issues such as occlusion become non-issues in that occlusion only exists for viewer centered descriptions. The advantages of a 3D iconic data structure are that it is neither viewer centered, nor is it object centered. It simply attempts a literal model of the surfaces present in the world, and may be used directly as a basis for navigation, collision avoidance or, following invocation, recognition itself.

MACHINE LEARNED CONTROL OF DYNAMIC SYSTEMS

Conventional approaches to the control of dynamic systems, such as robot arms, involves the modelling of the system dynamics followed by the use of the resulting inverse kinematics for control. Such methods clearly have problems whenever it is difficult to model the dynamics of the system.

One early attempt to provide effective control without knowledge of the system dynamics was performed by Michie and Chambers [7] in which a machine learning algorithm called BOXES was used to balance a simulated inverted pendulum. The only objective of the system was to avoid failure, which was reported when the angle of the pole exceeded a certain angle or the cart displacement reached either end of a finite length track. The experiment consisted of applying this control strategy in an attempt to keep the system from failing as long as possible. Learning occurs after each failure when the learning algorithm alters the control strategy, then experimentation is continued until the pole can be successfully balanced for an indefinite period.

The control strategy described is used to control an inverted pendulum and cart, a complex dynamic system, albeit a dynamic system which can be effectively controlled using classical means, e.g. a proportional plus derivative controller. So, in an attempt to increase the complexity of this particular dynamic system bang-bang control was used, and the learning algorithm was supplied with limited sensory data, the pole angle only. The analytical solution of the problem shows that four state variables: the cart position; the cart velocity; the pole angle and the pole velocity describe the state of the inverted pendulum and cart at any instant.

The state space is filled with four dimensional boxes, one dimension each for the four variables described above. In each box the variables fully describe one state and, based on the information contained within a given box that state decides whether the cart should be pushed left or right. If the pole is balanced successfully then the controller has become expert in balancing the pole, in that part of the state space only. However, in another part of the state space, one in which the controller has no experience, any attempt to balance the pole could meet with sudden failure. Thus, control strategies for many experiments are required in order to learn how to balance the pole, this experimental knowledge-base is then used to construct a generalised controller for the complete state space.

When the generalised controller was constructed, the box structure could be simplified to the point where human readable rules can be extracted, Figure 4, these control rules were of the form:

if the angular velocity of the pole is less than a given threshold then push left
else if the angular velocity of the pole is greater than a given threshold then push right
else if the angle of the pole is less than a given threshold then push left
else if the angle of the pole is greater than a given threshold then push right
if the velocity of the cart is less than a given threshold then push right
else if the velocity is greater than a given threshold then push left
else if the position of the cart is less than a given threshold then push right
else if the position of the cart is greater than a given threshold then push left

The rule above, derived by Sammut [12] from simulation, is similar to the Makarovic [11] rule, which was derived from examination of the equations of motion. The major

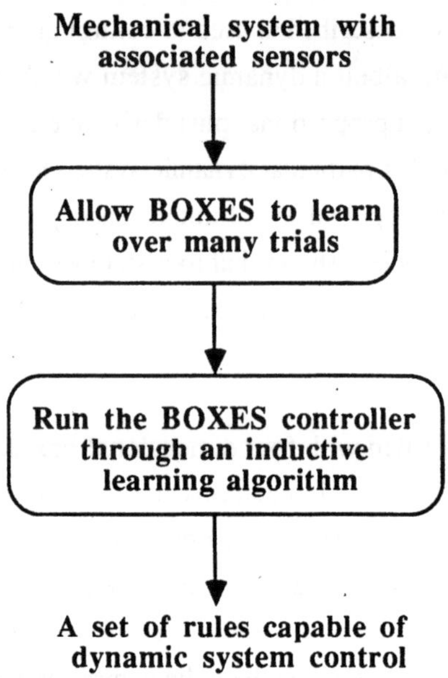

Figure 4 Using BOXES for generating a rule-based controller

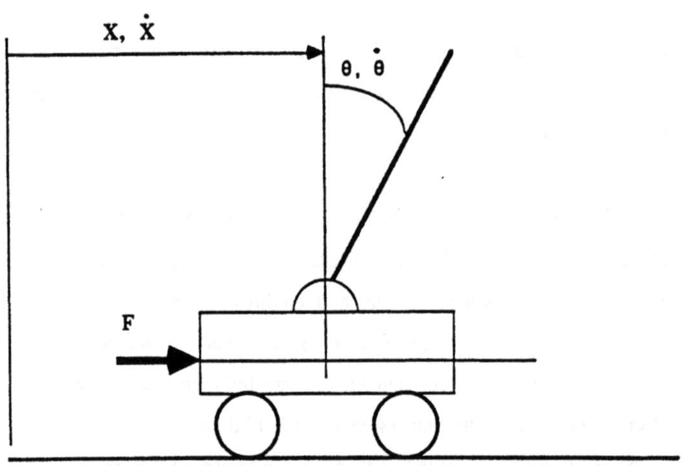

Figure 5 The inverted pendulum

difference between these two separate approaches was that whereas Sammut's rule was derived from knowledge of the system it was controlling, Makarovic learned only from examining the response of the system to a limited set of actions.

In the original work by Michie and Chambers [7], the algorithm worked by first partitioning the four dimensional problem space into a finite state space consisting of 255 states. Each state offered an, initially, random control solution (push left or push right), whose value was modified by a simple credit assignment algorithm. It is this initial partitioning exercise that is one of the domains being worked on in a new phase of this work. It is percived that automatic partitioning, rather than human partitioning, might lead to more effective and efficient learning, paricularly in the start-up phase. Automatic partitioning also becomes a necessity where uncertainty factors arise after learning is completed. For example, if a sudden change occurred in the system dynamics , e.g. a part falling off a satellite, giving a resulting change in inertia, it would be necessary for the controller to retrain itself. It is also useful for handling noisy signals or for cases where a state continually lands on a boundary.

This simple BOXES algorithm offered an effective control strategy which has been extended by recent work at the Turing Institute. After first replicating the original work an apparatus was constructed of the type shown in a diagrammatic form in Figure 5. This has been used to show how signals produced from a single sensor, an imposed uncertainty condition, can help to partition the state space and lead to effective control under dynamically unstable conditions. Using a rule-based control algorithm developed by Makarovic [11], that is in fact a derivation of BOXES, The pendulum balanced for 31 seconds. More recently however a rule-based algorithm developed at the Turing Institute has balanced the pendulum for 90 seconds.

Having successfully demonstrated the effectiveness of rule-based algorithms as an adaptive controller, the next phase of the project is the implementation of BOXES to other machine learned control domains. Further, it has been shown that the machine learned model for control has broad generalisibility. For example, Sammut [12] has shown that the rule base may be used directly for the control of a satellite. Here, the goal was to control the attitude of a robotised satellite in low Earth orbit, a satellite that has an on-board manufacturing capability operating in micro-gravity. Internal and external maintenance is undertaken by a robot arm. Working within these constraints, and in addition having to cope with the problem of remote sensing, the satellite had to function effectively with

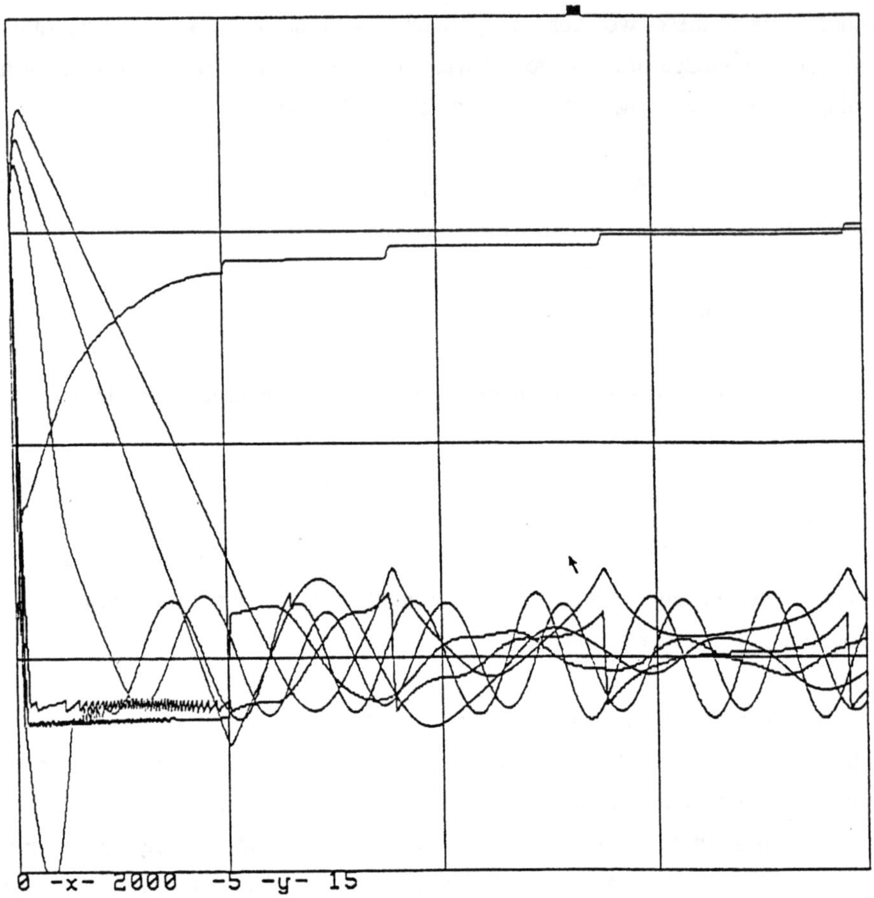

Figure 6 Satellite simulator controlled by BOXES rule-base

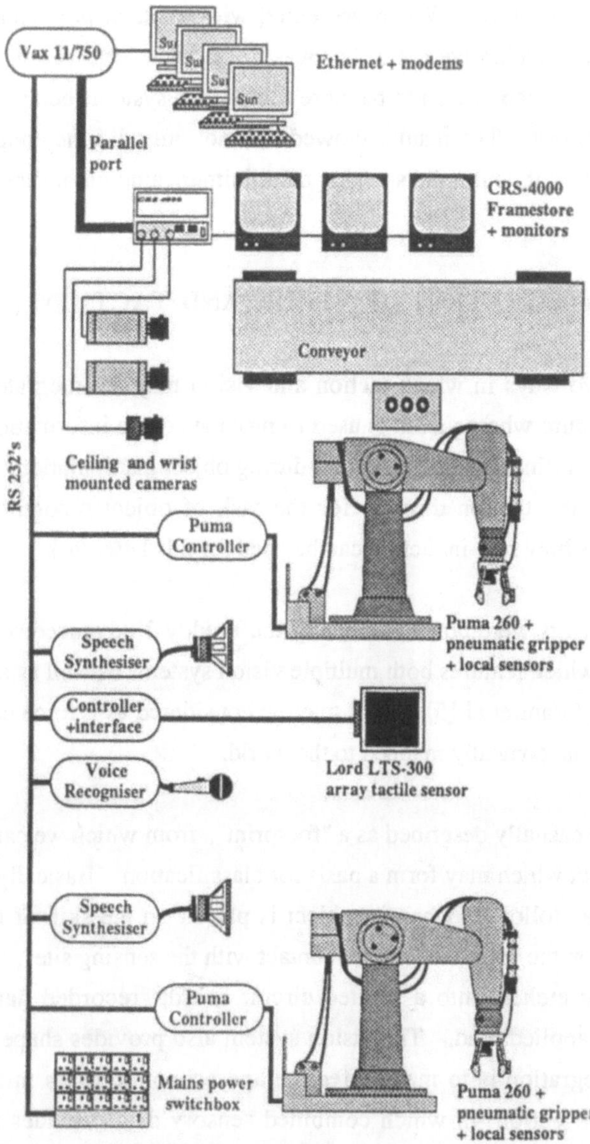

Figure 7 The Freddy 3 advanced robotics research testbed

minimum human intervention. When presented with a 'black box' simulation of this particular space craft an adaptive controller was developed, one based on AI methods described previously, which proved to be more robust than systems being developed from traditional control theory. The results showed that not only did the controller keep the satellite in its desired orbit, but it did so using the minimum amount of fuel, Figure 6.

RULE BASED INTEGRATION OF VISION AND TACTION

There are two ways in which taction and vision may be integrated. First, in a coarse-to-fine procedure where vision is used to provide coarse information for hand/eye control which is then refined by tactile signals during object manipulation. Second, in the use of both vision and taction directly for the task of object recognition. The work described here shows how rule-induction can be used for this latter task.

The experiments were conducted using the Freddy 3 advanced robotics research test-bed, Figure 7, which features both multiple vision systems as well as a Lord LTS-300 array tactile sensor, Grant et al [5]. This may be considered as a force camera in that it produces data which is iconically mapped to the world.

This may be casually described as a "footprint", from which we can recover shape attributes of an object which may form a basis for classification. Basically, the operation of the LTS-300 is as follows, when an object is placed on the skin of the sensor, the conductive coating on the underside makes contact with the sensing sites. These sites, an 80 x 80 array, are etched onto a printed circuit board, recorded data readings are proportional to the applied load. The vision system also provides shape attributes. The role of sensory integration is to make effective and economic use a minimal subset of attributes (vision or taction) in which combined sensory data provides a classification solution which is simpler and more accurate that using just a single sensor.

Separate areas of uncertainty were observed throughout the integration experiment. The most serious were the uncertainties inherent in the tactile sensor sensor itself, the product of poor specification and design. Not only did the sensor generate noise during its calibration experiments, Figure 8, it also gave hysteresis and creep curves that could be problematic in classification applications. Fortunately, all the work was undertaken in an environment that allowed the development of knowledge-based filters and tools to achieve

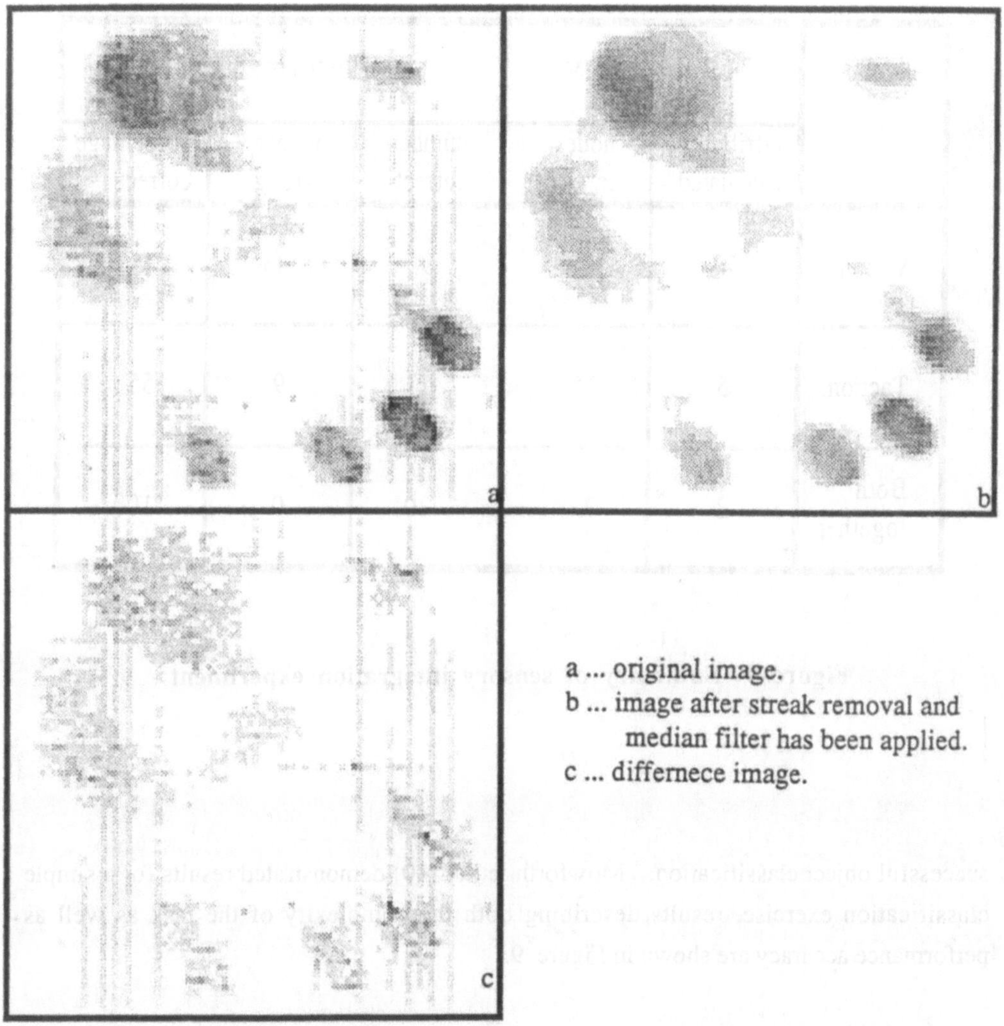

a ... original image.
b ... image after streak removal and median filter has been applied.
c ... differnece image.

Figure 8 Composite filtering

	Complexity		Performance		
	attributes calculated	nodes in tree	number correct	number wrong	percentage correct
Vision	4	15	9	11	45
Taction	5	15	11	9	55
Both together	2	7	20	0	100

Figure 9 Summary of sensory integration experiment

successful object classification. Mowforth et al [10] demonstrated results for a simple classification exercise, results describing both the complexity of the task as well as performance accuracy are shown in Figure 9.

ACKNOWLEDGEMENTS

The author thanks the SERC, National Engineering Laboratory, Westinghouse Corporation, Hunting Engineering and Lord Corporation for funding the research and/or their support. Thanks also to Peter Mowforth my co-researcher, Jonathan Shapiro and Paul Grant for the TOspace work, Claude Sammut and Paul Gisbey for the pole balancing and Stephen Rutherford for assistance with the sensory integration project.

REFERENCES

[1] Baird, s. and Lurie, M. *Precise robot assembly using vision in the hand.* In: Pugh, A., editor, Robot Sensor, Vol 1, IFS, pp 85-94.

[2] Brady, M. *Artificial intelligence and robotics.* Artificial Intelligence, 26(1), 1985, pp 79-121.

[3] Brost, R.C. *Automatic grasp planning in the presence of uncertainty.* In: Proceedings 1986 IEEE International Conference on Robotics and Automation, 7-10 April, 1986, 3, San Francisco, CA, pp 1575-1581.

[4] Durrant-Whyte, H.P. *Uncertain geometry in robotics.* In: Proceedings 1987 IEEE International Conference on Robotics and Automation, 31 March-3 April, 1987, Raleigh, NC, 2, pp 851-856.

[5] Grant, E., Mowforth, P.H., Rutherford, S. and Wagstaffe, N. *An array tactile sensor and its value in automation.* In: Proceedings of the 18th International Symposium on Automotive Technology and Automation, , ISATA, Florence, Italy, May 1988, pp 234-240.

[6] Harmon, S.Y. *Robot mobility: which direction and how far ?* In: Robots 11, SME 17th International Symposium on Industial Robots, 26-30 April, 1987, Chicago, ILL, pp 17-1 - 17-11.

[7] Michie, D. and Chambers, R.A. *BOXES: an experiment in adaptive control.* In: Dale, E. and Michie, D., editors, Machine Intelligence 2, Oliver and Boyd, Edinburgh, 1968, pp 137-152.

[8] Mowforth, P.H. and Gillespie, L. *Edge detection as an ill-posed specification task.* TIRM-87-026, November 1987, The Turing Institute, George House, 36 North Hanover Street, Glasgow G1 2AD, Scotland.

[9] Mowforth, P.H. and Grant, E. *Three examples of sensing for intelligent control.* In: Proceedings 1988 IEEE 3rd International Symposium on Intelligent Control, 24-26 August, 1988, Arlington, VA.

[10] Mowforth, P.H., Grant, E. and Rutherford, S. *Rule based integration of vision and taction.* TIRM-87-023, June 1987, The Turing Institute, George House, 36 North Hanover Street, Glasgow G1 2AD, Scotland.

[11] Makarovic, A. *Pole balancing as a benchmark problem for qualitative modeling.* IJS Delovno porocilo 4953, Jozef Stefan Institute, Ljubljana, Yugoslavia, December 1987.

[12] Sammut, C. *Experimental results from an evaluation of the algorithms that learn to control dynamic systems.* In: Laird, J., editor, Proceedings of the 5th International Conference on Machine Learning, Morgan Kauffman, San Mateo, CA, June 1988, pp 437-443.

Part II

Vision Algorithms and Architectures

Robotic Vision Knowledge System

Andrew K.C. Wong
PAMI Lab, Systems Design Engineering
University of Waterloo
Waterloo, Ontario, Canada

This article presents a robotic vision knowledge system based on some current sensor and machine intelligence methodologies, quite a large portion of which has recently been developed by the PAMI Group at the University of Waterloo.

In this system, there are two major sources of image input: grey tone images from CCD cameras, and range data images from synchronized laser scanner or from structured lighting scheme coupled with CCD cameras. The grey tone images are used for model generation and object recognition, location and tracking. The range data are used for surface profile measurement, gauging, and recognition. From range profile and discontinuity, edges can be detected and regions can be grouped into hyperpatches or enclosed surfaces of distinct geometry. Special techniques are used to relate the local profile information point features which form the object reference. In 3-D shape synthesis using grey tone images, a system integrating a fast feature extractor with a domain knowledge guided local feature filtering and geometric reasoning scheme is used. Synthesized shapes are then represented in the form of 3-D random graphs and attributed hypergraphs which can be translated into procedural knowledge in the form of rule network for real time object recognition and location using hypothesis refinement search strategies.

A research prototype of the vision knowledge system is currently under development for a multi-agent intelligent robotic workcell and for a project related to the Mobile Servicing System on Space Station. Preliminary experimentations already yield encouraging results.

1 Introduction

The general goal of computer vision is to construct, from images of physical objects, explicit and meaningful descriptions which can be processed and inferred by computers for various purposes. The way visual information is extracted by an imaging system and transformed into quantitative and symbolic representations crucially affects the effectiveness of a vision system. We assess this effectiveness by judging the performance of the three major phases of the system: 1) the 3-D measurement of an object; 2) the transformation of the measured data into a representation of the object and 3) the use of the representation for object recognition and location as well as 3-D scene interpretation. Hence, the major issues concerning many researchers in robot vision today are: 1) the effective acquisition of visual information from objects and scenes; 2) the recovery of two- and three-dimensional information from the acquired images; 3) the representation of the geometrical and spatial information in a suitable quantitative and/or symbolic form for inferences.

Because of the variable nature and complexity of the real world, the major problems encountered in a vision system are: 1) intensive computation and 2) susceptibility to various sources of noise. It is computation intensive partly because of the huge amount of data to be processed in a sizable image and partly because of the levels of data transformation required to compress data into a coherent interpretable form. Due to the sensitivity of various analytical techniques, often the analytical results highly depend on the technique used or the parameters set. Hence to maintain acceptable consistencies and reliability in the final interpretation of the results by a computer vision system is a difficult and usually frustrating task. Though considerable efforts have been devoted to filtering or eliminating noise at various processing levels, the results are

NATO ASI Series, Vol. F 66
Sensor-Based Robots: Algorithms and Architectures
Edited by C. S. George Lee

still not encouraging. One of the observations is that most of the current systems are devised to transform information from lower levels to higher levels and at each level of the transformation, noise could be introduced. But, often in a vision task, what mostly needed are certain visual and geometric information and constraints of the objects and the scenes. Hence, one feasible and realistic approach to computer vision is to use high level knowledge or highly redundant joint events obtained from visual and/or spatial aspects of the scene to tolerate or to bypass the effects of the low level noise. In this article, we present a machine intelligence and knowledge based approach to real world and real time 3-D vision for robotics applications based upon this philosophy. The system is largely evolved from current sensor and machine intelligence methodologies, quite a large portion of which has been recently developed by the PAMI Group at the University of Waterloo.

In this article, we first give an overview of a computer vision system and a brief survey of recent research activities in three-dimensional vision. Next we describe the robotic vision knowledge system developed at Waterloo. Finally, we present and discuss the results of some of the industrial and space applications.

2 Computer Vision: An Overview

Figure 1 gives a schematic of the basic processes and their relations in computer vision. Blocks linked by solid arrows represent processes generally adopted in existing systems. Those linked by dotted arrows describe activities related to some new trends and developments.

After the preprocessing phase, the image is usually transformed into a parametric form in terms of specific local feature codes or binary codes representing the foreground and background of the image [39]. Features may range from edges, streaks, to any local features such as corners and bright spots. In 3-dimensional range images, range, edges, surface orientation and curvature could be obtained through the use of smoothing and interpolation techniques [62].

The extracted features or grey level values obtained for different locations can be used to segment the image into regions or other coherent components such as borders, ribbons, curves, blobs, etc. This process is goal-oriented. In a more complex scene analysis, it may be necessary to separate different types of regions—regions with certain grey level or color, or regions with homogeneous texture content. In addition to segmenting a region based on its local features, tracking or reinforcing curves or borders of regions are generally included in the region segmentation and description process. An important phase in furnishing a region with low order structured information is texture analysis. The texture information can be used to segment regions [35]. Through the use of special classification and clustering techniques, the texture content of the regions can be represented and classified [46].

Once distinct regions are segmented, described, and labeled (or parameterized), the next phase is to represent the geometry of shape and the structural (or textured) content of distinct regions by special mathematical models for quantitative analysis, classification and comparison [61]. Though various mathematical and geometric models have been proposed today, their applicability is generally task dependent and justified largely by experimental results.

At the higher level of scene representation and analysis, relational structures are used. In this representation, both the shape and content (or attribute values) of each component as well as the type of relationship between components, will be included. Mathematical or algorithmic models have been developed for comparing and inferring relationships [62,2].

Robot 3-D vision requires shorter image processing and analysis durations. The 3-D scenes, in general, are more complex in a factory environment. To follow a comprehensive process of data transformation and analysis at various levels is often beyond the task requirement. A new trend

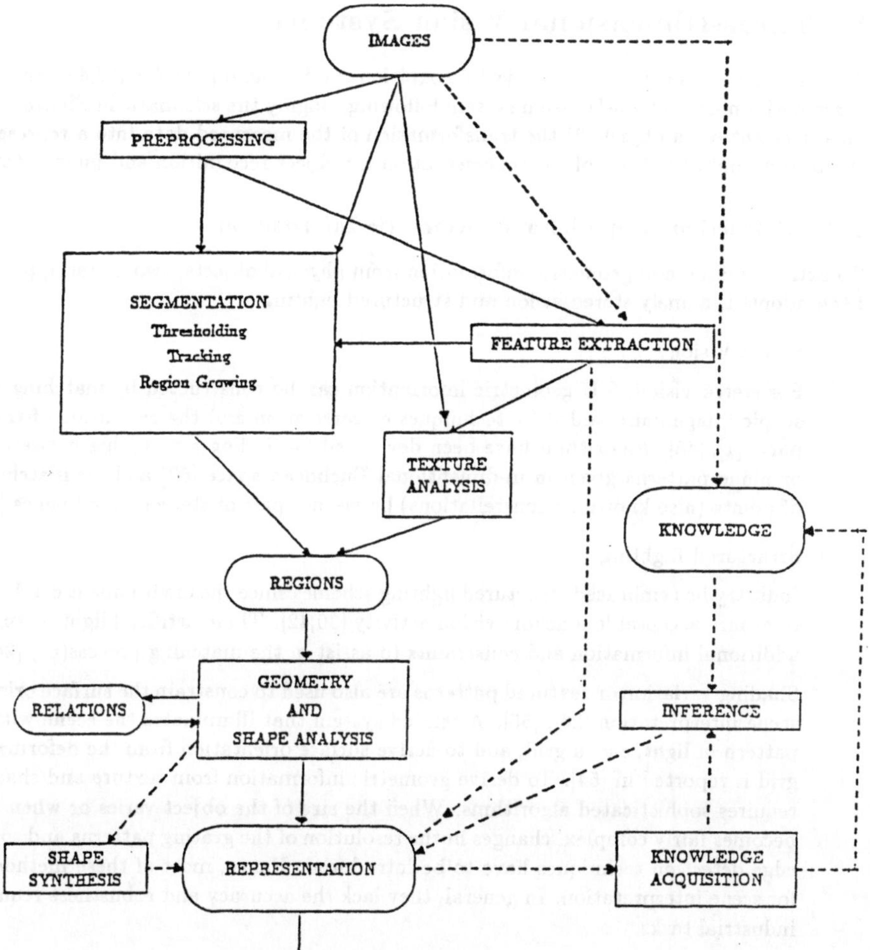

Figure 1: A Schematic of Computer Vision Activities

is to bypass some of the intermediate processes (dotted arrows on the right side of Figure 1). This is made possible if specific knowledge about the objects are used to tolerate preprocessing or early processing noise. To cope with the complex representation of the real world, automated shape synthesis and knowledge acquisition for object representation, recognition and location have been developed [61,19].

3 Three-Dimensional Vision Systems

A comprehensive survey on 3-D object recognition can be found in [8,7,27,16,42]. In general, the three major phases of a 3-D vision system following roughly the schematic in Figure 1 are 1) 3-D measurement of an object; 2) the transformation of the measured data into a representation of the object; and 3) the use of the representation for object recognition and interpretation.

3..1 Extraction of Spatial and Geometric Information

To extract spatial and geometric information from physical objects, two major approaches have been adopted, namely stereo vision and structured lighting.

1. Stereo Vision

   For stereo vision, 3-D geometric information can be constructed by matching the stereoscopic image pairs, aided by techniques of correlation and the resolution of the occluded parts [1], [48]. Algorithms have been developed for finding a matching between two point or plane patterns given in m-dimensional Euclidean space [52] and for matching subsets of points (also known as constellations) between a pair of stereoscopic images [63].

2. Structured Lighting

   Industry has embraced structured lighting schemes since the environment can be controlled to ensure acceptable machine vision activity [30,42]. These artificial light features provide additional information and constraints to assist in the matching process(es) [24].

   Shading variation or textured patterns are also used to constrain the surface orientation for scene interpretation [26], [55]. A related system that illuminates the scene with a regular pattern of light, e.g. a grid, and to derive surface orientation from the deformation of the grid is reported in [54]. To derive geometric information from texture and shade patterns requires sophisticated algorithms. When the size of the object varies or when the surface becomes fairly complex, changes in the resolution of the grating patterns and sophisticated edge detection techniques have to be introduced. Hence, most of these methods are used for scene interpretation. In general, they lack the accuracy and robustness required by the industrial tasks.

   Laser scanning [62] is another structural light method. Several commercial products that utilize laser scanning to obtain a depth map or depth image are now available. A synchronized structured lighting scheme in the form of stripes has recently been developed in the PAMI Laboratory at the University of Waterloo. It has the same capability as the laser scanning method. From the range data, spatial location and the $2\frac{1}{2}$-D and 3-D information of the object can be derived. The depth image contains the encoded spatial location and distance information [60,58].

3..2 Object Representation

An object representation scheme is generally examined under two criteria [23]:

1. descriptive adequacy, *i.e.* the ability of a representational formalism to capture the essential visual properties of objects and the relationship among objects in the visual world; and

2. procedural adequacy, *i.e.* the representation scheme's ability to support efficient processes of recognition and search.

Literature review of object representation schemes can be found in papers such as [7,37]. They include: 1) generalized cone or sweep representation [44,47]; 2) multiple 2-D projection representation [53]; 3) characteristic-views technique [14,48]; 4) skeleton representation [20,50]; 5) generalized blob representation [31]; 6) spherical harmonic representation [43]; 7) overlapping sphere representation [32]; 8) wire-frame representation [7,37]; 9) constructive solid geometry (CSG) representation [7,37]; 10) spatial-occupancy representation which include: a) voxel representation, b) octree representation [29], c) tetrahedral cell decomposition representation [7], d) hyperpatch representation [13]; 11) surface boundary representation [7]; 12) attributed graph and hypergraph representation [62].

Another approach uses directly the procedural knowledge organized in a form of rule network for object recognition and location from a perspective image of the scene [38]. It uses a hypothesis refinement strategy to direct the search. This approach has been demonstrated to be fast and reliable and well suited to domains where the number, type and basic geometric characteristics of the objects are known. It has the distinct advantage of requiring only one camera (one image).

4 Vision Knowledge System Configuration

4.1 An Overview

Figure 2 gives an overview of our configuration of the robotic vision knowledge system.

The system uses two major sources of image input: grey tone images from the CCD cameras and direct close-up range data images from the synchronized laser scanner (or structured lighting scheme coupled with CCD cameras). The grey tone images are used for model generation as well as for object recognition, location and tracking. The range data are used for surface profile measurement, gauging, and recognition.

To extract three dimensional information from range data, vectors normal to the surface at different points are computed; edges are detected and regions or enclosed surfaces of distinct geometry are grouped into hyperpatches [61]. The procedures include: 1) extracting profile and edges; 2) analysis and merging data to form hyperedges (or faces); 3) constructing an AHR of the scene; 4) performing a database search for subset of potential candidate AHR's; 5) conducting AH morphism operations (including AHR synthesis whenever necessary) for object recognition.

For fast object recognition, location and tracking, the grey tone images are more appropriate. Special local features are extracted for 3-D shape synthesis to form an AHR or a 3-D random graph [60]. The knowledge acquisition can also be directed by "experts" through their interpretation of the object and scene or by the use of 3-D dimension information from CAD data. This process can be coupled with an autonomous rule generation procedure under development.

The point features of an object can be fused with surface profile information, when the same CCD camera is used to acquire range information on the surfaces of an object. A special procedure has been developed which relates the surface to a local reference derived from the position information of a subset of well-defined conspicuous points. The combined information can be organized and integrated into an AHR. Once the declarative knowledge of 3-D objects or scenes are represented in various forms of AGR and/or AHR, the representations can serve as models. They will be integrated into the vision knowledge system for fast retrieval and compression in the object recognition and location phase.

For recognition and location of objects using CCD cameras, the knowledge-directed search methods based on hypothesis refinement are used. The method provides reliable and efficient scene interpretation by means of incremental refinement of the 3-D scene interpretation. Any domain knowledge that could be exploited will be input to the rule network to reduce the context

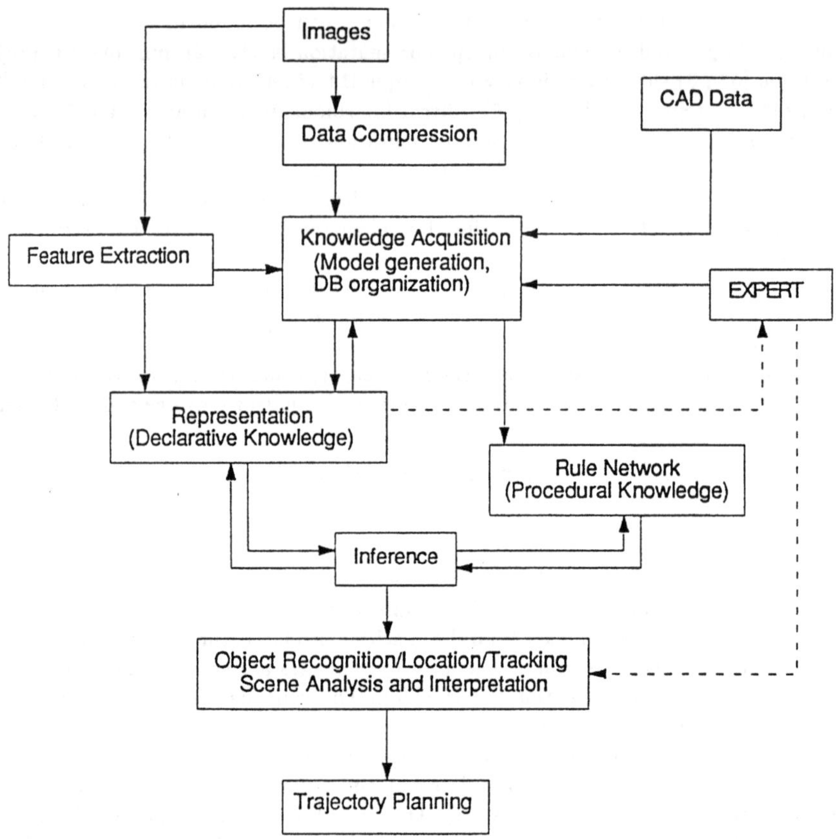

Figure 2: The Conceptual Configuration of the Vision Knowledge System

and to impose more specific constraints. The constraints provided by the scene interpretation guide the selection of key features used to establish object position accuracy. The vision system has been applied to robotic assembly tasks, scene interpretation for robot vehicle and visual inspection.

In case of conducting guaging or measurement on a 3-D surface profile, first the local reference of the objects of the scene relative to the camera or laser scanner has to be established. Then the local references obtained from the 3-D objects can be correlated with those stored in the knowledge base. Once the coordinate correlation has been established, direct comparison of corresponding points obtained from the range data with those in the model enables the system to measure the deviation of the observed surface from the model surface. Direct measurements of the surface relative to the local reference points of the object can also be conducted in the similar manner.

With the basic image analysis techniques and knowledge representation built into the system, on-line inferences for profile measurement and guaging, object shape synthesis and model generation, object recognition, location and tracking as well as scene interpretation can be achieved. Once the position information in the Euclidean space is acquired and inferred from the vision knowledge system, trajectory planning can be conducted to compute trajectories in the configuration space for various joints of the manipulator such that the entire posture of the manipulator can avoid collision with itself and any nearby objects while accomplishing an assigned manipulation or inspection task. In this section we shall describe some research activities leading to the development of such a system.

5 Knowledge Representation and Inference for Vision Systems

A representation of knowledge is a combination of data structures and interpretative procedures (*i.e.* inference) which when taken together will lead to *knowledgeable* behaviour [4]. Traditional AI thinking has divided knowledge representations into two major classes: declarative and procedural. Since almost no representation scheme is entirely one or the other, arguments about the relative value of one representation over the other are now of the past. But, the categorization does provide us with a way of grouping various representation schemes. The declarative representations encode knowledge as a collection of related facts. Procedural knowledge is embodied in the inference mechanisms which operate on these sets of facts. One group of declarative representations that will be looked at in detail is that which uses a certain type of graphs to capture the structural information in a domain. The emphasis of the declarative representation is the provision of a framework to store structural information in a most general, natural and comprehensive fashion. In principle, we wish the representation to be invariant. To utilize specific domain knowledge for effective inference, a certain form of procedural knowledge can be derived directly or indirectly from the objects or from their declarative knowledge.

5.1 Declarative Knowledge of Structural Representation: Graphs

In this section we shall focus on the knowledge representation of structural or relational models. The types of structural descriptions to be considered are: 1) semantic networks [4]; 2) graphs [9] including attributed graphs [56,2], hypergraphs [5,60], random graphs [33,64,57].

5.1.1 An Early Form: The Semantic Network Representation

The semantic network representation was originally proposed by Quillian [34] and Shapiro [45]. A semantic net represents information as a set of nodes interconnected by labeled arcs that represent relationships among the nodes. All semantic networks share the following features: 1) a data structure of nodes which represent concepts (generally in the form of a hierarchy of nodes connected by the ISA relationship) and other property links; and 2) specialized inferential procedures which operate on the nodes with the inheritance of information from the top levels of the hierarchy to the level of ISA links.

While semantic networks have been a successfully used knowledge representation in different application domains, some critical issues [12] still remain. For instance, how could information about classes be distinguished from information about instances of classes, or how could exception be handled. It has been found that various proposed *uniform default-style* representations still have serious flaws.

5.1.2 A More Basic Form: Attributed Graph and Hypergraph Representations

A graph can be seen as a more generalized form of the semantic network. A graph G is defined to be an ordered pair $(V(G), E(G))$ consisting of: 1) a nonempty set of vertices $V(G)$; 2) a set of edges $E(G)$ which is disjoint from $V(G)$ and 3) an incidence function Ψ_G which associates with each edge in $E(G)$, an unordered pair of vertices from $V(G)$ [9]. This type of structural description can be used to encode the low-level information about objects such as points, lines (edges), curves, and surfaces. It can also be used to encode relationships between parts of objects and between the objects themselves, for example: adjacency, intersection, union, and containment.

To furnish more specific graph structures for structural pattern representation and for effective inference, attributed graph representation (AGR), attributed hypergraph representation (AHR) and random graphs (RG) are introduced [61,56,64,57]. The mechanism for inference on graph representations is based on the retrieval, matching, recognition and transformation of graph and subgraph patterns. Usually, graph morphism algorithms are used for comparing graphs. A *graph morphism* between two graphs is defined as a one-to-one mapping from the vertex set of one graph onto that of the other, preserving their edge-to-vertex incidence relations. Such mappings can be defined to accommodate various constraints and optimality criteria according to the need and the nature of the problem.

An *attributed graph* is a graph $G_\alpha = (V_\alpha, A_\alpha)$ where $V_\alpha = v_1, v_2, ..., v_n$ is a set of attributed vertices and $A_a = ..., a_pq, ...$ is a set of attributed edges. The edge a_pq connects vertices v_p and v_q with attributed relation.

To represent 3-D objects or model, elementary area attributed and primitive block attributed graphs are introduced. An *elementary area attributed graph* $G_e = (V_e, A_e)$ is an attributed graph for representing a face or hyperpatch bounded by distinct and well-defined edges, where 1)V_e is a set of attributed vertices representing the boundary segments of the face and 2) A_e is the set of attributed edges representing the geometric relation between the segments. For example an angle of intersecting edge segments can be the relation between them. A *primitive block* of an object is a block bounded by surfaces such that there is no concave angular relation between any pair of the surfaces in the block. Hence a pyramid, a wedge, a cylindrical block can be a primitive block of an object. Then, a *primitive block attributed graph* is an attributed graph $G_p = (V_p, A_p)$ which represent a primitive block of an object. The attributed vertex set V_p represents the faces and the attributed edge set A_p represents the the geometric relations between faces.

In order to enable primitive or elementary features of an object to be grouped and organized in a hierarchical yet flexible manner, hypergraph representations are introduced into structural pattern recognition. A *hypergraph* [5,61] is defined to be an ordered pair $H = (X, E)$ where $X = x_1, x_2, ..., x_n$ is a set of vertices and $E = e_1, e_2, ..., e_m$ a set of hyperedges such that: 1) $e_i = \phi(i = 1, ..., n)$; 2) $Ue_i = X$ where X consists of a set of attributed vertices X_o and a set of hyperedges E_o. Each vertices is associated with an elementary area attributed graph representing a face (or hyperpatch), and each hyperedge is associated with a primitive block attributed graph representing a primitive block.

The AHR of a 3-D object can be constructed either manually using a CAD system or through a hypergraph synthesis process from the attributed hypergraphs derived from the range data of different views 3. When constructing the AHR for 3-D objects or their images, we can proceed in three stages: 1) construction of the elementary area attributed graph for each surface or hyperpatch; 2) construction of the primitive block attributed block attributed graph for each component block and 3) construction of the AHR for object(s) or image(s) by considering each surface as a vertex and each set of vertices associated with a primitive block as a hyperedge, the primitive block attributed graph being the attributed value of the hyperedge.

Using primitive blocks such as polyhedra or blocks with complete or partial cylindrical, conical or planar surfaces, we can construct complicated objects. In our system, each primitive block corresponds to a primitive block graph. The vertices, representing the faces of the primitive block, then form a hyperedge e_i. Here the complete AHR of an object model is given in Figure 3.

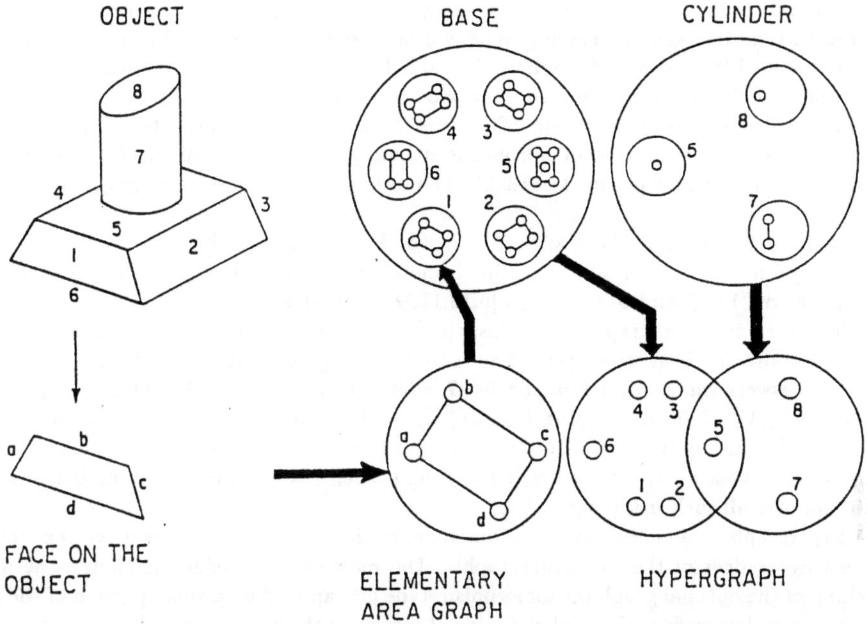

Figure 3: Attributed hypergraph representation of an object model.

For the recognition and description of objects in images with range data, a special form of AHR, known as Edge Feature Hypergraph Representation (EHR), the vertices of which are made up of edges or curves or line segments (observed or derived), is introduced [56]. For the derivation of 3-D information from objects and location of 3-D objects from 2-D images, another form of AHR, known as Point Feature Hypergraph Representation (PHR), the vertices of which consist of point features, has also been introduced. The PHR can take into consideration of the key point features as attributed vertices (real or projected features such as intersecting point between edges) and their 3-D spatial relationships as attributed edges. These feature points can serve as local references where surface profile information acquired from other structured lighting schemes or laser scanners can be integrated into the representation. The PHR can be directly represented by a parametric 2-D image in which the point features are treated as vertices, and distances or edge properties between points become the relation attribute values. Such a PHR is called an implicit image graph. The advantage of such representation is that it can be directly obtained from the parametric image without an additional symbolic representation construction process.

5.2 Graph Morphisms as an Inference Mechanism for Comparing Structures

Graph morphism can be described as the recognition that a graph, or one of its subgraphs, is embedded in another graph. These graphs will be designated as $G = (X, U)$ and $H = (Y, V)$ where G is called the *domain* of the morphism and H is called the *range*. Formally, a morphism, or mapping, of G onto H is denoted as a pair of mappings: $f = (\alpha, \beta)$ where α is a vertex morphism and β is an edge morphism. Therefore, $f : G \rightarrow H$ is equivalent to $\alpha : X \rightarrow Y$ and $\beta : U \rightarrow V$. There are several kinds of morphisms including graph isomorphism [17,18,25], subgraph isomorphism or monomorphism [15,51,2,22,21] and the largest common subgraph isomorphism [59].

Graph isomorphism is the mapping $f = (\alpha, \beta)$ is an isomorphism of $G = (X, U)$ onto $H = (Y, V)$ if and only if for all i and i' in X and $e_{ii'}$ in U there exists j and j' in Y and $e_{jj'}$ in V such that $\alpha(i') = j'$ and $\beta(e_{ii'}) = e_{jj'}$ [6,10,17,18,25,36,41,49].

A largest common subgraph isomorphism [59] is a one-to-one mapping between the subgraphs of two generally non-isomorphic graphs such that the largest number of incidence relations are preserved between them. More formally, let $G = (X, U)$ and $H = (Y, V)$ and $G_a = (X_a, U_a) \subseteq G$ and $H_a = (Y_a, V_a) \subseteq H$. A mapping $f_a = (\alpha_a, \beta_a)$ is the largest common subgraph isomorphism of G onto H if and only if for all i and i' in X_a and $e_{ii'}$ in U_a there exists j and j' in Y_a and $e_{jj'}$ in V_a such that $\alpha_a(i) = j$ and $\alpha_a(i') = j'$ and $\beta_a(e_{ii'}) = e_{jj'}$ where the number of $e_{ii'}$ is a maximum over all such mappings.

A hypergraph monomorphism [61] is based on maximal incidence preserving vertex matching between hyperedges of the two hypergraphs. The maximal hyperedge matching is actually a matching of the optimal graph monomorphism type [65] applied on a small portion of the graph represented by hyperedges. Instead of aiming at searching the entire sets of elements, it attempts to find component-component correspondence or subgraph isomorphism for both of the entire graphs. The hypergraph monomorphism provides a natural means to reduce the number of comparisons. For an object hypergraph $H_o(X_o, E_o)$ and a model hypergraph $H_m(X_m, E_m)$, there exists a monomorphism of H_o onto H_m if the following necessary conditions are satisfied:

1. each v_i in H_o is matched or partially matched by some vertices $v_{j'}$ in H_m in such a way that the elementary attributed graph G_{v_i} is monomorphic to G_{v_j};

2. each hyperedge in H_o and H_m is associated with a primitive block graph and for G_{e_i} associated with e_j in H_o, there is a monomorphism of G_{e_i} onto a primitive block graph $G_{e_{k'}}$ associated with $e_{k'}$ in H_m.

Figure 4 gives an example of 3-D object recognition based on attributed hypergraph monomorphism.

Finding the isomorphism between a graph and a subgraph of another graph belongs to the class of NP-complete problems. For the case of attributed graph, the average complexity can be greatly reduced if the contextual information and the structural relation between AGR's could be exploited. The specificity of the attributes and the attributed relations usually reduces the search space if vertex ordering [17], pruning [18] and branch-bound heuristics are appropriately applied. However the average complexity is still a function of the sixes of the graph. In a situation when there are many identical attributed vertices in an object, the vertex ordering method can do little for reducing the time complexity. One of the reasons of introducing the attributed hypergraph representation and monomorphism is to organize the graph into appropriate components such that part of the morphism finding process can be restricted to the comparison of subgraphs depicted by hyperedges.

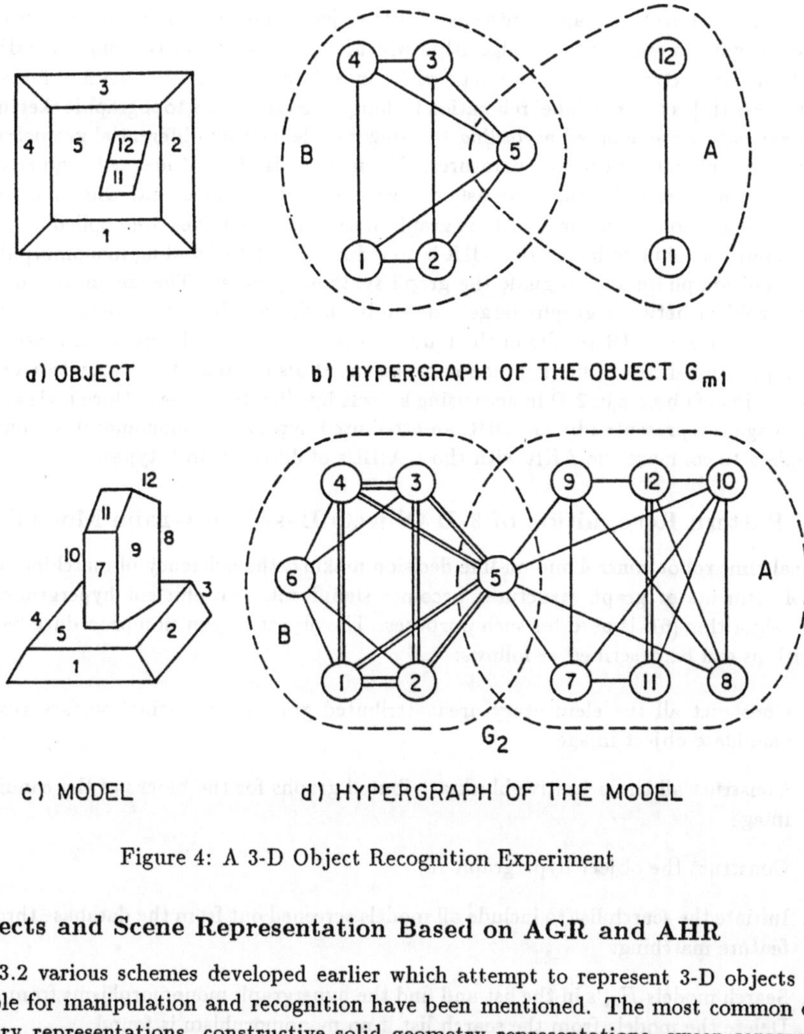

Figure 4: A 3-D Object Recognition Experiment

5.3 Objects and Scene Representation Based on AGR and AHR

In Section 3.2 various schemes developed earlier which attempt to represent 3-D objects in a form suitable for manipulation and recognition have been mentioned. The most common ones are boundary representations, constructive solid geometry representations, sweep representations and decomposition representations. Most of these schemes are feasible for acquiring the geometric information from the object image, yet they lack the flexibility for effective recognition if the orientation of the object varies, or certain parts of the object are occluded, or when the class of prototypes is large. Furthermore, in most of these schemes, the knowledge of the prototype object has to be input by the users.

In our system, we adopt attributed graph and hypergraph representations of 3-D objects (AGR and AHR) [61,56]. This data structure renders a very general representation of objects or scenes with high complexity. It provides a means to group components or parts of an object according to the relation that induces the hyperedges. Thus, the same object can be described in different ways depending on how the hyperedges are formed without affecting the basic primitive and primitive relations. This representation satisfies the requirement of the descriptive adequacy.

With respect to the shape synthesis of 3-D objects and models, we have developed a 3-D object recognition and synthesis algorithm which is capable of constructing an AHR based on the information extracted from an image with range data [61]. Another method for representing 3-D surfaces [11] is to use a label relaxation technique to estimate a topographic sketch consisting of surface patches segmented according to categories defined by differential geometry operators such as the gaussian and mean curvatures. The method is global (does not require the use of a local operator for classification), robust to noise and easy to implement. Once the segmentation is done, recognition procedures such as graph morphism techniques are applied.

The purpose of introducing the AHR is to reduce the cost of finding monomorphisms during the recognition phase and to guide the graph synthesis process. The complexity of finding the monomorphism between graphs largely depends on the number of vertices and arcs in these graphs. The use of AHR results in the reduction of the number of vertices and arcs, and hence the computational cost of finding monomorphisms. It also enables the recognition of the spatial configuration of objects in 2-D images using knowledge-directed search. Once a view of an object in an image is represented by an AHR, an attributed hypergraph monomorphism algorithm can be applied to compare the AHR with those AHR's of different prototypes.

5.4 Pattern Recognition of 3-D Objects Based on Graph Morphism

For real-time robot control and on-line decision making, the efficiency of matching between two sets of elements or graph structures becomes significant. An efficient hypergraph monomorphism algorithm [61] is used for such purposes. The object recognition procedure based on such algorithms can be described as follows:

1. Construct all the elementary area attributed graphs for distinct surface regions of the candidate object image.
2. Construct all the primitive block attributed graphs for the blocks of the candidate object image.
3. Construct the object hypergraph H_0.
4. Initiate the search list to include all models screened out from the database through special feature matching.
5. Search models H_m's in the list and find the hypergraph monomorphisms from H_0 to H_m's. Delete the models from the search list if no monomorphism is found.
6. If only one model is found then the object is recognized.
7. If monomorphisms from H_0 are found for more than one H_m's, obtain the hypergraph from the image of another view of the object and synthesize it with the previous H_0 (the hypergraph synthesis method is provided in Section 5.5).

5.5 Hypergraph Synthesis

From the image of each view of an object, we obtain an AHR which represents the geometric structure of only those edges and faces of that object visible from the vantage point of the laser scanner. We call that hypergraph an Image View Hypergraph (IVH) of the object. To gather more information, several images obtained from different views of a 3-D object should be used. We have developed a method by which AHR's obtained from different views can be combined (synthesized) to form a single AHR that yields an AHR of the entire object.

(a) The laser images for the object from several views.

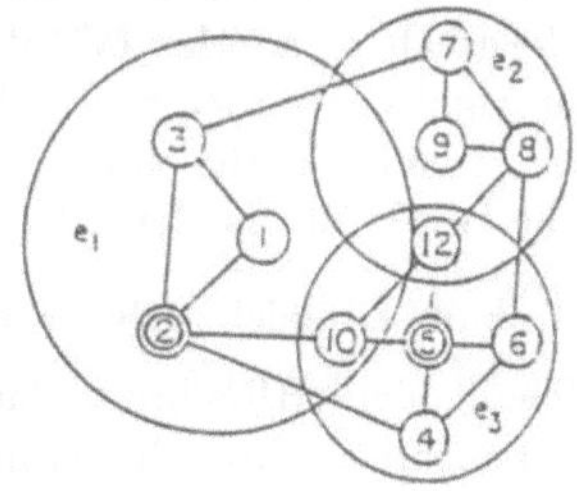

(d) SYNTHESIZED AHR FOR IMAGES IN (b)

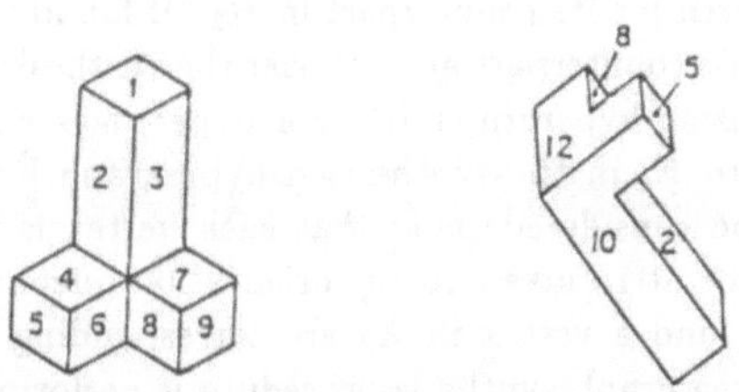

a) IMAGE 1 b) IMAGE 2

(b) Two different images with face labels.

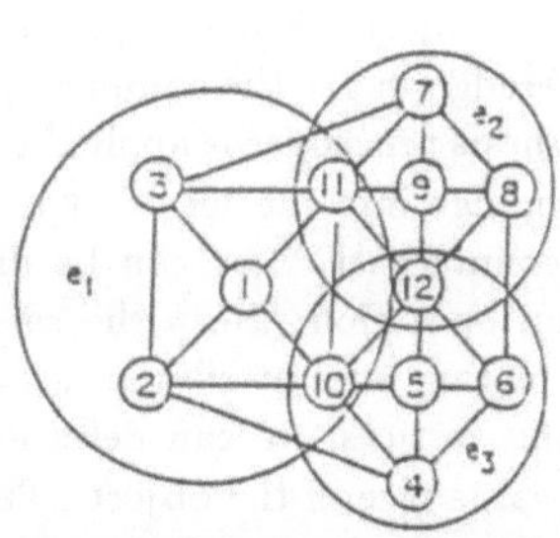

(e) SYNTHESIZED AHR FOR THE OBJECT

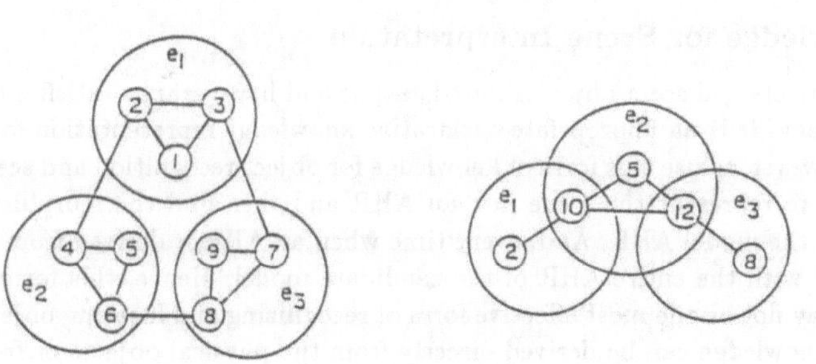

(c) IMAGE VIEW GRAPHS FOR IMAGES IN (b)

Figure 5: Hypergraph synthesis experiment.

We introduce the hypergraph synthesis for the following two purposes:

1. To combine two IVH's of a candidate object image in the object recognition process.

   When an IVH of a candidate object is monomorphic to two or more model hypergraphs in the database, another IVH should be obtained for further recognition in order to resolve the ambiguity. We then synthesize the two IVH's into one which contains all the information obtained about the object from both views. The comparison of the AHR synthesized from the IVH's with model AHR's in the database may yield a unique monomorphism.

2. To build a model AHR for a new object in the learning phase.

   The model AHR's in the knowledge base could be constructed from direct physical measurements or through learning using information derived from images of different views. Thus, several IVH's of the new model can be derived and synthesized into a model AHR.

The hypergraph synthesis procedure can be briefly described as below.

For two hypergraphs $H_1(X_1, E_1)$ and $H_2(X_2, E_2)$, the synthesis process is organized in two stages to obtain the synthesized hypergraph $H_r(X_r, E_r)$. First, two sets of hyperedges are considered. Let $e_p \in E_1$ and $e_q \in E_2$. If e_p and e_q correspond to an identical primitive block of the object, then the primitive block graph synthesis can be applied to them. For each hyperedge in H_1 the comparison is performed to search for its counterpart in H_2. If found the synthesis procedure is applied to that hyperedge and its counterpart and transfer the synthesized hyperedge into the set of hyperedges E_r in the resulting hypergraph H_r. For hyperedges with no counterpart, they can be directly transferred into E_r in the synthesized hypergraph by a union operation. Next the sets of vertices are to be considered (note that each vertex is an elementary area attributed graph). By comparing the attributes and the orientation reference of the vertices, we can determine if a vertex in X_1 and a vertex in X_2 are corresponding to the same face of the object. Then the elementary area graph synthesis procedure is performed on those vertices, and as a result the synthesized vertex can be transferred into the resulted hypergraph H_r. Finally a new adjacency matrix for the resulted hypergraph is constructed. Figure 5 demonstrates a hypergraph synthesis experiment given in [61].

5.6 Procedural Knowledge for Scene Interpretation

The representation of 3-D objects and scenes by attributed graphs and hypergraphs satisfies the criteria of descriptive adequacy. It is an appropriate declarative knowledge representation for a computer vision system. However, to use this form of knowledge for object recognition and scene interpretation, one has first to represent the scene into an AHR and then find the morphisms between the scene AHR and the model AHR. And, every time when an AHR is derived from an image, it has to be matched with the entire AHR of the candidate model. Hence, this form of knowledge representation may not be the most effective form of recognizing and locating objects in real time. If procedural knowledge can be derived directly from the physical objects or from the model AHR's of objects, they can be used to guide the search and inference process in a more focusing and direct manner. In this section, we present a robot vision system which uses a special form of procedural knowledge.

The term *procedural* implies that this representation is chiefly concerned with encoding the knowledge of *how* to do some actions or to execute certain procedure. It captures information which is not very easily described as a set of facts. It then provides a control mechanism for more effective search and inference. Control tasks for a vision system could include:

1. recognition - coordination of information and inference sources

2. planning - organization of system behaviour
3. calculations - organization of information needed to calculate object orientation, location and range

There is an increasing need to use procedural knowledge in the robot vision system. Emerging interest in vision guided robotics and autonomous robot vehicles requires effective analysis of complex scenes within the limitations imposed only by the natural constraints of the environment. Attempts to provide effective vision must deal with missing or extraneous features due to glare or shadow, partial or total occlusion by other objects or the robot arm itself, and the fundamental ambiguity resulting from projection of the original 3-D scene onto a 2-D image. Many of the traditional approaches to machine vision perform poorly under such conditions and will generally fail if the image representation is too impoverished (too few visible object features) or if features are concealed by noise.

The solution to providing effective vision is to adopt a problem oriented approach in which specific domain knowledge is used extensively for scene interpretation such that the use of the knowledge of the spatial relationship and the topology of the model and image features is exploited. The key to the success of such an approach depends on the ability of a vision system to:

1. have the various types of knowledge represented effectively, and
2. integrate the different knowledge representation schemes together such that the use of the various types of knowledge is effectuated.

In the vision knowledge system, we integrate the declarative knowledge with the procedural knowledge. Based on the spatial and geometric relation and visual features, AHR's are constructed and used as the basic data structure to represent declarative knowledge. A rule network is employed to represent procedural knowledge for shape synthesis, feature extraction, object recognition and scene interpretation. The scene interpretation process is a step by step hypothesis refinement process represented as a path through a subset of nodes in the network. The refinement of a new hypothesis is dependent on the declarative and procedural knowledge associated with the rule node of the search path. The entire process from image acquisition to hypothesis refinement is guided by the rule network and is referred to as knowledge-directed search.

5.7 Model Generation

Due to the flexibility of the AGR and AHR in representing 3-D objects, a shape synthesis process has been developed. In both the object recognition and the knowledge acquisition processes, often several images obtained from different views of a 3-D object are used. By this process, AGR's or AHR's obtained form different views can be combined (synthesized) to form a single AGR or AHR that yields a graph representation of the entire object.

Using the graph or hypergraph synthesis process, a unique model graph can be obtained for an object. This approach is superior to the alternative approaches which require, for a simple object, several representations corresponding to its various views. As a result, it is difficult for them to determine whether or not the representation is complete.

The AHR based on the structural decomposition view of an object would furnish a systematic and comprehensive way of representing objects and models. However, this representation may not render an economical way for determining graph morphisms or for providing a natural guide for the construction of the object recognition rule network. To reduce the complexity of the graph morphism problem, appropriate decomposition of AHR's is desirable. Since in each of the comparison processes only one characteristic view of the object features will be involved, disregard whether the comparison method is based on knowledge-directed search or on hypergraph morphism; it would be possible to introduce a new set of hyperedges which will be induced based on various "optimal" characteristic views. We call these subsets of attributed vertices characteristic view hyperedges. Thus, the model generated can be represented by two concurrent sets of hyperedges, one based on structural decomposition and the other based on characteristic decomposition. The intersection of these two hyperedges will yield another set of hyperedges, each of which is of even smaller size. Hence, the complexity of the graph morphism problems will be further reduced. The characteristic hyperedges of an object can provide another piece of domain knowledge in the autonomous construction of the rule network for the knowledge directed search.

5.8 Hypothesis Refinement and Knowledge-directed Search

Image analysis or interpretation may be viewed as a search of scene interpretation hypotheses which can be expressed as the possible location and orientation of modelled objects within the 3-D scene. Hypothesis refinement attempt to reduce the search space by conducting all search within the context implied by the current scene interpretation. The search context defines the natural constraints on object position relative to one or more surfaces in the 3-D environment.

Scene interpretation consists of two processing stages given below.

1. Establish the initial scene interpretation and camera viewpoints

   The position of each camera relative to the coordinate system of the scene interpretation is established through object recognition under unconstraint viewing conditions or may be obtained through a priori knowledge of the camera position relative to the support surface.

2. Refine scene interpretation with known camera viewpoint

   (a) Select (i) a support surface of the current scene interpretation; (ii) an object model and (iii) an appropriate support condition for the given task to limit the orientation of the object relative to the support surface.

   (b) For an object model feature, select corresponding image features by filtering according to the visibility characteristics defined by the camera viewpoint and support conditions.

   (c) For each model-image feature correspondence calculate the resulting constraints on object position. Repeat until object position is completely constrained.

   (d) Test each object position hypothesis by predicting the location of image features and testing for a valid match within the position tolerance of the corresponding model feature.

   (e) Update the scene interpretation and obtain the new support surfaces provided by the identified models.

The success of an analysis depends to a large extent on the ability to efficiently filter features in the image. Filter capability is determined by the nature of the model and the level of constraint provided by the support conditions.

Calculation of object position is dependent on the quality of the image features and the nature of the constraints of support conditions. The available constraints simplify calculation of object position and permit the use of feature with very limited information content (for example, under restricted support conditions, a single edge feature can completely constrain object position).

The procedure for testing a model hypothesis accommodates for the measurement accuracy of each feature of the object model and can directly access image data to bypass errors in low level feature detection. The object model also defines the conditions necessary for creating new support surfaces.

The process of exhaustively searching all model position hypotheses under all possible support conditions is maintained through the use of a control network. Each stage of the analysis is guided by the *search context* which defines the current scene interpretation and assumed search constraints.

Figure 6 illustrates how the support condition of an object could impose possible constraints of model features. Figure 7 demonstrates how such constraints can be used to define a filter which is able to extract the desired features under the hypothesized view point of the camera and the support conditions of the objects.

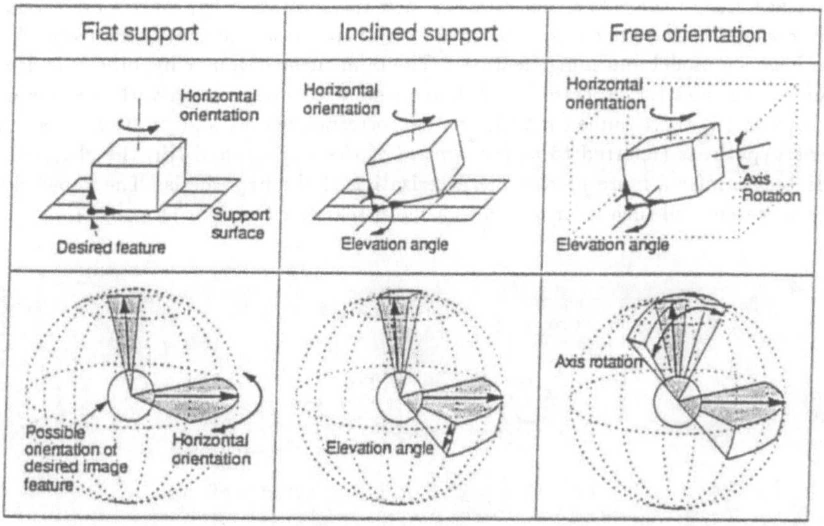

Figure 6: Use of support conditions to select image features.

The process of refining an interpretation hypothesis can generally be described by an ordered sequence of well-defined steps or stages. In actual practice, processing is more complex. Input to the systems may consist of multiple images or viewpoints, and additional workspaces may be defined with multiple object models and refinement strategies. An analysis of the broad range of possibilities is managed through a *knowledge-directed* search. The search is directed by the knowledge or context of the task or environment. This is similar in spirit to the "knowledge-directed image analysis" of Ballard, Brown and Feldman [3] in which image understanding was directed by a search query from the person using the system. The knowledge-directed search is guided by a rule network consisting of a network of *rule nodes*. Each step by step refinement process, or refinement strategy, is represented as a path through a set of nodes in the network.

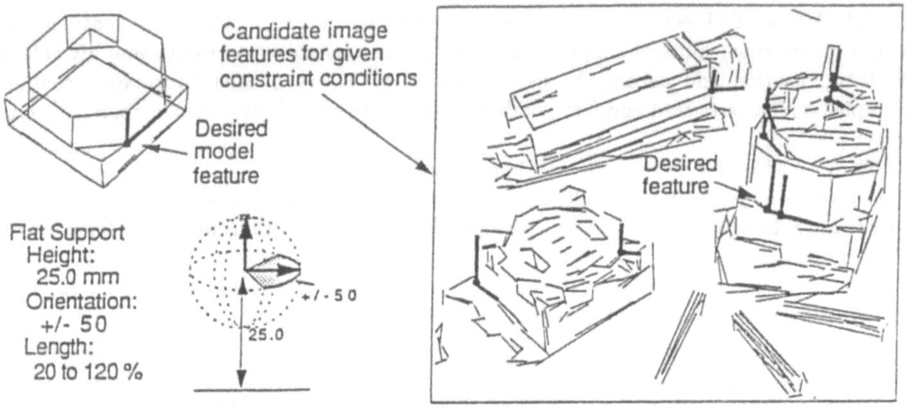

Figure 7: Selection of candidate image features.

At each step of the refinement strategy, new interpretation hypotheses are produced. Such hypotheses determine the constraints on the object position and specify an assumed correspondence between model and image features. The refinement of a new hypothesis is dependent on (i) the search rules (or processes) and domain knowledge associated with the rule node of the search path, (ii) the assumed model, feature correspondences, and position constraints of the current hypothesis (referred to as the *context* of the search), and (iii) the observed data that is selected to infer a more precise characterization of the hypothesis. The knowledge-directed search represents all such information as a set of search activations (Figure 8).

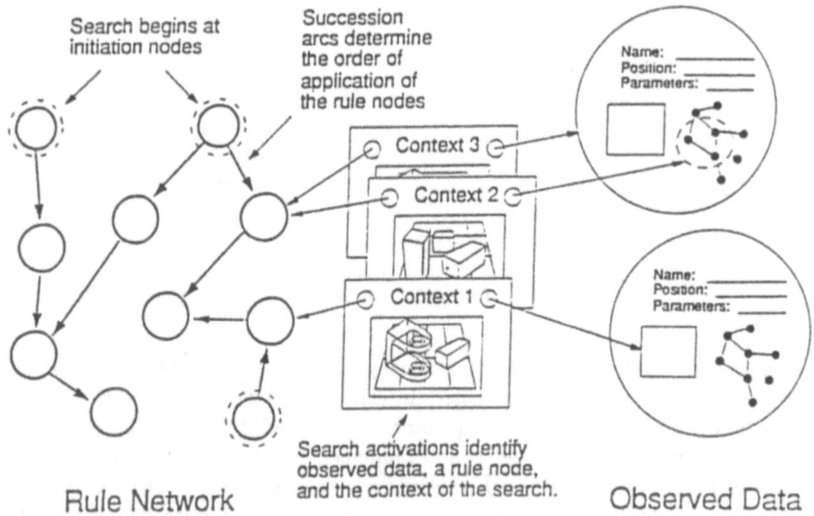

Figure 8: Rule Network and Search Activations

Each *search activation* is a relation that uniquely associates a node of the *rule network*, a set of *observed data*, and the *context* of the search.

Each node is an instance of a class of rule nodes which perform a specific task (e.g. acquire an image, detect image features or specify an object model, constraint and refinement strategy). The rule node classes employed in the present system are listed in Table 1.

Image Acquisition	**Obtain image from specified source (camera or disk file) and set camera parameters (focal length, imaging plane, focus setting).**
Line Detection	**Obtain line features according to specified limitations (line length, contrast, straightness, resolution).**
Camera Position	**Obtain camera position and orientation relative to the position reference.**
Feature Detection	**Obtain image features (line end points and corner features).**
Model Definition	**Define 3-D object model and interpretation constraints.**
Hypothesis Refinement	**Refine object position and orientation to verify possible hypotheses.**
System Control	**Select appropriate search strategies and object models. Construct a consistent world model. Interact with other processes.**
Model Editor	**Edit object models and examine the current scene interpretation. Obtain visual measurement of 3-D object dimensions. Interpret constraint conditions and define image filters for 3-D features.**

Table 1: Rule Node Classes

Each class has an associated set of processes and defined data structure for the input search context. Each instance of a rule node acts on the input context and observed data to produce any number of output context records. In addition, each instance of a rule node has a private internal memory to record specific parameters of the instantiation of the rule node or to compile a history of results or information acquired.

As context information is transmitted through the network, additional information is supplied by each node to eventually provide an interpretation of the world viewed by the various cameras of the vision system.

The general structure of the rule network is illustrated in Figure 9. The network layout is flexible. Any number of instances of each class of node may be defined. For example, there may be any number of image nodes, each image can be assigned specialized edge or feature detectors, and various models or refinement strategies may be defined.

5.9 System Implementation

PAMI Group's current vision system based on the knowledge-directed search provides analysis of single or multiple perspective images to locate parts visible in the workspace of a robot workcell. In its present configuration, the system consists of a SUN 3/160 workstation, a Matrox video

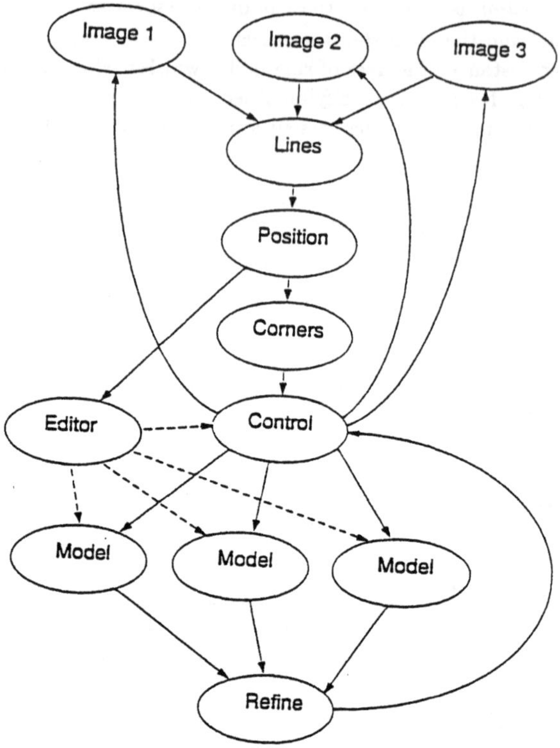

Figure 9: Structure of Rule Network

frame grabber, and from one to three SONY CCD video cameras. The system has access to a PUMA 260 and a Universal Machine Intelligence RTX robot arm. Each digitized video frame can be directly accessed by programs running on the SUN workstation. Approximately ten 512x480 images may be acquired each second. No additional special purpose hardware is required. Image processing, feature detection, and analysis are implemented in software in portable 'C' code and can be run on various computer systems (without modification on systems supporting the X Window graphics environment [40]). The system is capable of processing multiple camera/viewpoint input in a few seconds.

6 Applications

In this section, we present a number of experiments which were set up to demonstrate the potential and technical feasibility of the developed robot vision knowledge system in manufacturing and space environment. Most of these results are taken from [61,28].

The first example illustrates the use of visual guidance for robotic assembly. In this stage of the assembly, the arm must place the top casing of the motor onto the core winding and over the motor shaft. To achieve this task the system has been supplied with models of the motor casing and the partially assembled motor. Manipulation of objects is specified relative to the coordinate systems of the object models without concern for the absolute position of each object. For example, the required assembly task is expressed by the command shown below.

```
pickup ( from ((motor top)) to (motor base) offset (0 0 5 0 1 0) )
```

The offset specifies the position and orientation of the motor top relative to the motor base. The method used to grasp the motor top is automatically derived from the model definition.

Figure 10 (a) is the initial view of the scene. The motor components may be placed anywhere within the scene, although it is assumed that they have the correct vertical alignment for assembly. Figures 10 (b) and 10 (c) show the scene interpretation from the original viewpoint and as it would appear viewed from above the workspace (a view normally obscured by the position of the arm).[1]

Figure 10 (d) illustrates the actual assembly of the motor. Figure 10 (e) is the corresponding interpretation of the scene verifying correct assembly of the motor. Note the narrow tolerance between the motor shaft and upper casing—position error must be less than 2mm for successful assembly. Figure 10 (f) demonstrates the interpretation of a scene resulting from an assembly error. Detection of each object requires a few hundred milliseconds of processing time with 1.3 seconds required to complete analysis of the initial scene. Edge detection and calculation of camera position require an additional 2.5 to 3 seconds of processing time for each image. Note that errors in the absolute range due to inaccurate estimate of the camera focal length or other factors do not have an impact on the accuracy of assembly as all measurements are *relative* to the reference cross pattern.

The second example illustrates integration of multiple camera input shown in Figures 11 (a) and 11 (b). A mirror placed in the workspace provides an additional view of the scene. The mirror is purposely inclined relative to the table surface so that reflections of symmetrical objects are not mistaken for objects behind or 'inside' the mirror's surface. Although not shown, the position target was initially displayed to calibrate the positions of each camera. Consequently, it is possible to integrate the scene interpretations. (Note that the target provides the reference for the mirror image—it is not necessary to identify the actual mirror which produces the image.)

The interpretation of each image as shown from the original viewpoints are illustrated in Figures 11 (c) and 11 (d). Note the detection of the block hidden by the stacked block and cassette case and the detection of the transparent cassette case partially hidden by the box and only clearly visible in the mirror image. Figures 11 (e) and 11 (f) show an overhead view of the scene interpretation and a view from the opposite side of the workspace (as viewed from 'behind' or 'through' the mirror). Note that the mirror itself is not detected as no model of it is provided.

The final example is concerned with an error recovery task in printed circuit board assembly. An image of a printed circuit board is examined to test for improperly inserted components. Since the board is viewed from the side, as shown in Figure 12 (a), any insertion errors can be easily detected. Figures 12 (b) and 12 (c) show the scene interpretation from the original viewpoint and from a position above the board. Note that since flat support conditions are assumed, the position error appears as a change in horizontal orientation (shown more clearly in Figure 12 (d)). Given the available image resolution, the actual orientation cannot be accurately determined (the assumed hypothesis matches the model features precisely); however, the precise location could be obtained from knowledge of the true horizontal orientation and use of inclined support conditions.

[1] The object model is superimposed on the detected edges and identified by the object model name. All lines of the object model are shown giving the effect of a wireframe or transparent object.

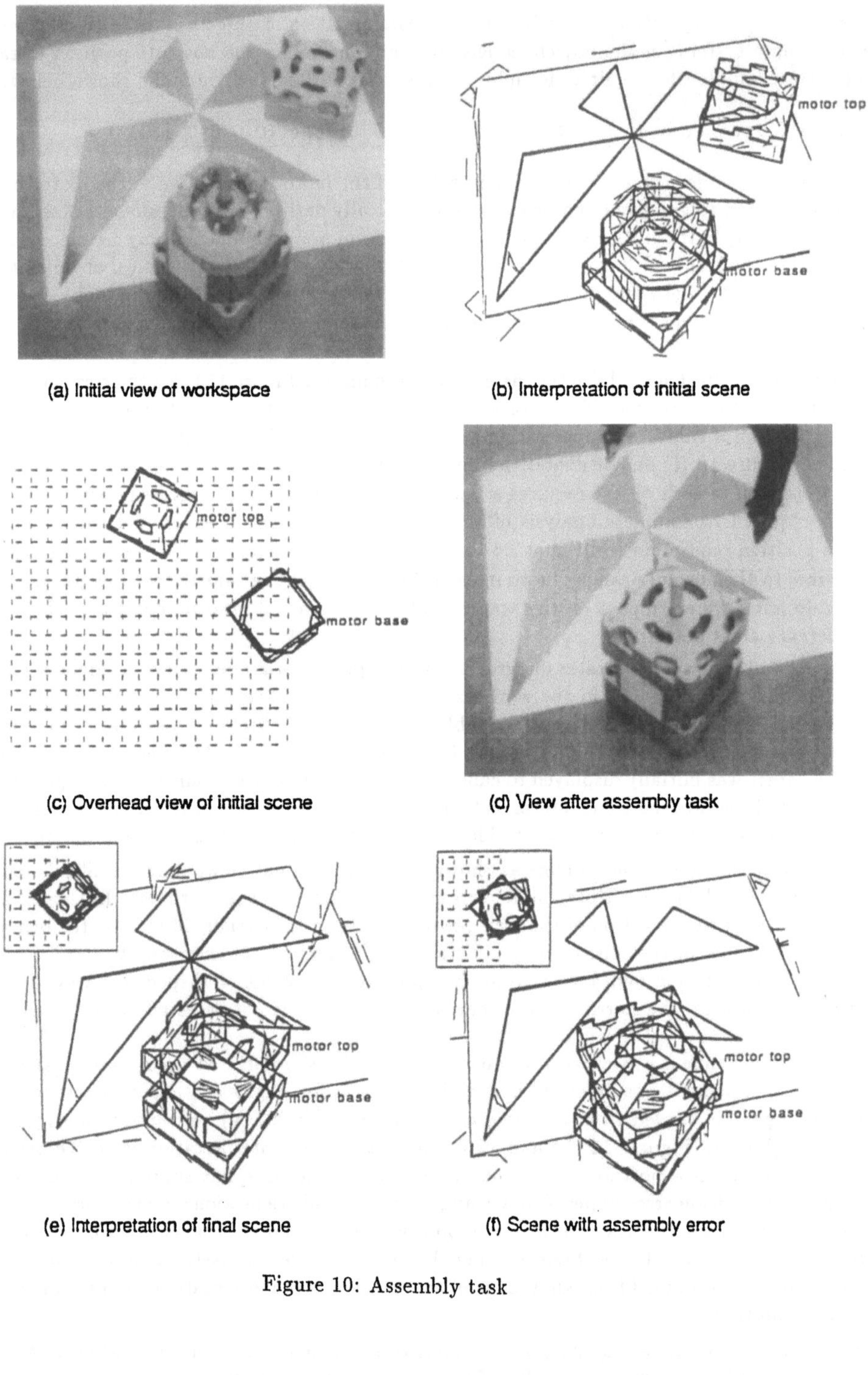

(a) Initial view of workspace

(b) Interpretation of initial scene

(c) Overhead view of initial scene

(d) View after assembly task

(e) Interpretation of final scene

(f) Scene with assembly error

Figure 10: Assembly task

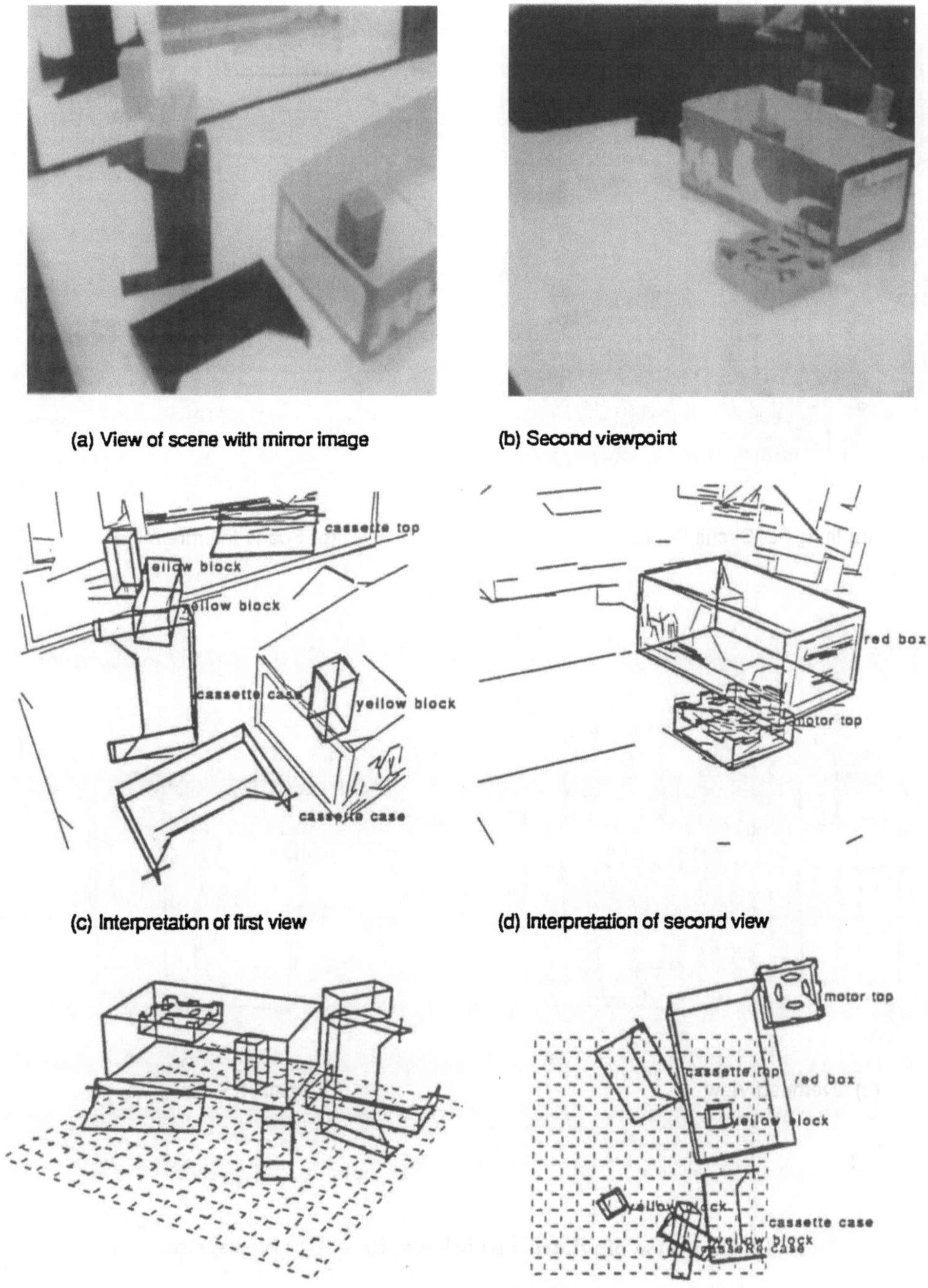

(a) View of scene with mirror image
(b) Second viewpoint
(c) Interpretation of first view
(d) Interpretation of second view
(e) View 'through' mirror
(f) Overhead view

Figure 11: Integration of multiple viewpoints.

(a) Image of Circuit Board

(b) Scene Interpretation

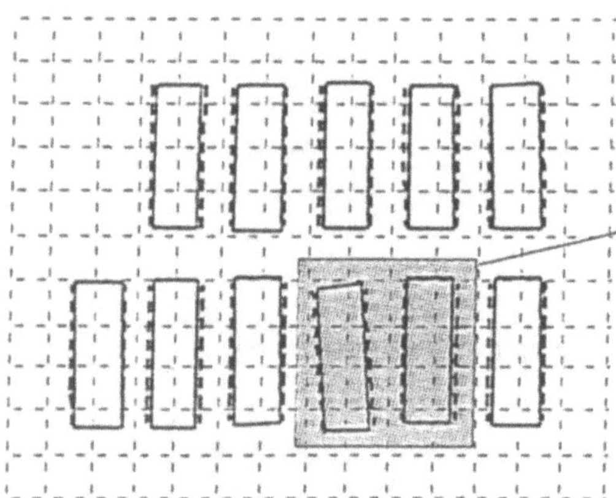

(c) Overhead View

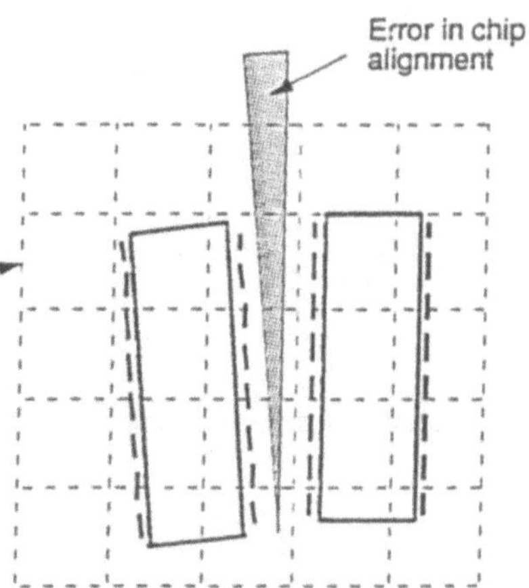

(d) Position Error

Figure 12: Inspection of printed circuit board by a 3-D vision system.

6.1 A Vision Based Local Path Generation

In this example, we demonstrate how the computer vision system is involved in generating a local path for a robot manipulator with obstacle and singularity avoidance capabilities [28]. The path generation system achieves singularities avoidance by establishing proper bounds for the rate of change of the Jacobian matrix representing the transformation between the joint speeds and the end effector Cartesian speed. These bounds become additional constraints for an optimization problem formulated to obtain the optimal path for the robot manipulator.

The identification and precise location (Figure 13) of industrial parts, components of an A.C. motor, is accomplished successfully by building up the correct 3-D models using search strategies to recognize and locate the motor top and the motor base. An unsymmetrical frame structure (Figure 14) is also built and placed in the workcell as an obstacle in order to provide an extra test to the path planning algorithm.

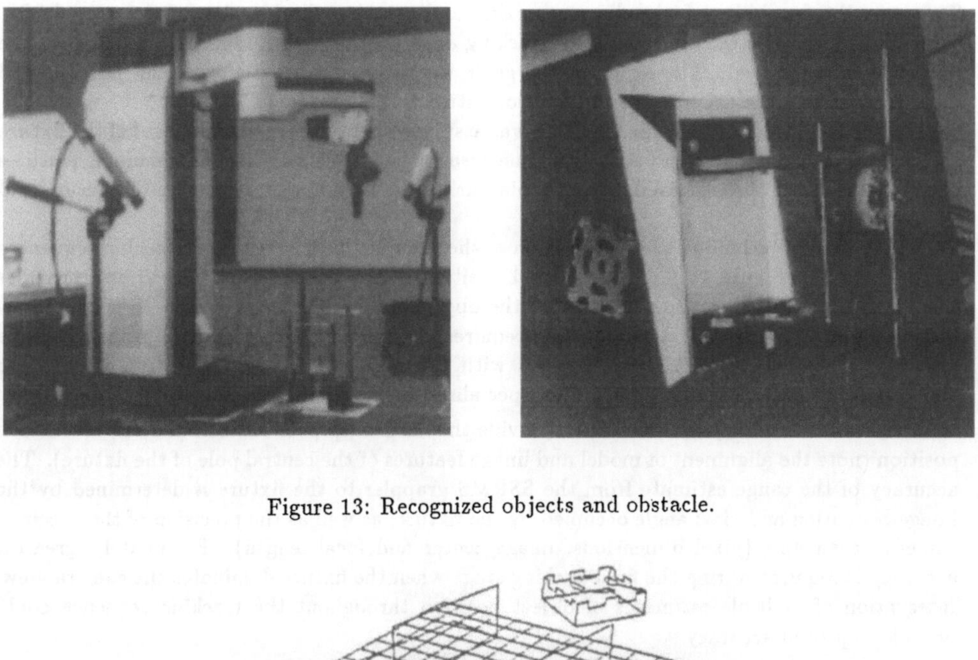

Figure 13: Recognized objects and obstacle.

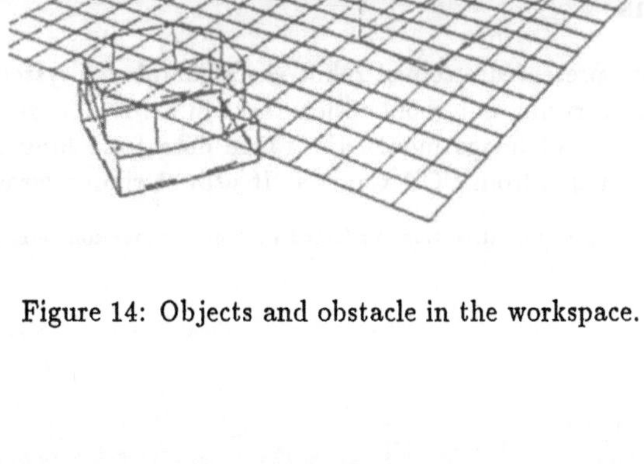

Figure 14: Objects and obstacle in the workspace.

The experimentation is designed with the model top and motor base located randomly on the opposite side of the reference pattern which is approximately placed in the centre of the RTX robot work space. After the reference pattern is located, the frame structure is set between the motor top and the motor base. The local path planning program gets the position of these parts and evaluates the path for transporting and placing the motor top on the motor base bypassing the frame structure. Currently, the obstacle avoidance is implemented under the assumption that a Cartesian path with enough clearance from the obstacle would result in an obstacle free path in configuration space. This assumption is only valid for a specific class of task environments. The experimentation so far shows that a real time robotic local control package in a computer based environment is readily available while a real time vision guided robotic local control package in which the obstacle avoidance trajectories of all joints are properly evaluated is still in development.

6.2 Tracking of the Grapple Fixture for a Space Robot Arm

To assess the feasibility of visually recognizing and tracking the location of the grapple fixture used by the SSRMS, simple visual tracking experiments with a non-flight mockup of the fixture were conducted. The experimental goals were to determine the ease and reliability of identification and the accuracy of the position estimate.

For the conditions of the experiment, it was assumed that only the visual target of the fixture (black rectangle with circle and post) would necessarily be visible and that the remaining features of the fixture could be obscured, for example during the final stages of mating the fixture and SSRMS grappler.

Applying this technique to identification of the grapple fixture results in tracking sequences as illustrated in Figure 15. The calculated position of the fixture is calculated and provides the appropriate transform to superimpose the object model over the detected object features. Identification of the fixture is reliable and requires approximately 2 seconds of processing time per image on the SUN 3/160 workstation with the majority of processing time devoted to detection of edge features (without use of specialized image processing hardware).[2]

Orientation accuracy is sufficient to provide the fixture tip position within 1 cm of the true position (note the alignment of model and image features of the central pole of the fixture). The accuracy of the range estimate from the SSRMS grappler to the fixture is determined by the image resolution and view angle occupied by the fixture, as well as the precision of the specified camera parameters (pixel dimensions, image center and focal length). Fortunately, greatest accuracy is required during the final docking stage when the fixture dominates the camera view. Integration of multiple estimates of object position throughout the tracking sequence could provide improved accuracy.

7 Conclusions

In this article, we have presented a robust robot vision knowledge system which demonstrates real-time capability in carrying out robot vision tasks in a robot workcell environment. The system accepts two types of image input: the range data from laser scanner or structured lighting and greytone images from CCD Camera. It uses attributed graph and hypergraph as

[2] Processing time, including feature detection is affected by overall image complexity.

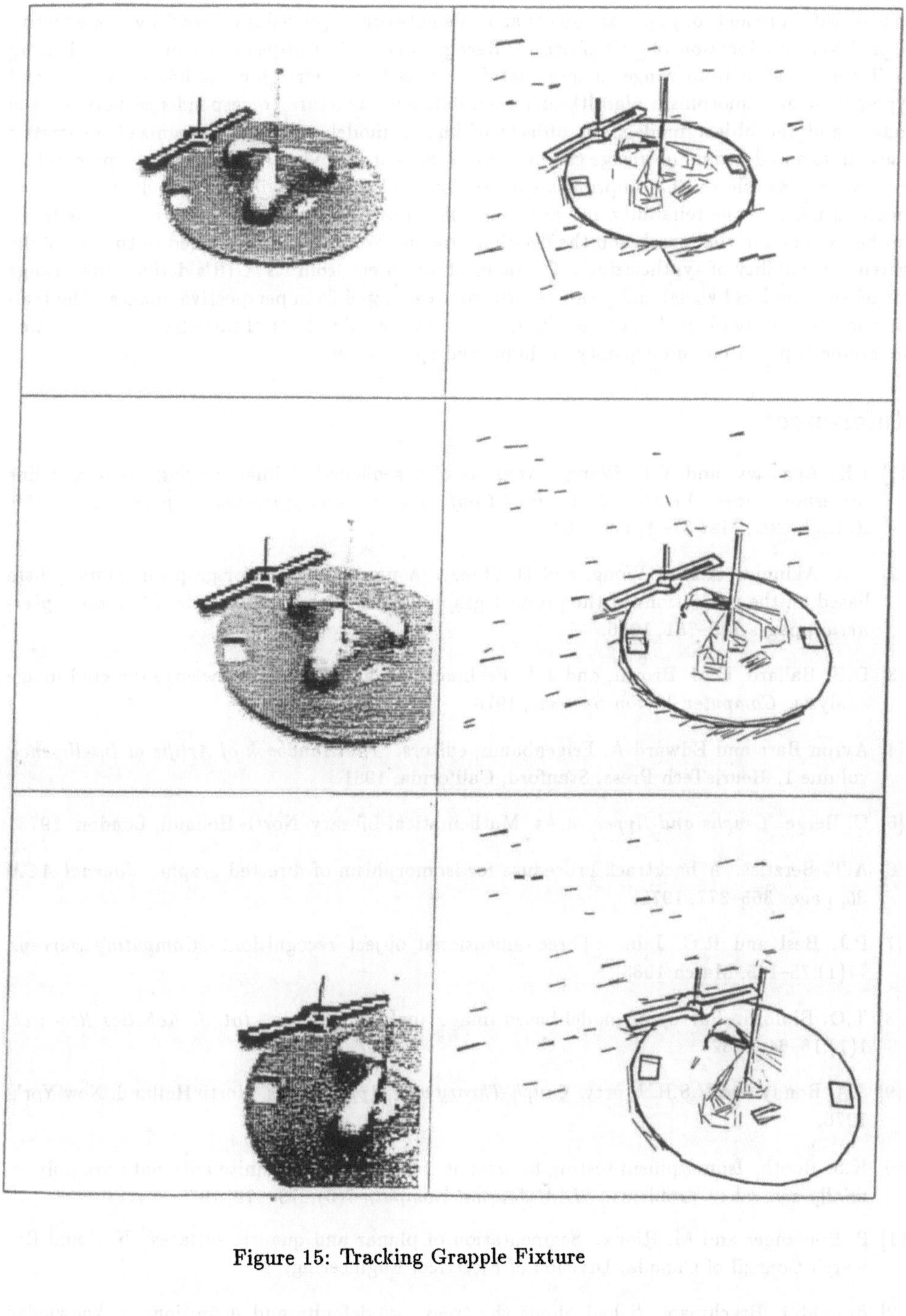

Figure 15: Tracking Grapple Fixture

representation models of physical objects and rule network as procedural knowledge for effective recognition and location of 3-D objects. Recognition of 3-D objects can be achieved using AHR construction from range images; database search for retrieving candidate models; and hypergraph monomorphism algorithms for establishing structure correspondence between the images and the object models. If objects of known model are to be recognized, geometric constraints and domain knowledge can be used in the search of visual features on the perspective images. We have shown that appropriate use of domain knowledge will greatly reduce the search time and increase the reliability and accuracy since noise from various sources at various levels can be excluded in the search and the decision process. We have also described in the paper the system's capability of synthesizing 3-D shapes of an object from its AHR's derived from range images or from local visual and geometric features extracted from perspective images. The real-time and reliable performance of the vision system to several industrial tasks have demonstrated the system's potential and capacity for industrial application.

References

[1] J.K. Aggarwal and Y.F. Wang. Analysis of a sequence of images using point and line correspondences. In *1987 IEEE Int'l Conf. Robotics and Automation*, pages 1275–1280, Raleigh, NC, Mar 31–Apr 3 1987.

[2] F.A. Akinniyi, A.K.C. Wong, and D. Stacey. A new algorithm for graph monomorphism based on the projections of the product graph. *IEEE Trans. on Systems, Man and Cybernetics*, pages 740–751, 1986.

[3] D.H. Ballard, C.M. Brown, and J.A. Feldman. An approach to knowledge-directed image analysis. *Computer Vision Systems*, 1978.

[4] Avron Barr and Edward A. Feigenbaum, editors. *The Handbook of Artificial Intelligence*, volume 1. HeurisTech Press, Stanford, California, 1981.

[5] C. Berge. *Graphs and Hypergraphs.* Mathematical Library. North-Holland, London, 1973.

[6] A.T. Berztiss. A backtrack procedure for isomorphism of directed graphs. *Journal ACM 20*, pages 365–377, 1973.

[7] P.J. Besl and R.C. Jain. Three-dimensional object recognition. *Computing Surveys*, 17(1):75–145, March 1985.

[8] T.O. Binford. Survey of model-based image analysis systems. *Int. J. Robotics Research*, 1(1):18–64, 1982.

[9] J.A. Bondy and U.S.R. Murty. *Graph Theory with Applications.* North Holland, New York, 1976.

[10] K.S. Booth. Isomorphism testing for graphs, semigroups, and finite automata are polynomially equivalent problems. *SIAM Journal Comput.*, 7(3):273–279, 1978.

[11] P. Boulanger and M. Rioux. Segmentation of planar and quadric surfaces. National Research Council of Canada, Division of Electrical Engineering.

[12] Ronald J. Brachman. 'i lied about the trees' or, defaults and definitions in knowledge representation. *The AI Magazine*, pages 80–93, Fall 1985.

[13] M.S. Casale and E.L. Stanton. An overview of analytic solid modeling. *IEEE Computer Graphics Applications*, 5(2):45–56, 1985.

[14] I. Chakravarty and H. Freeman. Characteristic views as a basis for 3-d object recognition. In *Society for Photo-Optical Instrumentation Engineers Conference on Robot Vision*, volume 336, pages 37–45, Bellingham, Washington, 1982.

[15] J.K. Cheng and T.S. Huang. A subgraph isomorphism algorithm using resolution. *Pattern Recognition*, 13(5):371–379, 1981.

[16] R.T. Chin and C.A. Harlow. Automated visual inspection: A survey. *IEEE trans on PAMI*, PAMI-4(6):557–573, Nov 1982.

[17] D.G. Corneil and C.C. Gotlieb. An efficient algorithm for graph isomorphism. *Journal ACM*, 17(1):51–64, January 1970.

[18] N.J. Deo, J.M. Davis, and R.E. Lord. A new algorithm for digraph isomorphism. *BIT*, 17:16–30, 1977.

[19] M. Earnshaw, C. Wong, and A.K.C. Wong. Map matching algorithm for points. Technical report, University of Waterloo, Department of Systems Design Engineering, Waterloo, Ontario, March 1988.

[20] G. Garibotto and R. Tosini. Description and classification of 3-d objects. In *IEEE 6th International Conference on Pattern Recognition*, pages 833–835, New York, 1982.

[21] D.E. Ghahraman, A.K.C. Wong, and T. Au. Graph monomorphism algorithms. *IEEE Trans. of SMC*, SMC-10(4):189–197, 1980.

[22] D.E. Ghahraman, A.K.C. Wong, and T. Au. Graph optimal monomorphism algorithm. *IEEE Trans. of SMC*, SMC-10(4):181–189, 1980.

[23] W. Havens and A. Mackworth. Representing knowledge of the visual world. *IEEE Computer*, pages 90–98, Oct 1983.

[24] S.W. Holland. Consight-i: A vision-control robot system for transferring parts from belt conveyors. *Computer Vision and Sensor Based Robots*, pages 81–97, 1979.

[25] J.E. Hopcroft and R.E. Tarjan. *Isomorphism of Planar Graphs*, pages 131–152. Plenum, 1972.

[26] T.S. Huang, editor. *Image Sequence Processing and Dynamic Scene Analysis*. Springer, Berlin, 1983.

[27] R.P. Kruger and W.B. Thompson. A technical and economic assessment of computer vision for industrial inspection and robotic assembly. In *IEEE 69*, number 12, pages 1524–1538, Dec 1981.

[28] R.V. Mayorga, K.S. Ma, and A.K.C. Wong. A local path generation method for robot manipulators in a computer vision based environment. In *Ninth Symposium on Engineering Applications of Mechanics Conference*, London, Ontario, May 1988.

[29] D.J. Meagher. Geometric modelling using octree encoding. *Computer Graphics and Image Processing*, 19(2):129–147, 1981.

[30] R.K. Miller. 3-d machine vision. *SEAI Technical Publications*, 1984.

[31] P.G. Mulgaonkar, L.G. Shapiro, and R.M. Haralick. Recognizing three-dimensional objects from single perspective views using geometric and relational reasoning. In *IEEE Conference on Pattern Recognition and Image Processing*, pages 479–484, Jun 1982.

[32] J. O'Rourke and N. Badler. Decomposition of three-dimensional objects into spheres. *IEEE trans on Pattern Analysis and Machine Intelligence*, PAMI-1(3):295–305, 1979.

[33] Edgar M. Palmer. *Graphical Evolution: An Introduction to the Theory of Random Graphs.* Wiley-Interscience Series in Discrete Mathematics. John Wiley and Sons, New York, 1985.

[34] M.R. Quillian. *Semantic Memory*, pages 354–402. MIT Press, Cambridge, Mass., 1968.

[35] H. Raafat and A.K.C. Wong. A texture information-directed region growing algorithm for image segmentation and region classification. *Computer Vision, Graphics, and Image Processing*, 43:1–21, 1988.

[36] R.C. Read and D.G. Corneil. The graph isomorphism disease. *J. Graph Theory*, 1:339–369, 1977.

[37] A. Rosenfeld. Image analysis: Problems, progress, and prospects. *Pattern Recognition*, 17(1):3–12, 1984.

[38] K.C. Rueb and A.K.C. Wong. Visual part identification and location in a robot workcell. *International Journal of Machine Tools and Manufacture, Special Supplement on Robotics and Artificial Intelligence*, 28(3):235–249, 1988.

[39] P.K. Sahoo, S. Soltani, and A.K.C. Wong. A survey of thresholding techniques. *Computer Vision, Graphics and Image Processing*, 41:233–260, 1988.

[40] R.W. Scheifler and J. Gettys. The x window system. Technical report, Massachusetts Institute of Technology, Cambridge, MA, Jul 1986.

[41] D.C. Schmidt and L.E. Druffel. A fast backtracking algorithm to test directed graphs for isomorphism using distance matrix. *J. ACM*, 23:433–445, 1976.

[42] H.E. Schroeder. Practical illumination concept and technique for machine vision applications. In *SME Robots 8*, pages 27–43, Detroit, MI, Jun 1984.

[43] R.B. Schudy and D.H. Ballard. Model-detection of cardiac chambers in ultra-sound images. Technical report, Computer Science Department, University of Rochester, Rochester, NY, 1978. ref no. TR-12.

[44] S.A. Shafer and T. Kanade. The theory of straight homogenous generalized cylinders and taxonomy of generalized cylinders. Technical report, Carnegie-Mellon University, Pittsburgh, PA, 1983. ref no. CMU-CS-83-105.

[45] S.C. Shapiro and G.H. Woodmansee. A net-structure-based relational question answerer. In *Proc. IJCAI*, pages 325–346, Los Altos, Calif., 1971. Morgan Kaufmann.

[46] H.C. Shen and A.K.C. Wong. Generalized texture representation and metric. *Computer Vision, Graphics and Image Processing*, 23:187–206, 1983.

[47] B.I. Soroka and R.K. Bajcsy. A program for describing complex 3-d objects using generalized cylinders as primitives. In *Pattern Recognition and Image Processing Conference*, pages 331–339, New York, 1978.

[48] C. Thorpe and S. Shafer. Correspondence in line drawings of multiple views of objects. In *International Joint Conference on Artificial Intelligence*, pages 959–965, 1983.

[49] W.H. Tsai and K.S. Fu. Error correcting isomorphisms of attributed relational graphs for pattern analysis. *IEEE Trans. on SMC*, SMC-9(12):757–768, 1979.

[50] K.J. Udupa and I.S.N. Murthy. New concepts for three-dimensional shape analysis. *IEEE trans on Computers*, C-26(10):1043–1049, 1977.

[51] J.R. Ullmann. An algorithm for subgraph isomorphism. *J. ACM*, 23:31–42, 1976.

[52] S. Umeyama and T. Kasvand. A new algorithm for point and plane pattern matching. *National Research Council Canada*, (28455), Nov 1987.

[53] T.P. Wallace and P.A. Wintz. An efficient three-dimensional aircraft recognition algorithm using normalized former descriptions. *Computer Graphics Image Processing*, 13:96–126, 1980.

[54] P.M. Will and K.S. Pennington. Grid coding: A preprocessing technique for robot and machine vision. *Artificial Intelligence*, 2:319–329, 1971.

[55] A.P. Witkin. Recovering surface shape and orientation from texture. *Artificial Intelligence*, 17:17–45, 1981.

[56] A.K.C. Wong. Knowledge representation for robot vision and path planning using attributed graphs and hypergraphs. *NATO Advanced Study Institute Series: Machine Intelligence and Knowledge Engineering for Robotics Application*, pages 113–144, 1987.

[57] A.K.C. Wong. Structural pattern recognition: A random graph approach. *Pattern Recognition Theory and Applications, NATO ASI Series, Computer and Systems Sciences*, 30:323–346, 1987.

[58] A.K.C. Wong. Trends and developments in computer vision. In *IEEE Asian Electronic Conference 1987*, pages 707–714, 1987.

[59] A.K.C. Wong and F.A. Akinniyi. A net based algorithm for largest common subgraph isomorphism. In *1983 International Conference on Systems, Man and Cybernetics*, pages 197–201, 1983.

[60] A.K.C. Wong and S.W. Lu. Representation of 3-d objects by attributed hypergraphs for computer vision. In *1983 International Conference on Systems, Man and Cybernetics*, pages 49–53, 1983.

[61] A.K.C. Wong and S.W. Lu. Recognition and knowledge synthesis of 3-d objects based on attributed hypergraphs. *IEEE Trans. on PAMI*, 1989.

[62] A.K.C. Wong, S.W. Lu, and M. Rioux. Recognition of 3-d objects in range images by attributed hypergraph monomorphism and synthesis. In *First IFAC Symposium on Robot Control*, pages 389–394, Barcelona, Spain, November 1985.

[63] A.K.C. Wong and R. Salay. An algorithm for constellation matching. In *Eighth International Conference on Pattern Recognition*, pages 546–554, Paris, France, October 1986.

[64] A.K.C. Wong and M.L. You. Entropy and distance of random grahps with application to structural pattern recognition. *IEEE Trans. on Pattern Analysis and Machine Intelligence*, 7(5):599–609, September 1985.

[65] M.L. You and A.K.C. Wong. An algorithm for graph optimal isomorphism. In *Seventh International Conference on Pattern Recognition*, pages 316–319, 1984.

Algorithm for Visible Surface Pattern Generation - a Tool for 3D Object Recognition

J. Majumdar, P. Levi, U. Rembold

Forschungszentrum Informatik, Haid- und Neu-Straße 10-14,
D-7500 Karlsruhe, F.R.G.

Abstract

This paper describes useful algorithms, developed for model-based object recognition, which happens to be one of the basic problems in the area of Robot Vision research. The important vision-oriented functions derived through these algorithms are: (i) the generation of the3D convex hull of an object to calculate its feasible stable positions, (ii) the determination of the pattern of the visible surfaces in the orthographic projection and finally (iii) the extraction of characteristic features invariant to the object rotation. The parameters will be used for a consequent matching phase. The feasibility of the algorithm is demonstrated through several sample objects.

1 Introduction

In order to realise a flexible assembly system, an industrial robot must be able to interact with its environment through visual information. This paper describes the development of a CAD-based machine vision system. The present system, in principle, generates several vision-oriented functions which enable one to recognise the pattern of visible surfaces of an object in the orthographic projection for each of its stable positions on the worktable.

A flexible system for 3D object recognition necessitates the development of an effective modelling tool suitable for a general description of 3D objects. Geometric modelling is used in CAD/CAM research for easy model generation, graphic visualisation and for the support of the vision analysis.

Advantage of using CAD modellers are:

- an unambiguous representation of 3D objects is possible

NATO ASI Series, Vol. F 66
Sensor-Based Robots: Algorithms and Architectures
Edited by C. S. George Lee

- object models generated by a CAD modeller, after proper transformation, may be used for the evaluation of pictures of an object
- the recognition and analysis of an object is possible without physically scanning the real object.

A CAD-based machine vision system thus provides information required for the automatic analysis of camera data.

A survey of various model-based recognition systems is reported in [1]. Baumgart [2] developed a 3-D geometric modelling system (GEOMED) for the application to computer vision. Hermen[3] established the three-dimensional structure of the visible surfaces of an object from a single view by assuming the object in the shape of a polyhedron which has perpendicular adjacent faces and edges. Koshikawa et al [4] have used a solid modeller (GEOMAP) for finding the stable position of the object from observed surface normals. Bolles et al [5] have designed a CAD model based vision system (3DPO) for advanced research on Robot Vision. Henderson et al [6] have used a Computer Aided Geometric Design (CAGD) system for visual recognition and manipulation of an object.

In the present work, the objects are modelled using a CAD system called ROMULUS, which offers the facilities of defining and manipulating models of 3D objects. Bodies are created in ROMULUS using standard shapes such as cubes, blocks, wedges, cones, cylinders etc. A complex object is constructed by binary operations such as union, subtraction or intersection between the bodies already defined. The transformation is provided to the system through a translation, rotation or change of the size of a body. The modeller is equipped with a CAD/CAM interface whose format is called FEMGEN (Finite Element Mesh GENerator). It is used to access the geometrical data from the CAD system. A description of the FEMGEN format is available in [7]. The geometrical data in the FEMGEN format consists of vertices, edges and faces of the object in the 3D space.

The vertex points of the object in the 3D space are given by the coordinates (x,y,z) with respect to a world coordinate system. Each linear edge of the object is described as a pair of points e=[p1,p2], where p1 is referred to as the starting point and p2 as the end point of the edge. Each curved edge of the object is described as a circular arc consisting of three points e=[p1,p2,c], where p1 is the starting point, p2 is the end point and c is the center point of the arc. Each surface of the object is described as a sequence of edges f=[e1,e2,e3,e4] which defines a closed, non-self intersecting chain of linear and curved line segments.

The generated vision-oriented functions with the above mentioned data structure as input are:

1. Generation of the 3D convex hull
2. Generation of the visible surface pattern

3. Extraction of the characteristic features

2 Algorithms

The entire algorithm for generating these vision-oriented functions are subdivided into two parts - the preprocessing part and the main part as shown in Fig. 1.

In the preprocessing part, the input data (vertex, edge and face list of the object) is read and the boundary and the internal regions of each of the object are identified. Thereafter, the curved edges and faces are interpolated to polygonal form. In the main part, the 3D convex hull of the object is first calculated and consequently the list of stable positions is obtained from the face list of the convex hull. For each stable position of the object on the worktable, the pattern of visible surfaces for an orthoghaphic projection is generated and the characteristic features are extracted from this pattern for use in a subsequent matching stage.

Preprocessing part

1: [make_regions] Construction of a closed polygonal chain of the faces and identification of the boundary and internal regions.

2: [plane_shapes] Conversion of the body into a polygonal form by interpolation of the curved edges and subdivision of curved faces into pieces of plane faces (see Fig. 2).

   2.1: Replace all the curved edges by piecewise linear segments.

   2.2: Transform the curved faces into polyhedra by joining the corresponding points representing the curved edges.

Main part

1: [convex hull_3D] Generation of the 3D convex hull of the object.

2: [stable_position] Generation of the list of stable positions.

3: For each stable position:

   [projection] Project all the faces of the object on the worktable and determine the resulting total face.

   [write_features] Determine the characteristic features of this total face in the output data.

The following sections describe the detailed algorithm for each task of the main part.

2.1 Algorithm 1 [convex hull-3D]

Input: Set of points defining the object in 3D space.
Output: Convex hull defined by its faces and edges.

1: [initial_face] Select the initial face F (see Fig. 3).

1.1: Search for a point A with the maximum value of z-coordinate from the given set of points.

1.2: Define a point H on the plane z = z max ($H \neq A$).

1.3: Rotate the plane z = z max about the axis AH until it hits the first point B from the given set of points.

1.4: The plane containing A & B is now rotated about AB until it hits a third point C which also belongs to the given set of points.

1.5: Determine all the points of the set which lie on the plane containing the points A, B & C.

1.6: At this stage, apply the 2D convex hull algorithm proposed by Andersen [8] to generate the resulting convex hull (polygonal chain) for all the points of the set lying on this plane.

1.7: The convex hull so obtained is the desired initial face.

2: For all the edges k of the face F, if k is not in storage, store k as unmarked, otherwise mark k.

3: [face_calculation] For all unmarked edges k′ of F, calculate a second face F′to k′and F (see Figs 4(a) & (b)).

3.1: Calculate the angle α between the face F and the other faces (eg., p1, p2, p′ is one of such face).

3.2: Select p′ so that α is minimum.

3.3: Calculate the equation of the plane containing the points p1, p2 & p′.

3.4: Determine all the points that satisfy the equation of this plane.

3.5: Apply the 2D convex hull algorithm for the resulting set of points

3.6: The convex hull so obtained is the new face F′.

4: Return to step 2 recursively.

The algorithm terminates only when marked edges are in storage, since in that case no more recursive call takes place. An edge is said to be marked when it is calculated from two different faces and both the faces containing this edge is found. Declaring all the edges as marked signifies that each edge is formed as an intersection of the two adjoining faces of the desired convex polygon. Thus the convex polygon is fully defined through its edges and faces.

2.2 Algorithm 2 [stable-position]

Input: List of the faces of the convex hull.
Output: List of stable positions of the object.

1: For each face of the convex hull on the worktable, use the transformation matrix (linear and rotational transformation) to calculate the new coordinates of all its vertex points.

2: Check for a stable position.
If the projection of the center of gravity lies within the bounded polygon of the convex hull face on the worktable, the position is stable and otherwise the position is not stable.

2.3 Algorithm 3 [projection]

Input: Object B in a stable position with the transformation matrix.
Output: Pattern of visible surfaces after projection.

This algorithm may be subdivided into the following steps:

1: [transform_object] Calculate the new coordinates of all the vertex points of the object using the transformation matrix corresponding to the stable position.

2: [orient_object]

2.1: For all the faces F of B, if the face F is at right angles to the worktable then:

- Remove the face F from the face list of B.
- Insert the face F in the list of the faces of B which are at right angles to the worktable.

2.2: For all the remaining faces with their internal regions:

- Determine the starting point (smallest y-coordinate which has smallest x-coordinate) and orient the polygonal chain of the face in a clockwise direction.
- Determine the maximum stretching of the face in the x- and y-direction.
- Determine the equation of all the edges of this face. An illustrative example is shown in Fig. 5.

3: [intersect_object]

3.1: Construct a linear list l with the boundary and the internal region of all the faces.

3.2: For all pairs of the faces F1 and F2 from the list l:

- If the maximum stretching of F1 and F2 intersect with themselves, test with each edge of F1 and each edge of F2, whether an intersection point exists. If so, subdivide the edge at the point of intersection (see Fig. 6).

4: [sort_altitude]

4.1: Set F2 = face list of b, set face list of B = empty.

4.2: So long F2 is not empty:

- Search the face F1 from the face list F2 so that no other face from the face list F2 lies above the face F1 (see Fig. 7).
  - (i) If a point v of F1 has the same value of x and y coordinate as the point w from F2 but has a higher value of z-coordinate, then the face F1 lies above the face F2 (Fig. 7(a)).
  - (ii) If a point v of F1 has a value of z-coordinate higher than that of a point w in face F2 but F1 lies within the boundary of F2, then the face F1 lies above the face F2 (Fig. 7(b)).
  - (iii) If a point v of F1 has a higher value of z-coordinate than that of a point w in face F2 but F2 lies within the boundary of F1, then the face F1 lies above the face F2(Fig. 7(c)).
- Remove the face F1 from the face list F2 and insert F1 as the first element of the face list of B.

5: [visible_faces] Replace all the faces by its visible parts through replacement of their visible portions only when the other parts are covered by the overlapping faces (see Fig. 8).

5.1: With all the faces F2 from face list of B:

- With all the faces F1 which are successor of F2:
  - (i) If F1 lies within the boundary of F2, remove F1 from the face list and proceed to the next face.
  - (ii) Replace the boundary of the face F1 and all its internal regions by its visible parts through replacement by their visible portions only when the other parts are covered by the overlapping faces.
  - (iii) If F2 lies within the boundary of F1, insert F2 as the internal region of F1, remove F2 from the face list and proceed to the next F2.

6: [unit_faces] Join all the visible parts of the faces from the cluster into a total face (see Fig. 9).

6.1: With all pairs of face F1, F2 from the face list of B:

- If the faces F1 and F2 have one common edge, then perform the following steps:
  - (i) Insert the boundary of F1 to the boundary of F2 and the internal regions of F1 to the internal region of F2.
  - (ii) Remove F1 from the face list.

6.2: Set F1 = face list of B without the first face. Set boundary of the total face = boundary of the first face. Insert the first face as the internal region of the total face.

6.3: So long F1 is not empty:

- Search a face F2 from the face list F1 so that it touches the boundary of the total face.

- Insert the boundary of the face F2 to the boundary of the total face.
- Insert the face F2 as the internal region of the total face.
- Remove F2 from the face list of F1.

2.4 Algorithm 4 [write-features]

Input: Visible surface pattern after projection.
Output: List of the characteristic features extracted.

1: Determine the perimeter of the total face and calculate the length of each edge in a normalised form with respect to the perimeter.

2: Output the features of the total face.

- Determine the lenght of all the edges.
- Determine the face area .
- Merge the edges which are in contact with each other.
- Determine the type of face (linear, curved or mixed).
- Determine the perimeter of the face.
- Output the type and the length of all the edges.
- Output the features of all the internal regions.

3 An application Example

The above algorithm was implemented in C language in a Micro VAX II workstation. The results, depicted in Figs 10 to 13, show the convex hull, the possible stable positions and the visible surface pattern corresponding to different stable positions of some sample objects.

Acknowledgement

This work was carried out at the Institute of Realtime Computer Systems and Robotics (Prof.-Dr.-Ing. U. Rembold and Prof. Dr.-Ing. R. Dillmann) of the University of Karlsruhe, and at the research group "Technic Expert Systems and Robotics" (Prof. Dr.-Ing. U. Rembold and Prof. Dr. P. Levi) of the Forschungszentrum Informatik, Karlsruhe. The work was supported by the Deutsche Forschungsgemeinschaft.

References

[1] Chin, R. T. and Dyer, C. R., "Model-Based Recognition in Robot Vision", Computing Surveys, Vol. 18, No. 1 (March 1986).

[2] Baumgart, B. G., "Geometric Modelling for Computer Vision", Technical Report AIM-249, STANCS-74-463, Computer Science Department, Stanford University (October 1974).

[3] Hermen, M., "Representation and Incremental Construction of a Three-Dimensional Scene Model", Carnegie-Mellon University Report, CMU-CS-85-103 (1985).

[4] Koshikawa, K. and Shiral, Y., "A 3-D Modeler for Vision Reserach", Proceedings of the International Conference on Advanced Robotics, Tokyo, Japan (September 1985).

[5] Bolles, R. C., Horaud, P., "3DPO : A Three-dimensional Part Orientation System", Robotics Research, Vol. 5, No. 3 (1986).

[6] Henderson, T., Hansen, C., Samal, A., Ho, C. C., Bhanu, B., "CAGD-Based 3D Visual Recognition" Proc ICPR, Paris, pp.230-232 (1986).

[7] Romulus Users Manual, Shape Data Ltd. (1984).

[8] Anderson, K. R., "A Reevaluation of an Efficient Algorithm for determining the Convex hull of a Finite Planar Set", IPL, Vol. 7, p. 53 (1978).

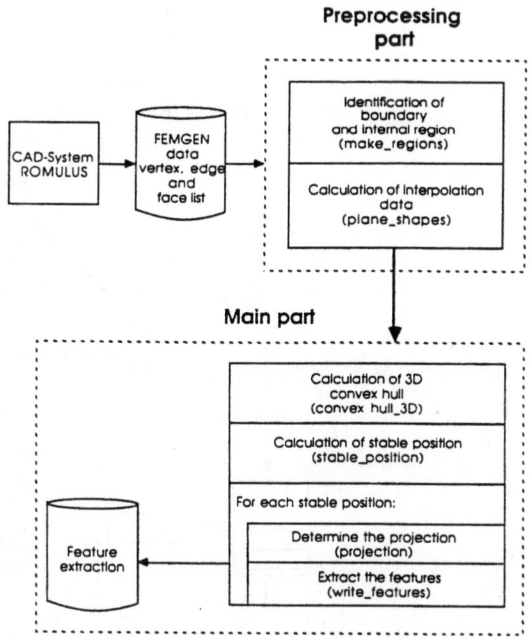

Fig. 1 **Algorithm for generating the vision-oriented functions**

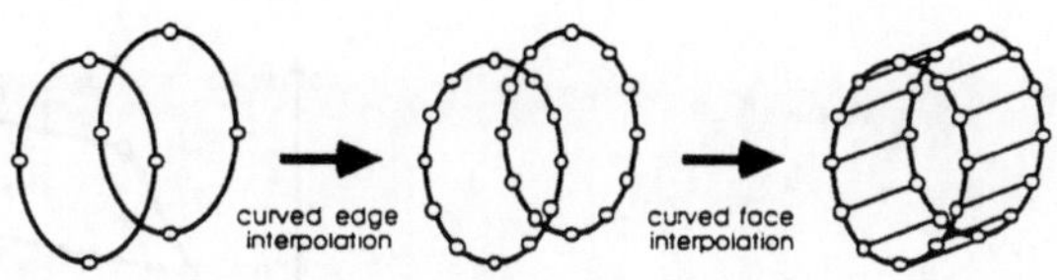

Fig. 2 **Interpolation of the curved face**

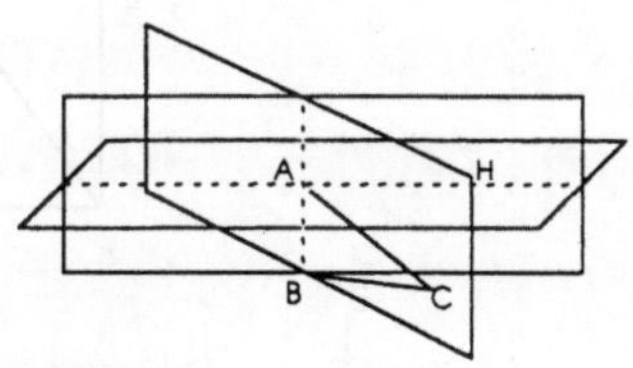

Fig. 3 **Calculation of the initial face**

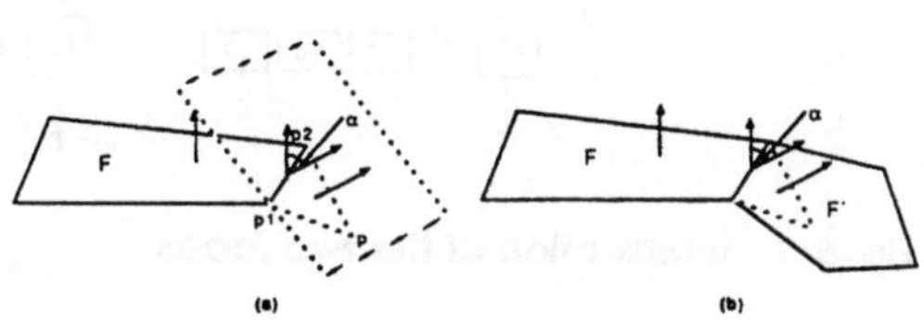

Fig. 4 **Calculation of the face**

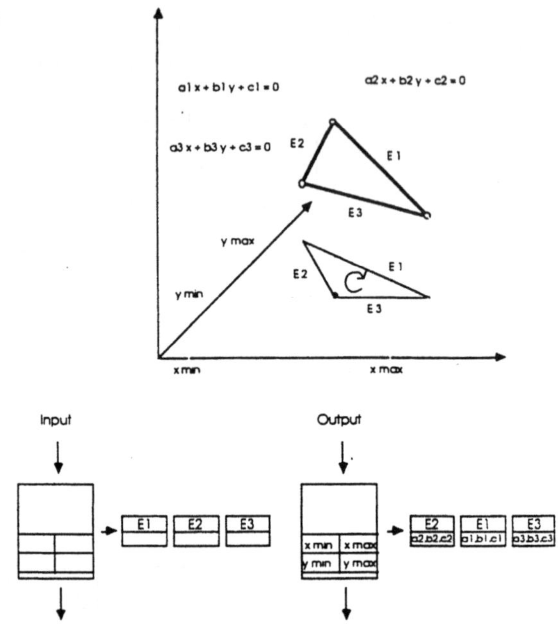

Fig. 5 **Orientation of the face**

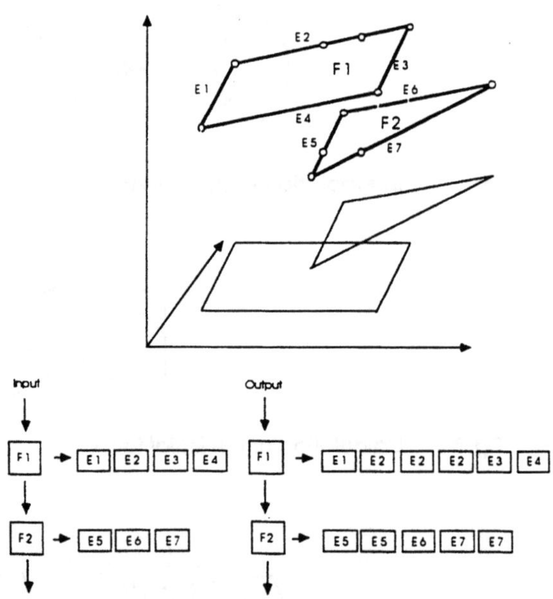

Fig. 6 **Intersection of the two faces**

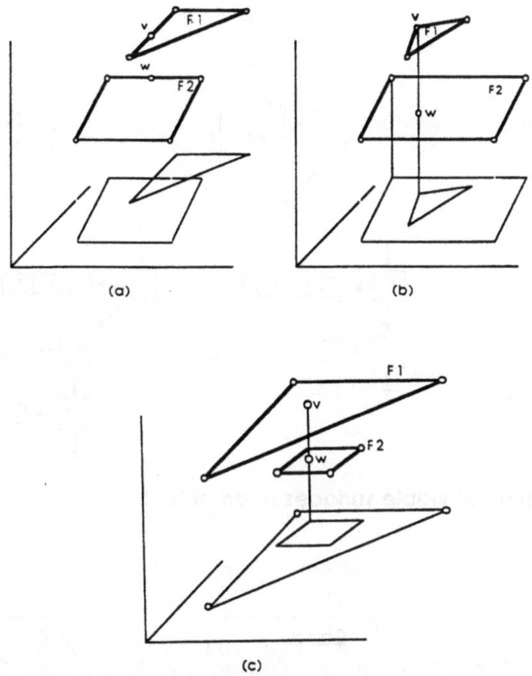

Fig. 7 Sorting the faces of the object according to their height from the worktable

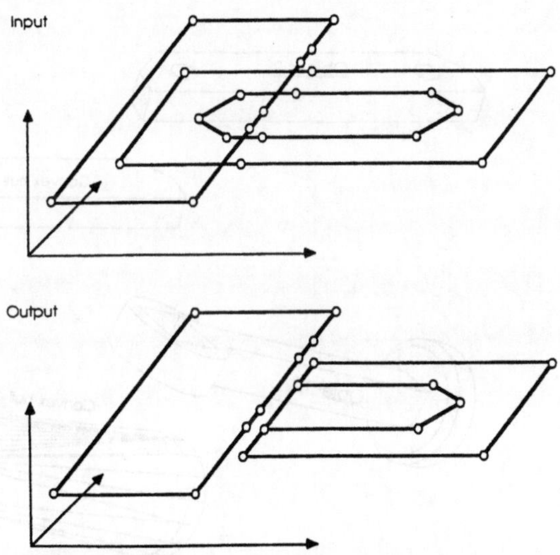

Fig. 8 Replacement of the faces by their visible parts

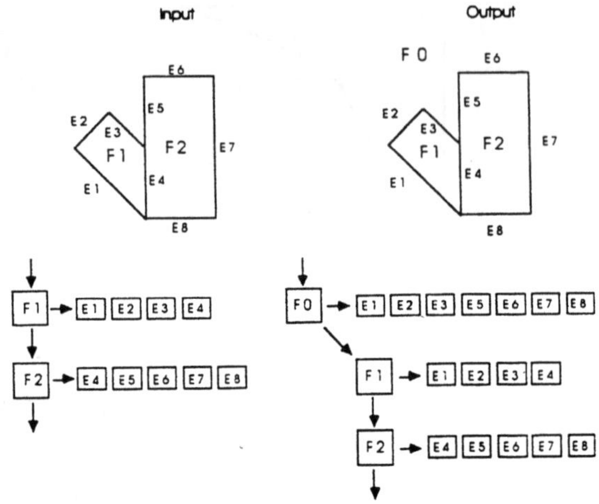

Fig. 9 Union of visible surfaces of an object

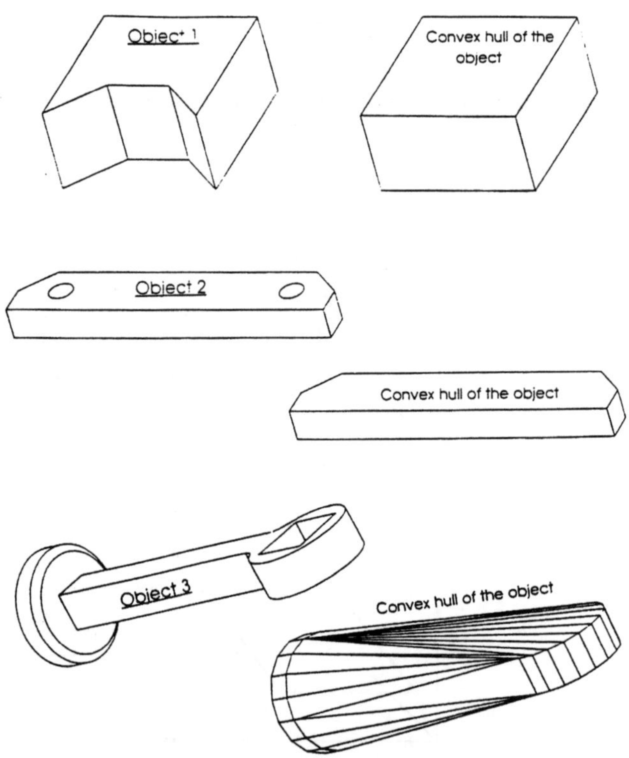

Fig. 10 Some sample objects with their convex hull

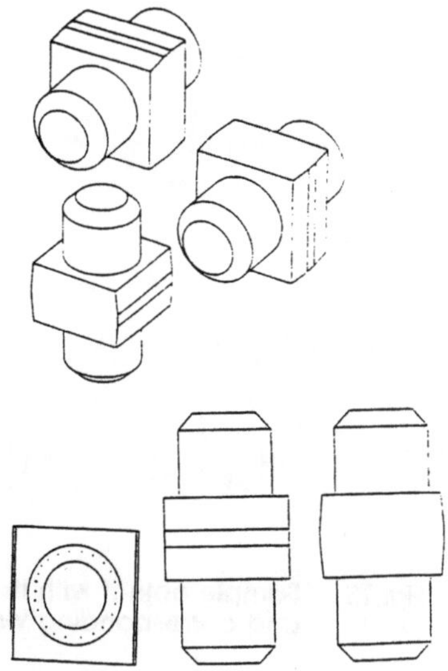

Fig. 11 Sample object with its possible stable positions and corresponding visible surface pattern

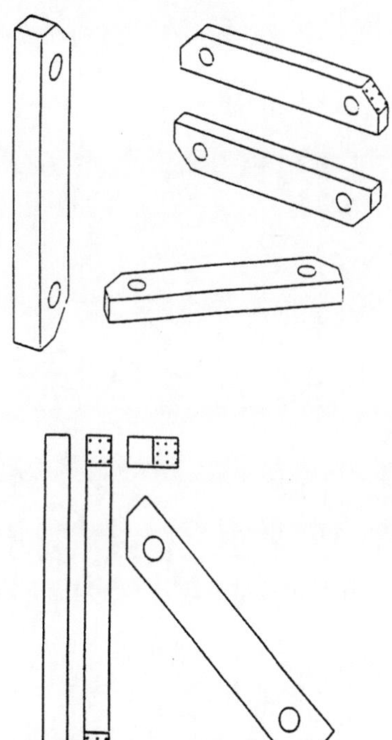

Fig. 12 Sample object with its possible stable positions and corresponding visible surface pattern

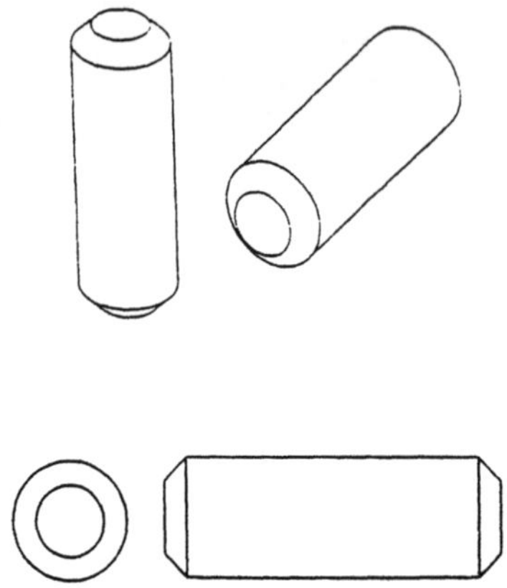

Fig. 13 Sample object with its possible stable positions and corresponding visible surface pattern

KNOWLEDGE-BASED ROBOT WORKSTATION: SUPERVISOR DESIGN

Robert B. Kelley
Electrical, Computer, and Systems Engineering Department.
Rensselaer Polytechnic Institute
Troy, NY 12180

I. INTRODUCTION

There are several problems currently inhibiting the growth of automation in industry. In particular, the growing interest in the application of robots to assembly tasks is being limited by the way such tasks are programmed and executed. Current robotic assembly systems force an exact, detailed description of the task to be executed. To perform a given assembly task, the detailed actions of the robot, as well as the successive positions and orientations of the gripper, must be specified. In addition, the work environment of the robot and the state of the objects in it must be completely controlled for the robot to successfully accomplish its task. These requirements could be reduced by the use of a variety of sensors and, in this way, allow the degree of uncertainty in the environment to be increased. Nevertheless, the use of sensors alone could also make the programming phase more difficult.

The basis of this problem stems is the lack of integration of the workstation environment model into the system design which forces the states of objects, manipulators and grippers to be completely controlled. Such an approach often prevents the system from recovering from unplanned events or errors that might occur during task execution. Further, the task plan is entered either through a *teach-box* or a robot programming language such as VAL. The entry requires considerable effort and skill on the part of the programmer. If any execution steps must be modified, this tedious programming task must be completely repeated in most cases.

This leads naturally to the consideration of intelligent, *knowledge-based robot workstations* which have integrated planning and sensing capabilities and allow for both automatically programming the robot and successfully executing the assembly task in spite of uncertainties in the task environment. Such a robotic system would interface with the user at a level that allows task plans to be entered in terms of parts to be moved and mated. It would also allow organized, automatic recovery from unexpected, non-deterministic deviations encountered during execution through the maintenance of a current environmental model. A system possessing these capabilities would greatly promote flexible knowledge-based automation because it would provide a very high level user interface and would not require frequent operator intervention. Systems attempting to solve some of these problems are described in the literature [Chochon and Alami 1986; Vijaykumar and Arbib 1987; Angermuller and Hardeck, 1987; Clermont, Hermant and Gaspart 1986]. However, most of the proposed intelligent robotic systems are

NATO ASI Series, Vol. F 66
Sensor-Based Robots: Algorithms and Architectures
Edited by C. S. George Lee

made specific in the early stages of the planning process in the sense that they are tied to some of the characteristics of the robot and other elements that will carry out the assembly.

In contrast, the goal of a cooperative research effort involving the Institut de Cibernética of the Universitat Politécnica de Catalunya and the Robotics and Automation Laboratories of the Rensselaer Polytechnic Institute is to build an automatic programming and supervision robotic knowledge-based system which is able to transform a symbolic description of an assembly task and a set of pieces, into a complete, physical assembly. It is a hierarchical knowledge-based planning and execution system that will allow manipulators and sensors to intelligently perform industrial assembly tasks in a variety of robotic workstations. Incorporated in this system will be a sensor-based model of the workstation environment so that deviations from the expected world model can be sensed and corrected by automatic on-line recovery mechanisms.The remainder of this article is on the design, implementation and successful testing of one level of this system, the Supervisor. The Supervisor is responsible for managing the on-line execution of a given task plan in a specified workstation. The presentation begins with a summary of the operating framework and objective followed by an overview of the knowledge-based robotic system.

II. Framework and Objectives

The development of such a system requires the identification and solving of major gaps in current technology and the bringing together of a variety of scientific and technical domains. Among these are programming and description languages, modeling, planning, motion and force control, computer vision, sensor fusion and error recovery. Some contributions in these fields are outlined below and can be found in more detail in some recent papers [Basañez et al., 1988; Juan and Paul, 1985, 1986; Ilari, 1987; Ilari and Reyna, 1986; Kelley and Bonner, 1985; Kelley, 1986; Thomas and Torras, 1988].

The output from this system will be the execution of the desired assembly task in the given robotic workstation. A variety of manipulators may be used in order to complete the task. For feedback purposes, the environment will be continuously monitored by sensors which may include 2D and/or 3D vision systems, gripper based proximity sensors and tactile arrays, grasping force sensors, wrist force/torque sensors, etc. By combining manipulation with abundant sensing, deviations from desired actions and locations can be detected and corrected.

The underlying design approach of this knowledge-based system is the Principle of Least Commitment. Under this principle, choices are deferred to lower levels of the hierarchy until a decision is forced to be made in order to continue successful planning and execution of the task. For example, the initial planning level does not consider specific characteristics of the particular robotic workstation, but relegates these decisions to lower levels in the hierarchy. Also, once the assembly is mapped to a particular environment, the execution order of the tasks is not defined until the on-line levels receive the list of tasks to execute. Analogously, the decision about which sensor to use to get some information at execution time can be postponed, in some cases, until the moment this information is needed. This design approach provides the system with greater flexibility and allows it to make a more efficient use of the available resources. The hierarchical programming and execution system being developed will map user-specified three dimensional part assembly tasks into various target robotic workstations, and will execute these tasks efficiently using the manipulators and sensors available in the workstation.

III. Knowledge-Based Robotic System Overview

The hierarchical knowledge-based system being designed is presented in Figure 1. This system was first reported by [Kelley and Bonner 1985], and refined descriptions were presented in [Moed and Kelley, 1987; Basañez et al., 1988; Moed and Kelley 1988]. As seen in Figure 1, along with the forward planning and execution system, there is a feedback path from each level to replan or re-execute an assembly if unexpected environmental deviations occur. A brief description of each component of the system follows:

A. Databases.

DAM: Dynamic Assembly Model contains the geometric description of the current state of the parts being assembled.

SAD: Static Assembly Database provides the descriptions of parts being assembled in terms of geometric constraints and sensor based information.

DEM: Dynamic Environment Model provides the current state of all objects in the workstation in terms of sensor based features.

SED: Static Environment Database contains the initial state of the workstation.

B. Assembly Planner.

The user input to this system is a part-centered description of the desired assembly. This input describes which parts are to be mated and includes specifications on how the parts should be joined, in terms of surfaces and features, as well as other geometric relationships between the objects. One method for achieving this assembly description is through the use of a Computer Aided Design (CAD) representation of the parts involved in the assembly. By using a CAD tool as the user interface, an assembly designer would be able to visually manipulate the desired parts displayed on a video screen to describe the desired assembly. It would no longer be necessary to tediously program the task in a conventional manner, using a language such as LAMA [Lozano-Perez and Winston, 1977] or AUTOPASS [Lieberman and Wesley, 1977].

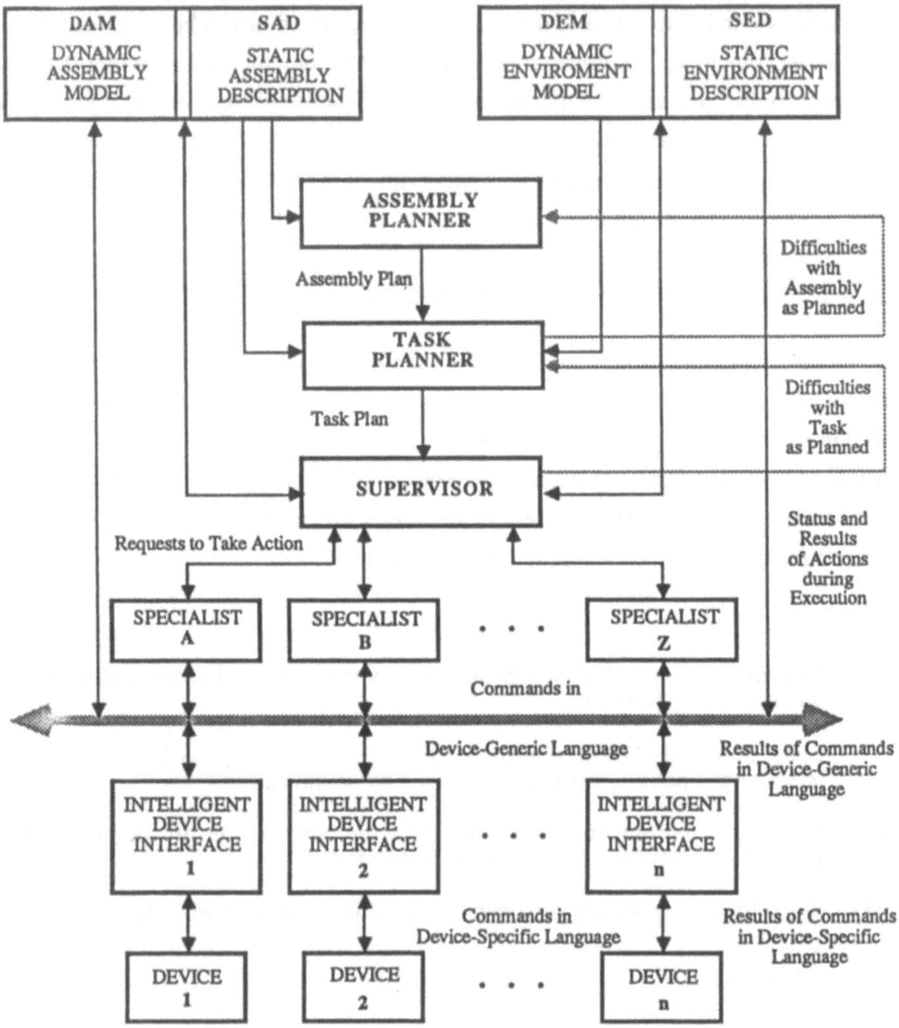

Figure 1. Knowledge-based robot system architecture.

The assembly planner is an off-line automatic program which geometrically transforms 3D objects, and plans *what* must be done to achieve the physical assembly of a part based on part constraints, and is not concerned with *how* the actions are accomplished. Using the SAD, this input can be translated to describe the specific task in terms of known part features. If a severe error occurs during task execution which prevents replanning or reexecution by the lower levels of the hierarchy, the assembly planner can consult the DAM, which contains the current state of the assembly, to replan the transformation of parts and try another attempt at successful execution.

The assembly planner implements a constraint-based model of the tasks, in which object-centered planning consists of the progressive refinement of the initial high-level assembly description through the successive application of constraints inherent in the different assembly operations. Three types of constraints are considered:

1. *Shape-matching constraints* between the mating parts of the workpieces to assembled (complementary shape and similar parameters);

2. *Constraints on the degrees of freedom* (degrees-of-freedom) that define the relative positions and orientations of workpieces (aligned, coplanar, etc.) [Herve, 1978]; and

3. *Constraints of non-intersection* between workpieces.

It must be noted that accessibility constraints, those limiting the possible sequences in which an assembly can be built, are subsumed under the third type above, since they can be handled by detecting intersections between moving workpieces. Three separate operators have been developed to deal with the three constraint types, their activity being coordinated by means of a message-passing control structure. A detailed description of each operator, as well as of the control structure, is provided in Thomas and Torras [1988]. The procedure sketched is essentially based on that proposed by Ambler and Popplestone [1975], but conveniently simplified and refined to increase its efficiency and to permit a uniform treatment of some special cases.

Non-intersection constraints are used to discard those relative poses of workpieces which, despite satisfying all shape-matching constraints and constraints on the degrees-of-freedom, lead to interferences. If the degrees-of-freedom that cause the interference are those characterizing an assembly operation, then the above constraints should be more precisely named *accessibility* constraints, since they deny accessibility for a given assembly sequence. Two ways of dealing with this type of constraint have been explored. The first requires the explicit construction of Configuration Space (C-space). The second is based on an implicit representation of this space. More concretely, the former involves carrying out a uniform sweep of C-space, which leads to a concise representation of this space as a list of ranges (forbidden areas) of one variable. The functions and predicates used to define C-obstacle boundaries are those described in Canny (1986). On the other hand, the implicit way of dealing with C-space is based on the use of local experts to move along C-boundaries, as proposed by Donald (1984).

A very simple *message-passing control structure* has been devised to coordinate the activity of the three operators described and adequately combine the limitations on the possible solutions imposed by each of them. Essentially, it works by pruning a tree of alternatives through the application of the most restrictive constraint at each stage. The usual cyclic sequence of operator application that results is first shape-matching constraints, next degree-of-freedom constraints and finally non-intersection constraints.

This off-line planner produces an assembly plan which consists of the object transformations which are required to produce the desired assemblies.

C. Task Planner.

The task planner revises the assembly plan by mapping the object transformations onto specific available manipulators and sensors which are under the control of *on-line* Specialists. The off-line task planner produces a detailed task plan which describes the specific manipulation and sensing actions which are needed to achieve the geometric assembly. The task planner also has access to the SED which provides a model of the nominal workstation environment. If an unrecoverable error occurs during execution of the task, the task planner can consult the DEM, which contains the current state of the environment, to remap transformations onto other manipulators. If this is impossible, the task planner requests a new assembly plan from the assembly planner.

Three basic operations on the pieces in an assembly task have been considered: picking, movement, and insertion. For each of these operations, a corresponding *off-line* Specialist has been designed: the Grasping Specialist, the Trajectory-Finding Specialist, and the Insertion Specialist.

From a description of the assembly operation to be performed, the Grasping Specialist generates a list of grasping sites on the workpieces to be assembled. The list is ordered according to a measure of the grasping site qualities (stability, tolerance to uncertainties, etc.). The list generation process starts by determining the portion of the workpiece surface that will remain accessible after the assembly operation has been completed; this is done by performing some specific subtraction operations on the boundary-based representation of objects provided by the CAD database. Next, a search for certain types of grasping sites (two parallel faces, face-edge, face-vertex, etc.) is carried out on this surface, guided by an evaluation function reflecting the *a priori* quality of each site. Finally, sensor strategies to verify the correct execution of each grasp, together with recovery procedures to apply in case of slippage, rotation of the workpiece within the gripper and other anomalous situations, are included for each site.

Currently, the Trajectory-Finding Specialist solves a simplified version of the general problem of generating collision-free trajectories for a 6 degree-of-freedom manipulator, namely that involving a non-articulated mobile body in a 2D environment. This is analogous to the motion of a gripper on a table. The approach followed combines an initial global search, based on the R-MAT model of free-space [Ilari, 1987], with a subsequent local search in C-space [Ilari and Reyna, 1986]. The R-MAT model is a subset of the medial axis transform whose underlying graph is a minimal deformation retract of the Voronoi diagram that can be reached from every point in free-space through a straight-line motion while preserving the clearance existing at that point. Depending on the cost function defined upon this graph, different global paths will be retrieved by a best-first search process. The underlying idea is to supply to the subsequent local search process the global path most likely to lead to a C-space solution path. Three heuristics have been built into the local C-space search, one of them guides the evolution of the translation degrees-of-freedom of the mobile body and the remaining two guide the evolution of its rotational degrees-of-freedom. A detailed description of the model, the search procedure and the heuristics implemented can be found in Ilari [1987].

The Insertion Specialist determines the fine motion strategy to put the objects in the final position in the assembled product. Two phases have been distinguished: the approach phase, before physical contact between the objects is made, and the interaction phase, in which contact forces and torques are generated. In this latter phase, not only the trajectory, but also active compliance must be planned. To do this, a generalization of C-space that makes use of an elastic model of the objects surfaces is proposed. From the generalized C-space, the insertion specialist determines the fine motion of the gripper, the compliance center and frame, and the force and torque control references.

The output of the task planner is a set of workstation specific instructions in an AUTOPASS-like language which specifies the positioning, mating, and sensing of objects in the workstation. This output is called the task plan.

D. Supervisor.

The Supervisor disburses the task plan in an organized manner to a set of Specialists. The supervisor is responsible for the real-time management and monitoring of the assembly process, the on-line coordination of these specialists, and for error recovery. The Supervisor is provided with the task plan and delegates the specified actions to a group of Specialists. The Supervisor is responsible for managing the resources of the workstation, which in this case are manipulators and sensors, but may be extended to include the work volume that a task requires. To optimize task execution, jobs are scheduled to execute in parallel as resources allow. The Supervisor contains experimentally determined error recovery routines for inconsistencies that cannot be handled by the individual Specialists. The Supervisor is also responsible for maintenance of the DEM and DAM as the execution process continues. If an unrecoverable error occurs during this execution, the Supervisor requests a new Task Plan from the Task Planner.

E. Specialists.

Each Specialist is an on-line, independent process that is expert in performing one type of task. Each specialist may utilize a variety of devices such as a 6-degree-of-freedom robotic arm, a stereo vision system, and so forth to accomplish its specific task. Examples of Specialists are: Gross Motion, Pick, Place, Grasp, 2D Vision Object Detection, 3D Vision Ranging.

Each on-line specialist is a separate process running in a multiprocess environment. This allows the concurrent parallel execution of several distinct specialists. The execution of the on-line specialists is monitored and managed by the supervisor as discussed above. Upon receiving detailed information about the specific task at hand from the supervisor, the on-line specialists control the necessary manipulators, sensors, and peripherals to complete the operations. Upon completion, each on-line specialist returns a message which identifies the state of objects in the environment that have been transformed. If an error occurs, the message contains information detailing the error conditions, if known. Some local error recovery routines may be executed by the individual on-line specialist if the deviation is sufficiently simple.

The Specialists communicate with devices (such as sensors and manipulators) in a device generic language. This language passes though an Intelligent Device Interface (IDI) where it is translated into device specific commands. By using a generic language, the system can be transported from one environment to another and the Specialists will not have to be modified. It is the function of the Intelligent Device Interfaces to translate the generic vocabulary into commands that the actual workstation device can understand. This method of interfacing allows different physical devices to be substituted without redesigning the upper levels of the hierarchy.

With this overview of the complete system description, the features and functions of the Supervisor can now be examined in depth.

IV. DESIGN OF THE ON-LINE SUPERVISOR

This section presents an initial design of the On-Line Supervisor which provides a structural base from which a robust knowledge-based system can be derived. In many ways, an on-line Supervisor of robotic tasks is similar to the Operating System of a typical computer. Both must be able to manage a host of jobs executing in real time with a finite set of resources. An Operating System must be able to schedule tasks, allocate resources, and provide communication links. Other added features of an Operating System might include parallel job execution, fault tolerant design and error reporting. Similarly, the Supervisor design can be separated into five main functional blocks: Resource Management; Concurrency Detection; Task Scheduling; Error Recovery; and Interprocess Communication. For the five functional blocks to work together cohesively, a background automaton was developed which drove the assembly execution to completion. Also, an interface was designed which provided the ability to create an off-line task plan.

A. Background Automaton

The workspace of the robotic workstation is separated into various objects each having a specified set of attributes. As these objects are moved around the workspace, they are transformed from one state to another. Under normal circumstances, the features of an object after it has been transformed is the end state that was desired. However, errors may sometimes occur during execution which must somehow be corrected. For this purpose, an automaton was created to keep track of the object states, changes, and errors.

Workspace objects can be either parts, tools, platforms, or manipulators. The state of an object reflects its position and orientation and specifies features which relate to the object in question. An instance of an object transformation specifies the part/manipulator/tool which is to be changed in terms of its position, orientation and features as well as the desired end conditions for this object. Included in the information for manipulators is the motion through which the object will be transformed.

Objects can be in one of three states, Valid, Unvalidated, or Invalid. A Valid state is assigned to an object when its position and features match the desired end conditions specified in the transformation entry. An Invalid state is assigned when there is a deviation from the desired end conditions after a transformation. An Unvalidated state is assigned to an object while the object is being transformed. This reflects the fact that during the transformation, the location of the object is indeterminate.

The automaton in Figure 2 describes the process through which objects in the workspace are transformed from one state to another. Initially, all objects are in known positions, so each object is in a Valid state. When an instance of an object transformation is encountered, the object moves from a Valid state to an Unvalidated state. The object remains in an Unvalidated state until the transformation is done. When the transformation is completed, sensors determine whether the object is in the desired location. If the object is located where expected, the object enters a Valid state once again. If the objects position, orientation or features deviate from the desired, it enters an Invalid state. If the object is in an Invalid state, one of a set of error recovery routines is executed in order to correct the object. When one of these routines is executed, the object is once again transformed and enters an Unvalidated State. This process continues until either the object enters a Valid state, or all relevant recovery routines are exhausted. If the latter occurs and the object is still in an Invalid state, the Supervisor requests assistance from the Task Planner to replan execution from the current workstation state.

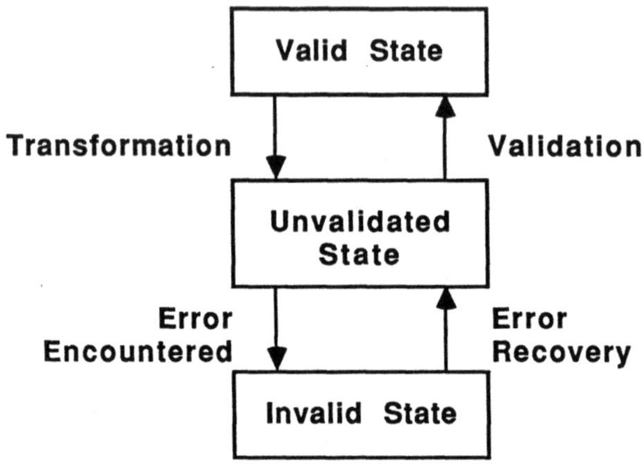

Figure 2. Background transformation state automaton.

B. Task Planner Interface

The Supervisor requires a list of tasks to run in order to begin execution so a simple *task plan language* was developed. This language segments tasks into two types: Observations and Transformations. Observations are tasks which require only sensing, and no motion or manipulation of objects in the workstation. Transformations are tasks that require manipulation of objects in the workstation, but may include sensing as well as the manipulation. Finding the location of an object using a 2D vision system is an example of an Observation task. Grasping a part with a force of 3 kg is an instance of a Transformation task. Each entry in the task plan contains the name of the task to be executed, parameters specifying the desired location and features for a Transformation, or simply Specialist specific parameters required for an Observation. The task plan entry also includes the objects in the workspace that will be Transformed/Observed in the process of execution. This last piece of information is required later by the Concurrency Detector. Figure 3 presents a sample task plan for a Pick/Insert Task.

Upon startup of the current assembly system, the Supervisor reads in the list of tasks which describe the workstation specific Transformations and Observations that must occur. First, it translates the entries into a form which will drive the automaton, and then assembly execution begins.

C. Resource Management

Associated with each task is a set of resources available in the workstation that are necessary for the successful execution of that task. Resources required by a task are manipulators and sensors that must be used to carry out the execution. For example, the execution of a PICK task requires the PUMA arm, Gripper, Force Sensor, and Proximity Sensors as resources. The table of resources required for defined tasks is named the Static Resource Table since the resources required by a task do not change over time. In general, given a set of Specialists S $\{s_i \varepsilon S; 1 \leq i \leq n\}$ and a set of Resources R $\{r_j \varepsilon R; 1 \leq j \leq m\}$, then S x R is an n x m matrix where each row i specifies the set of Resources E_i needed by s_i $\{E_i \varepsilon R\}$. For two Specialists s_i, s_j, if $E_i \wedge E_j = \phi$, then the two Specialists do not require the same resources.

The Resource Manager must also keep track of the resources being used during the execution of the Task Plan. Whenever a task begins execution, the Resource Manager adds the resources associated with that task to the Dynamic Resource List. Resources not on this list are available for assignment to tasks that are waiting to run. When a task has completed, its resources are deallocated from the Dynamic Resource List. Successful management of this Dynamic Resource List promotes task concurrency, since tasks requiring the same resources will be detected and prevented from executing simultaneously, thereby avoiding a conflict in the assembly process.

D. Concurrency Detection

Since the Task Plan does not provide an itemized list of jobs that can be run concurrently, an algorithm must be present in the Supervisor to perform this function. Using conventions developed in Operating Systems design, a method was created to detect execution parallelism in a set of tasks.

1) Conditions for Parallel Execution. Given a list of steps in a program, one can determine if two consecutive steps can be executed in parallel by using Bernstein's Conditions [Bernstein 1966]. These conditions state that given a variable A:

$$W_1(A) \wedge W_2(A) = \phi$$

$$W_1(A) \wedge R_2(A) = \phi$$

$$R_1(A) \wedge W_2(A) = \phi$$

where $R_i(A)$ is the Read Set of all variables A for Step i,
and $W_i(A)$ is the Write Set of all variables A for Step i.

	TASK	TYPE	VARIABLES
1.	FIND Cage A Vision C	Observation	Cage A
2.	FIND Cage B Vision D	Observation	Cage B
3.	PICK Cage A Arm E	Transformation	CageA Cage B
4.	GROSS MOTION Cage A, Cage B Arm E	Transformation	Cage A Cage B Card 1
5.	PLACE Cage B Arm E	Transformation	Cage B Card 1

Notes:

The task FIND uses 2D vision to locate objects in the workspace.

The task PICK removes a PC card from a card cage with a robotic manipulator.

The task GROSS MOTION moves the robotic manipulator from one location to another.

The task PLACE inserts a PC card into a given card cage.

Figure 3. Sample task plan for a Pick/Insert task

Given these constraints on execution concurrency and a list of programming steps to run, it is possible to extract steps which may run in parallel, and steps which are Data Dependent [Hwang and Briggs, 1984]. A line is data dependent on a previous step when one of Bernstein's conditions is violated by the two lines. In simple terms, data dependent lines both try to access the same variable. If one line changes the variable before another step can access it, the earlier step changes the data received by the later step when it does access the variable. This can cause problems if the old data was required by the second line, since the old data is now gone. By enumerating all the data dependencies between individual steps, non-dependent steps can be determined and executed concurrently, while dependent steps will execute in the proper order.

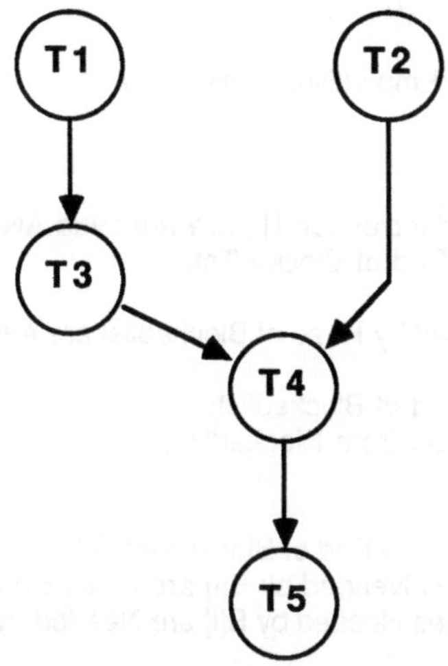

Figure 4. Concurency graph for example task plan.

Bernstein's conditions can be modified to apply to tasks in a robotic workstation by segmenting tasks into the two types described earlier, Transformations and Observations. These task types can be mapped onto Bernstein's conditions by associating Object Transformations with Write and Object Observations with Read. Objects such as parts and tools in the workspace become the variables of the data dependent enumeration. A concurrency graph is easily constructed for each Task Plan, where nodes represent tasks, and arcs are placed between dependent tasks. The graph for the Task Plan of Figure 3 is presented in Figure 4. As tasks execute in the workspace, other jobs are blocked from executing if they are dependent upon waiting or running tasks, as shown by the graph. When these dependencies are resolved, the job can be scheduled for execution.

E. Task Scheduler

It is the responsibility of the Task Scheduler to schedule the steps in the Task Plan for execution using the information provided by the Resource Manager and Concurrency Detector. The Scheduler selects a job from a list of waiting tasks, signals the Communication Center to begin execution of that job and calls the Resource Manager to update the Dynamic Resource List. Because of the possibility of process starvation, an algorithm was developed which schedules tasks that have been waiting for resources the longest before scheduling recently available jobs. This prevents "hogging" of resources by certain sections of the task plan. The algorithm makes use of a Blockledlist which contains tasks whose data dependencies have been removed, yet whose resources are not yet available. Task number i on the Blockedlist is named B[i].

```
WHILE Not Empty(T) OR Not Empty(Blockedlist) DO
    BEGIN
        FOR i := 1 to n
            BEGIN
                IF Data Dependencies for T[i] are removed AND T[i] is not on Blockedlist THEN
                    Add T[i] to End of Blockedlist.
            END;
            IF Resources Needed by Head of Blockedlist are Available THEN
                BEGIN
                    Execute Head of Blockedlist;
                    Remove Task from Blockedlist;
                END;
            ELSE
                FOR i:= Head + 1 to End of Blockedlist DO
                    IF Resources Needed by B[i] are Available AND
                    No Resources Needed by B[i] are Needed by Head of Blockedlist THEN
                        BEGIN
                            Execute  B[i];
                            Remove Task from Blockedlist;
                        END;
                    END;
                END;
            END;
        END;
    END;
```

This algorithm forces tasks waiting for resources to queue on the Blockedlist. The first element on the Blockedlist, called the Head, is the task that has been waiting the longest for resources. It is checked first for possibility of execution. If it cannot execute, only tasks which do not require any of resources needed by the Head can execute. Thus, the Head will only be blocked from executing while the already scheduled tasks are still running, and will be allowed to execute as soon as they complete. It is important to note that when a task is scheduled to execute, its resources are automatically placed on the Dynamic Resource List, and these resources are deallocated immediately upon completion. Also, a task only moves onto the list of completed tasks when it successfully terminates. If a task returns with an error, recovery mechanisms must be invoked before further steps can execute which are data dependent upon this task.

F. Error Recovery

The design of the Supervisor provides two systematic checks for errors that may occur during the execution of a Task Plan. First, when a Specialist is unable to complete a task that it has been issued by the Scheduler, the Specialist returns an error code which specifies the type of problem that it encountered during execution. Within the Supervisor there exists a set of error recovery routines for each Specialist which provide experimentally determined mechanisms for correcting an unexpected event based upon the error code received. The second check for execution errors occurs when a transformation task completes and an objects state must change from Unvalidated to either Valid or Invalid. This determination is made by using sensors available in the workstation to decide if transformed objects are at their desired locations and possess necessary features.

Depending on the type of error encountered, different types of recoveries are attempted to correct the situation. One routine simply reexecutes the task in error, to try to accomplish the task again. Other routines reexecute the same Specialist, but provide different parameters. A third set of error recovery routines may need several Specialists to execute. In difficult cases, the Task Planner is invoked to replan execution. When new tasks are required for error recovery, they are assigned a very high scheduling priority by being put on the Head of the Blockedlist. All tasks that were data dependent upon the error causing task are now made data dependent upon the last step of this error recovery routine.

G. Interprocess Communication

Since the Supervisor and Specialists are running concurrently in an on-line multiprocessing environment and communication is essential to the functioning of the hierarchy, a protocol was developed to allow high-speed data flow between the two levels. This protocol and associated communication mechanisms is housed in a functional block called the Communication Center.

The Communications Center is a separate running process in the multiprocessing system. Upon startup of the Supervisor, the Communicatons Center is also created and a bidirectional data path is established between the two. When an instance of a Specialist is created, it receives a bidirectional data path with the Communications Center as well. Each direction of the bidirectional path is in fact a queue which allows data to be held on it until it is required by the receiving process. By using a queue as the data line construction, one does not have to worry about losing data due to improper handshaking, since the data will remain on the line until read.

To allow for efficient, high-speed communications, the Communicatons Center was designed to be interrupt-driven. The Communicatons Center can handle interrupts from either the Supervisor or from any of the Specialists. Supervisor interrupts are required to start the execution of a Specialist, write to a Specialist, or to close a Specialist upon completion. When the Communicatons Center receives an interrupt from the Supervisor, it enters a service routine which receives the operation to be performed (open, read, write, or close) and the Specialist number to perform it on.

When a Specialist has data to transfer to the Supervisor, it issues an interrupt to the Communications Center. Since the Communicatons Center is constantly I/O-multiplexing the Specialists for interrupts, it jumps into an interrupt service routine as soon as one is encountered. The Communicatons Center then receives the data from the interrupting Specialist. When this data transfer is completed, the Communicatons Center interrupts the Supervisor and sends the data which is now prefixed with bytes identifying the Specialist that it originally came from. While this data transfer is taking place, interrupts from other Specialists are placed on a signal stack to be serviced after completion of this critical section.

V. One Experimental Implementation

The Assembly System Supervisor and four Specialists were implemented on a VAX 11/750 with the Unix 4.3bsd operating system under the RAL Hierarchical Control System. At the time of this implementation, the Task Planner and Assembly Planner were not fully designed, so a simple task plan was developed as input to the Supervisor. Also, the environment and object databases had not been created. However, for the assembly task at hand, it was possible to maintain an environmental model within the Supervisor.

A. Language Choices

Since Unix promotes inter-language communication, the different functions of the Supervisor were each written in the language best suited for their particular application. The languages chosen were OPS5, LISP, and C.

OPS5 is a forward chaining expert system shell [Brownstein et al., 1966; Forgy, 1981] capable of storing objects and their attributes, a feature which is essential for the current implementation of the Supervisor. Also, OPS5 under Unix permits event-driven data modifications, so the attributes of stored object entries can be changed dynamically by other processes on the system. This aspect fits well into the interrupt driven design that has been discussed above. Further, OPS5 is rule-driven, and thereby allows the user to encode experimentally determined results to be used in error recovery situations. Learning simply requires the addition of new rules to the rule base.

Since OPS5 is a declarative language, it does not provide the standard algorithmic structure that is required to easily implement the entire Supervisor structure. LISP [Winston, 1981] was selected to encode these algorithms for two reasons. First, many of the data structures used are list-based. Second, OPS5 easily communicates with LISP under Unix. Therefore, the Resource Manager, Task Scheduler, and Concurrency Detector were written in this language.

An important feature of the Supervisor design is the ability to execute tasks in parallel in a multiprocessing environment. The C language [Kernighan and Ritchie, 1978] was selected as the best language to achieve the high speed message passing necessary for interprocess communication. Unix 4.3bsd is written in C, and contains a host of routines which promote interprocess communication through message pipelines. Interrupt structures are also defined and available between executing processes. Therefore, the Communications Center was written in C.

B. Workstation Environment and Task Description

The workstation consisted of the following devices:

1. A PUMA 600 robot under cartesian control.

2. A pneumatically servoed force sensing gripper. The gripper is equipped with proximity sensors which reliably report the distance (up to 5 cm) of an object with known reflectivity. It is also equipped with a cross-fire sensor between the fingers and an overload sensor in the wrist.

3. A large work table supporting the following three pieces.

4. A large card rack which contained PC cards for picking.

5. A small card cage mounted on a 4 degree-of-freedom force detecting platform sensor in which PC cards were inserted.

6. A reflective metallic plate for Z calibration of the table plane.

The assembly task consisted of the following steps performed by the arm:

1. Calibrate the robot to the Z plane of the table by use of gripper-based proximity sensors.

2. Locate the large card rack using the proximity sensors.

   Note: The location of the card rack could be varied by 30 cm in the X and Y directions, and the orientation variable by ± 15 degrees.

3. Locate the card in the cage and grasp the card with sufficient force for retrieval.

4. Move Card to small card cage.

5. Fix the orientation of card cage using proximity sensors.

6. Insert card into small card cage and release when a given force is detected by the platform sensor.

C. Specialist Descriptions

To accomplish these tasks, a set of on-line Specialists was designed. These Specialists are:

CALZ employs the proximity sensors to determine the table normal vector and establishes a frame of reference in which the table is the Z boundary plane.

PICK detects the orientation of the card rack and aligns the gripper perpendicular. It finds the left or right edge of the rack and derives the center point, reaches into the rack a known distance, descends until the cross fire sensor is triggered, grasps the card with given force and then ascends to fixed height above rack.

GMOTION moves the arm along a linear trajectory between two set points.

PLACE employs similar techniques as PICK to determine location of the cage. It moves the arm to a set position relative to the cage and descends until a given force is detected by the platform sensor.

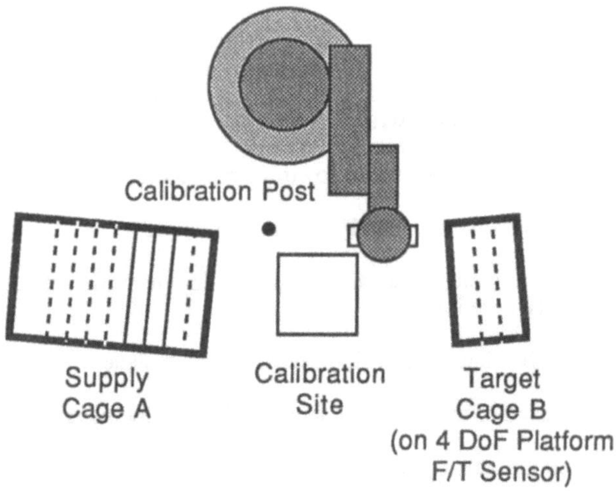

Figure 4. Experimental robotic workstation for Pick/Insert task.

D. Analysis of Experiment

The Supervisor was successfully able to execute the pick/insert assembly task described above. The Supervisor was able to keep a record of object locations and features in the workspace, but only minimal verification was available due to the absence of a vision system. The Resource Manager tracked resources and the Task Scheduler, under the longest-waiting-time-next algorithm, began Specialist execution in the desired order.

Further, a set of error recovery routines were encoded in the Supervisor which allowed the location of the large card rack and small cage to be varied. If the proximity sensors on the gripper could not find an object, the task was reattempted with a new gripper orientation, scanning these sensors along a different trajectory.

Since these was no inherent parallelism in this experiment, the concurrency graph was linear, and the tasks executed sequentially.

E. Supervisor Transportability: Another Experiment

As part of a cooperative research effort between the Robotics and Automation Laboratories and the Institut de Cibernética, the Supervisor, Specialists, and some associated hardware were transported to the Institut de Cibernética for further testing and demonstration [Moed, 1987]. The Supervisor software was modified to function in a VAX/VMS environment. Although these modifications were quite severe, the theoretical architecture of the Supervisor was not altered, and still functioned properly.

Equipment available at the Institut de Cibernética allowed the addition of another Specialist to the Assembly System. This Specialist performed 2D visual object identification from a camera overlooking the work cell. This Specialist provided the position and orientation of the large card rack and the small card cage in a similar pick/insert experiment. With this added information, the location of these objects could vary greatly without affecting task completion.

Experiments performed in the new environment demonstrated that the Supervisor still managed the execution of Specialists according to design. The Concurrency Detector was tested, since the VISION and GMOTION Specialists could run simultaneously. Unfortunately, due to equipment failure, Specialists requiring the Proximity Sensors could not be run. However, the success of the Supervisor in a new environment is an important indicator of the robustness of the initial design.

VI. CONCLUSIONS

A robotic assembly system has been presented which allows both the automatic programming of the robot and the successful execution of the assembly task, despite uncertainties in the task environment. To achieve this system, the Principle of Least Commitment is applied so that the successive stages from planning to execution become progressively more specific. The system is partitioned into off-line and on-line modules. The off-line modules comprise an automatic programming and planning system. The off-line modules are the assembly planner, the task planner, and off-line specialists. The assembly planner is responsible for determining *what* has to be done. The task planner maps the operations to a robotic environment and describes *how* the task is to be completed. The off-line specialists assist in planning the specific tasks.

The on-line modules are the supervisor, on-line specialists, and intelligent device interfaces. The supervisor manages the execution of the tasks in the workstation, and decides *when* the operations will execute. The on-line specialists control the physical manipulators, sensors and peripherals and execute the desired task. Since there can be a variety of unique devices in an workstation which can perform the same functions but require different programming languages, the intelligent device interfaces translate generic specialist commands into device-specific instructions.

The design, implementation, and testing of the Supervisor as part of an integrated hierarchical planning and execution knowledge-based robot workstation is presented in some detail. The Supervisor is responsible for on-line management of an assembly process. Using constructs developed in Operating Systems, the functionality of the Supervisor is separated into five main sections: Resource management; Concurrency Detection; Task Scheduling; Error recovery; and Interprocess Communication. The design presented here contains the necessary intelligence to recover from unexpected deviations in the workstation environment and provides one look into on-line robot task management. An experiment performed under Supervisor control demonstrates that the design provides a structural base for a robust on-line robotic workstation execution.

ACKNOWLEGMENT

The work reported here was performed by many researchers at the Institut de Cibernética of the Universitat Politécnica de Catalunya and the Robotics and Automation Laboratories of the Rensselaer Polytechnic Institute. The contributions of Luis Basañez (IC), Carme Torras (IC), Alberto Sanfeliu (IC), Michael Moed (RAL), and Susan Bonner (RAL) are especially acknowledged. This research was supported in part by the U.S.–Spain Joint Committee for Scientific and Technological Cooperation through grant CA8309-188.

REFERENCES

Ambler, A. P. and Popplestone, R. J.: "Inferring the Position of Bodies from Specified Spatial Relationships," *Artificial Intelligence*, vol. 6, 157-174 (1975)

Angermuller, G. and Hardeck, W.: "CAD-Integrated Planning for Flexible Manufacturing Systems with Assembly Tasks," IEEE Intl. Conf. on Robotics and Automation, Raleigh, NC, 1822-1826, 1987

Basañez, L., Kelley, R.B., Moed, M.C. and Torras, C.: "A Least-Commitment Approach to Intelligent Robotic Assembly," IEEE Intl. Conf. on Robotics and Automation, Philadelphia, PA, 1318-1319, 1988

Bernstein, A. J.: "Program Analysis for Parallel Processing," *IEEE Trans. on Electronic Computers*, vol. EC-15, no. 5, 757-762, 1966

Brownstein, L. et al.: *Programming Expert Systems in OPS5*, Addison-Wesley (1986)

Canny, J.: "Collision Detection for Moving Polyhedra," *IEEE Trans. on Pattern Analysis and Machine Intelligence*, vol. PAMI-8, no. 2, 200-209, 1986

Chochon, H. and Alami, R: "NNS, A Knowledge-Based On-Line System for an Assembly Workcell," IEEE Intl. Conf. on Robotics and Automation, San Francisco, CA, 603-609, .1986

Clermont, G., Hermant, M. and Gaspart, P.: "FIACRE: A Flexible and Integrated Assembly Cell for Research and Evaluation," 16th Intl.Symp. on Ind.Robots, Brussels, 1986

Donald, B.R.: "Global and Local Techniques for Motion Planning," M.Sc. Thesis, Massachusetts Institute of Technology, 1984

Forgy, C.L.: "OPS5 User's Manual," Department of Computer Science, Carnegie-Mellon Univ, 1981

Herve, J.M.: "Analyse Structurelle des Mecanismes par Groupe des Deplacements," *Mechanism and Machine Theory*, vol. 13, 437-450, 1978

Hwang, K. and Briggs, F.A.: *Computer Architecture and Parallel Processing*, McGraw-Hill (1984)

Ilari, J.: "Study of New Heuristics to Compute Collision-Free Paths of Rigid Bodies in a 2D Universe," Ph.D. Thesis, Universitat Politécnica de Catalunya, 1987

Ilari, J. and Reyna, J. LL.: "Some Experimental Result Using Heuristics for Solving the Find-Path Problem in C-space," IFAC Symposium on the Theory of Robots, Vienna, 1986

Juan, J. and Paul, R.: "Programming Automatic Assembly for Robots," 1st IFAC Symposium on Robot Control (SYROCO '85), Barcelona, 407-410, 1985

Juan, J. and Paul, R.: "Model for Automatic Programming of Fine-Motion in Assembly," IEEE Conf. on Robotics and Automation, San Francisco, CA, 1582-1587, 1986

Kelley, R.B.: "Vertical Integration for Robot Assembly Cells," IEEE Intl. Conf. on Robotics and Automation, San Francisco, CA, 585-590, 1986

Kelley, R.B. and Bonner, S.: "Understanding, Uncertainty, and Robot Task Execution," 1st IFAC Symposium on Robot Control (SYROCO '85), Barcelona, 331-335, 1985

Kernighan, B.W. and Ritchie, D.M.: *The C Programming Language*, Prentice-Hall (1978)

Lieberman, L.I. and Wesley, M.A.: "AUTOPASS: An Automatic Programming System for Computer Controlled Mechanical Assembly," *IBM J.of Res. and Devel.*, vol. 21, no. 4, 321-333, 1977

Lozano-Perez, T. and Winston, P.H.: "LAMA: A Language for Automatic Mechanical Assembly," Fifth Intl. Joint Conf. on Artifical Intelligence, Cambridge, MA, 710-716, 1977

Moed, M.C.: "An Intelligent Supervisor for a Robotic Assembly System," Master's Thesis, Rensselaer Polytechnic Institute, 1987

Moed, M. C. and Kelley, R. B.: "The Design of an Assembly Cell Task Supervisor." IEEE Intl. Conf. on Robotics and Automation, Philadelphia, PA, 652-653, 1988

Moed, M.C. and Kelley, R.B.: "An Expert Supervisor for a Robotic Work Cell," SPIE Conf. on Intelligent Robots and Computer Vision, Cambridge, MA, 496-501, 1987

Thomas, F. and Torras, C: "Constraint-Based Inference of Assembly Configurations," IEEE Intl. Conf. on Robotics and Automation, Philadelphia, PA, 1304-1305, 1988

Vijaykumar, R. and Arbib, M.A.: "Problem Decomposition for Assembly Planning," IEEE Intl. Conf. on Robotics and Automation, Raleigh, NC, 1361-1366, 1987

Winston, P.H. and Horn, B.P.: *LISP*, Addison-Wesley (1981)

Robot/Vision System Calibrations in Automated Assembly

F. G. King
Body & Assembly Operations

G. V. Puskorius, F. Yuan
R. C. Meier, V. Jeyabalan and L. A. Feldkamp
Research Staff

Ford Motor Company, Dearborn, Michigan

ABSTRACT

A vision guided robot for assembly is defined to be a robot/vision system that acquires robotic destination poses (location and orientation) by visual means so that the robot's end-effector can be positioned at the desired poses. In this paper, the robot/vision system consists of a stereo-pair of CCD array cameras mounted to the end-effector of a six-axis revolute robot arm. Automated calibration methodologies for local and global work volumes of the robot/vision system are described, including a perspective error transform calibration method for cameras. Multiple component assembly and robotic fastening has been demonstrated with the developed vision guided robot.

1. Introduction

For most vision guided robots, the vision modules are mounted remotely away from the robot, either on the ceiling or the floor, with the robot's work envelope within the field of view of the vision modules (see Figure 1). Recent advances in the miniaturization of cameras [16] and other optical systems has made the mounting of vision modules to the robot arm a practical strategy for robot guidance. This paper addresses calibration issues associated with mounting a stereo pair of CCD array cameras onto the end-effector of a six-axis revolute robot arm [8] as shown in Figure 2. The advantages and disadvantages of a robot-mounted vision system are discussed in the next section. Since a key to a useful robot/vision system lies in its accuracy, calibration of the robot/vision system is presented in detail in section 3. In section 4, an assembly strategy based upon the geometric relationships between various elements of the robot/vision system and its environment is outlined. A description of an assembly application using the robot/vision system is presented in section 5. Section 6 describes the use of the robot/vision system to infer global errors of the robot and the cameras. Section 7 discusses another method of calibrating the cameras.

2. Robot-Mounted Vision System

Robots guided by remotely mounted stereo cameras have been previously demonstrated [9]. This method of robotic guidance relies on having the robot's work envelope lie within the field of view of the stereo vision system. Most industrial robot arms have rather large work envelopes, consisting roughly of a hemisphere of 1.5 m radius. The planar resolution of a typical CCD array camera, even with sub-pixel calculations, is only on the order of one part in a thousand. Hence, for a robot arm of one and

NATO ASI Series, Vol. F 66
Sensor-Based Robots: Algorithms and Architectures
Edited by C. S. George Lee

a half meter radius, the planar resolution of a remotely mounted camera that views the entire work envelope is only about 3 mm, whereas the accuracy requirement for many assembly tasks in the automotive industry is on the order of 1.5 mm. This planar resolution limits the depth resolution of a pair of stereo cameras, the accuracy of which is a function of the separation of the stereo baseline and of the angle of convergence of the cameras. A common approach for increasing the resolution of such vision systems is to use multiple stereo camera pairs with overlapping fields of view.

An alternative approach for vision guided robots is to mount one or more cameras to the end-effector of the robot arm, thereby allowing the vision system to be used throughout the working envelope of the robot arm. Hence, the complexity of the system is reduced by having fewer cameras. Moreover, since this strategy allows for image acquisition throughout the work envelope of the robot arm, the field of view of the stereo camera system can be restricted, thereby increasing the resolution of the vision system. Another advantage of a robot mounted vision system is that, due to the mobility of the robot arm, the vision system can be moved to view around obstructions.

3. Robot/Vision System Calibration

With the cameras mounted to the end-effector of the robot arm, a general calibration of the robot/vision system consists of two steps: (1) determining camera model parameters with respect to a reference coordinate frame, and (2) determining a geometric relationship between the reference and robot end-effector coordinate frames. The result of the calibration procedure is a pair of camera calibration matrices and associated distortion correction terms, calculated relative to the end-effector coordinate frame. By using the camera calibration matrices in conjunction with the kinematic model of the robot arm, three dimensional coordinates with respect to the robot's base frame can be determined by the stereo vision system.

3.1 Camera Calibration Relative to Reference Frame

The calibration of a single camera is the estimation of camera model parameters that are used for the mapping of points expressed relative to a world coordinate frame to the vision system's pixel coordinates. Assuming a pin-hole model and the laws of Gaussian optics for the camera, the mapping can be shown to be a linear function of the world frame coordinates and is represented by a 3x4 calibration matrix in homogeneous coordinates [1,3,6]. This calibration matrix can be decomposed into the product of an internal and an external transformation matrix [5]. The external matrix is a 4x4 homogeneous transformation matrix relating the coordinate frame of the camera's center to an external reference coordinate frame. The internal matrix, which is also expressed in homogeneous coordinates as a 3x4 matrix, is a function of the camera's focal length, scaling parameters and coordinates of the image element's origin. Due to the homogeneous formulation, there are eleven unique elements in the calibration matrix. Thus, to determine the elements of the calibration matrix, it is necessary to measure the pixel coordinates of a minimum of 6 non-coplanar points whose coordinates are known with respect to the reference coordinate frame [1,14,15]. A linear least-squares procedure can be formulated to calculate the elements of the calibration matrix. In practice, a precisely machined calibration plate mounted orthogonally to a translation stage is used to establish the reference frame and the coordinates of known points.

The accuracy of the pin-hole camera model can be improved by accounting for geometric aberrations that are due to departures from the ideal conditions of Gaussian optics. The major geometric aberration that can affect the position of an image is radial distortion. The lowest order radial distortion component is a cubic function of the radial distance of the image from the optical axis [2]. For the CCD array cameras used in this study, the distortion correction to the camera model results in increasing the accuracy of calibration by approximately 50 percent over the nominal pin-hole camera model. Higher order polynomial corrections to radial distortion and polynomial corrections for tangential distortion were found to have a negligible effect on the accuracy of calibration. Similarly, the effects of image plane tilt with respect to the optical axis were also found to be negligible for small angles. With this calibration, an accuracy of better than 0.1 mm is achieved for the stereo vision system in a field of view of approximately a 125 mm cube.

3.2 Camera Calibration Relative to End-Effector Frame

In order to express the coordinates of imaged points relative to the robot's base frame, the camera calibration matrices are transformed to relate the camera-centered coordinate frames to the robot's end-effector frame. This transformation is expressed as a 4x4 homogeneous transformation matrix and is denoted by T_r^6. The relationship can be established by relating the coordinates of a minimum of four non-coplanar known points measured relative to these two coordinate frames. These data points can be obtained by manually translating the robot arm to touch a set of known points on the calibration block with a probe attached to the end-effector. The coordinates of the points relative to the end-effector frame are determined by the robot controller, and the coordinates relative to the reference frame are known a priori from the construction of the calibration block. This method is straightforward, but time consuming and may yield inconsistent results.

To determine this relationship between the end-effector and external reference frames automatically, a procedure using rotational and translational motion of the robot has been proposed [13]. The methodology to be described is based only on translational motions and uses a machined probe attached to the robot end-effector in the field of view of the vision system.

Let E denote the end-effector frame and r be the external reference frame to which the cameras are calibrated. Then the relation between E and r is expressed as T_r^E. Let

$$T_r^E = \begin{bmatrix} R & D \\ 0 & 1 \end{bmatrix} \quad \text{and} \quad T_E^r = \begin{bmatrix} R^{-1} & -R^{-1}D \\ 0 & 1 \end{bmatrix}$$

By determining R and D, one can solve for T_r^E.

Note that when the robot arm is moved from E0 to E1, there is a corresponding motion of r0 to r1 because the vision system is mounted on the robot arm (see Figure 3). The relations between all these reference frames are given as

$$T_{r0}^{r1} = T_r^E \; T_{E0}^{E1} \; T_E^r \qquad \text{and} \qquad T_{r0}^{E0} = T_r^E = T_{r1}^{E1}$$

If P is a point known with respect to r0 and P_{r1} is obtained from vision cameras, then

$$P_{r0} = T_r^E \; T_{E0}^{E1} \; T_E^r \cdot P_{r1} \qquad (1)$$

$$\text{Let} \quad T_{E0}^{E1} = \begin{bmatrix} R_0^1 & D_0^1 \\ 0 & 1 \end{bmatrix}$$

$R_0^1 = I$ and D_0^1 is known if the robot is moved by pure translations. So equation (1) becomes

$$R * D_0^1 = P_{r0} - P_{r1} \qquad (2)$$

R can then be solved by using the least squares method with at least 3 translational moves.

To solve for D in T_r^E, use a machined probe with tip P so that P_{E0} is known. Determine P_{r0} with vision cameras. Then $P_{E0} = T_{E0}^{r0} P_{r0}$ would give

$$D = P_{r0} - R * P_{E0} \qquad (3)$$

To summarize, the computation steps for T_r^E are

1) Determine stereo vision calibration matrices relative to base reference frame r.
2) Make n ($\geq$ 3) programmed translational moves and measure P_{ri}, i = 1,...,n.
3) Determine rotational part R by equation (2).
4) Pick up probe and determine probe tip by stereo vision cameras.
5) Determine translational part D by equation (3).
6) Form $T_r^E = \begin{bmatrix} R & D \\ 0 & 1 \end{bmatrix}$

3.3 Calibration Errors and Compensation

Numerous factors, e.g. incomplete camera model, noisy data, and other random or systematic errors, limit the overall applicability of the robot/vision system calibration as described above. In the inference of the transformation matrix T_r^6, a variety of measures are taken to minimize the effects of these errors. First, the motion of the robot arm is programmed so that the effects of gear backlash are eliminated. This is accomplished by actuating each joint of the robot in a fixed direction when approaching the desired pose. Another possible source of error is the relation of the probe center with respect to the end-effector frame. In fact, the components of the probe's location that are orthogonal to the sixth rotational axis of the robot arm are difficult to measure accurately. These components can be determined by viewing a point fixed in space while rotating the robot's sixth axis. This has the effect of calculating a correction to the matrix T_r^6. However, the resulting camera calibration matrices may still be related to a reference coordinate frame that is not precisely aligned with the end-effector coordinate frame.

Of even greater significance are the geometric and non-geometric robot errors which result in inaccurate estimates of robot end-effector poses. If real-time visual feedback control is not a feasible means of robot guidance, then some corrective strategy must be applied to use the vision system for robot guidance in an open loop mode. Local and global approaches for correcting these errors will be briefly discussed in the following sections.

Others [4] have used a remotely mounted vision system to increase the precision of a robot arm in a local volume defined by the field of view of the vision system. Correction parameters based on 6 dimensional cartesian coordinates are inferred from visual data and robot poses. This methodology should be directly applicable to a robot mounted vision system and can be extended to a larger work envelope by using numerous viewing points with associated local calibration volumes.

4. Assembly Strategy with Vision Guided Robot

There are five essential elements in vision guided robots for automated assembly. They are the workpiece, the parts or components to be assembled into the workpiece, the robot and its controller, the calibrated vision system and the gripper. Assume that the workpiece contains holes into which components are assembled. The fundamental goals of the robot/vision system are to determine the robot poses for the assembly operations of picking, inserting and fastening of the components into the workpiece, to execute the appropriate motion between the poses, and to perform the assembly operations. Geometric relationships and constraints associated with these five elements are discussed leading to a strategy for using vision guided robots in automated assembly. The discussion is restricted to a robot/vision system which performs the insertion of a part P into a hole H on the workpiece based on 3D visual data. The workpiece contains a pre-selected feature F which the vision system images so that the pose of the hole can be determined.

4.1 Use of Robot/Vision System

The first geometric relationship is that of the vision system and the robot. As described in the previous section, let C_A^r and C_B^r be the camera calibration matrices and T_r^6 be the transformation between the robot end-effector and the reference frame to which the vision system is calibrated. Then as depicted in Figure 4, postmultiplying the camera calibration matrices by the transformation matrix T_r^6 expresses the camera calibration matrices with respect to the end-effector coordinate frame. This relationship allows the vision system to be used for 3D vision guidance.

In a typical application, these 3D coordinates are transformed to the robot base frame through the kinematic model.

4.2 Fixed Viewing Application

Express the viewing pose and the target pose of the robot arm as transformation matrices ${}_vT_0^6$ and ${}_tT_0^6$, respectively. Let the geometric relationships between the imaging feature F with respect to the end-effector frame be denoted by T_6^F, between the hole H with respect to the feature F by T_F^H, and between the part P with respect to the end-effector coordinate frame by T_6^P. Both T_F^H and T_6^P may be derived from CAD data bases for the workpiece, part and the gripper. T_6^F is obtained from visual data while ${}_vT_0^6$, from the robot kinematic model. Combining these transformations, the target robot pose is given by

$${}_tT_0^6 = {}_vT_0^6 \cdot T_6^F \cdot T_F^H \cdot (T_6^P)^{-1}$$

This formulation assumes ideal conditions which, in practice, are usually violated. First, the poses described by the kinematic models of the robot arm may not accurately reflect the true robot poses. Second, the vision system may be calibrated relative to a reference coordinate frame that is slightly offset in location and orientation from the end-effector frame. Third, the geometric relationship of the gripper with respect to the end-effector frame may not be precisely known since the physical location of the end-effector frame is not measurable. Note that the various geometric errors are constant with the exception of those of the robot arm, which are functions of the joint variables. However, if the viewing position is fixed for a particular assembly task, then the error in the robot pose becomes a constant for that viewing pose. Thus, under this constraint, all the errors on the RHS of the above equation are constant. The error in the target robot pose, on the other hand, is not constant, since the target pose is a function of the location and orientation of the workpiece. However, it is nevertheless reasonable to approximate this error as a constant since the angular ranges of the six joint axes do not vary considerably under the local volume constraint imposed by fixing the imaging point. For small errors, the correct target robot pose can be approximated as the product of the nominal target robot pose with an error transformation:

$${}_tT_0^6 = {}_vT_0^6 \cdot T_6^F \cdot T_F^H \cdot (T_6^P)^{-1} \cdot (I + {}_t\delta T) ,$$

where I is a 4x4 identity matrix and ${}_t\delta T$ is an error transformation matrix [9], the elements of which are linear functions of all the above errors. The local calibration problem is then to determine the six components of ${}_t\delta T$. The methodology for increasing robot precision in a local volume [4] is easily revised for determining these error terms.

If, in the above equation, T_F^H and $(T_6^P)^{-1}$ are not known a priori, then the local calibration problem reduces to determining the 12 components of ${}_t(T_F^6)$ given by:

$${}_t(T_F^6) = T_F^H \cdot (T_6^P)^{-1} \cdot (I + {}_t\delta T)$$

For either case, the unknown correction parameters can be inferred by iteratively relating taught robot poses to the viewing pose and to the pose of the imaging feature as determined by the vision system.

5. Concept Demonstration of Instrument Panel Assembly

An accuracy of 1.5 mm is achieved with the above local calibration methodology, i.e. by fixing the imaging point of the robot/vision system. With such an accuracy and a local volume constraint of a 125 mm cube, vision guided robotic assembly of components into an automotive instrument panel (IP) is demonstrated. The robot-mounted stereo camera system is used in the open loop feedback scheme in which the pose of a randomly positioned IP is visually determined and communicated to the robot only once. The IP can be tilted or rotated by $\pm 15^{o}$ and can be positioned so that the imaging features are contained in the local volume. After receiving the pose information, the robot moves to pre-taught positions to pick up end-effectors and components and performs the assembly tasks. There are four components, clock, heater control unit, radio and speedometer cluster. Because of the variation in size and weight of the four components, a parallel jaw gripper and a suction cup end-effector are used. Each end-effector is capable of picking up two components, thus demonstrating the concept of multi-purpose grippers. For a laboratory environment, the use of a finger exchange unit, rather than a tool exchanger for the robot, is more adaptive and flexible. Since one of the end-effectors requires suction, the concept of vacuum exchange has also been shown. The components are presented in such a way that the concepts of just-in-time delivery and part kitting are assumed.

With two changes of fingers, the picking and insertion of the four components are accomplished in 65 seconds. To achieve such a cycle time, image processing has to performed while the robot is moving to pick up the first set of fingers and component. This is accomplished by storing the images of both cameras in buffers. By using three pre-painted dots or pre-punched holes as the imaging feature, the image processing and stereo correspondence time is less than 5 seconds. The pose information is transmitted to the robot before it has even picked up the first set of fingers. In comparison, a natural feature, such as the boundary of an air vent grill, requires a processing time that is 5 times longer due to the fact that the air vent can be in different configurations.

After inserting the components into the IP, the robot switches fingers to pick up a fastening tool. The above open loop feedback approach is not sufficient since (1) the tolerance between the screw and the fastening hole is less than 1.5 mm, (2) the screw in the fastening tool is not held rigidly and wiggles while the robot moves the fastening tool into position, and (3) the fastening tool may not be held rigidly with respect to the robot wrist since the force of the weight of the fastening tool is greater than the capability of the finger exchange unit.

To perform robotic fastening, an error correction to the open loop control is needed. After the robot moves to a position above the screw hole, the vision system is used again to determine the relative position between the screw and hole. This use of the vision system is feasible since both the screw and the hole are within the field of view of both cameras. A priori information about the location and orientation of the fastening tool tip is also required.

6. Global Calibration of Robot/Vision System

As indicated in section 4, one of dominant errors of the robot/vision system are the geometric and non-geometric errors of the robot. A global calibration procedure for the robot/vision system has been developed by using the vision system on the robot end-effector [11]. The calibration produces a single set of parameters which can be used throughout the entire work envelope of the robot arm. The methodology determines the geometric errors of the robot/vision system and the effects of joint compliance and gear backlash. The primary constraint employed by the methodology is that the coordinates of a single point in space, as measured by the stereo vision system, must not change with different robot joint angle configurations. A threefold improvement in the positioning accuracy of a robot arm can be obtained with this methodology.

6.1 Geometric Errors of the Robot

Let A_{i-1}^{i} be the 4X4 homogeneous transformation matrix representing the pose of the robot's i^{th} link coordinate frame relative to $(i-1)^{th}$ link. The differential errors of the link and joint parameters give rise to a corrected relationship

$$^{c}A_{i-1}^{i} = A_{i-1}^{i} * (I + \delta A_{i-1}^{i})$$

For N-axis robot, the corrected representation of the end-effector frame to the robot base fram is given by

$$^{c}T_{0}^{N} = \prod_{i=1}^{N} (A_{i-1}^{i} + \delta A_{i-1}^{i}) = T_{0}^{N} (I + \delta T_{0}^{N}) \quad (4)$$

by ignoring high order terms.

6.2 Non-geometric Robot Errors

In the robot that is being used, the most significant non-geometric errors are gear backlash and compliance. Backlash is modeled as

$$\Delta\theta_i' = B_{ii} \text{ sign } (\theta_i - (\theta_i)_{previous})$$

For backlash, induced by the third motor, of joint 2, which is coupled to joint 3 with a rubberized timing belt, the model is

$$\Delta\theta_2'' = B_{23} \text{ sign } (\theta_3 - (\theta_3)_{previous})$$

The balanced design of the links eliminated the need to compensate for joint compliance due to the weight of the links. For compliance due to the weight of the end-effector, it is modeled as a linear function of the cross-product of a vector parallel to the joint axis with a vector relating the mass position to the joint axis.

6.3 Algebriac Formulation of Geometric Relationships

From section 4, we have for camera J, $C_J^6 = C_J^r \; T_r^6$ where r is the reference frame to which the vision system is calibrated. Because of the errors in the robot, the corrected camera calibration matrix should be

$$^{c}C_{J}^{6} = C_{J}^{r} * \Delta T_{r}^{6} = C_{J}^{r} * (I + \delta T_{r}^{6})$$

Let ^{6}P denote the homogeneous coordinates of a target relative to the end-effector frame and ^{J}U, the homogeneous pixel coordinates of camera J. Then

$$^{J}U = C_{J}^{6} * \Delta T_{r}^{6} * {}^{6}P$$

Let $^{6}_{m}P = \Delta T_{6}^{r} * {}^{6}P$ (5)

Then $^{6}_{m}P$ is the target coordinates determined from the uncorrected camera calibration matrices, i.e. the measured target coordinates. The target point relative to the robot base is given by $^{0}P = {}^{c}T_{0}^{6} * {}^{6}P$ where $^{c}T_{0}^{6}$ is defined by equation (4).

Combining equations (4) and (5), one obtains

$$^{0}P = T_{0}^{6} * (I + \delta T_{0}^{6}) * (\Delta T_{6}^{r})^{-1} * {}^{6}_{m}P \qquad (6)$$

Since $(\Delta T_{6}^{r})^{-1} = (I - \delta T_{6}^{r})$, equation (6) becomes

$$^{0}P = T_{0}^{6} * (I + \delta T_{0}^{6} - \delta T_{6}^{r}) * {}^{6}_{m}P \qquad (7)$$

if high order terms are ignored.

The 43 unknowns in equation (7) include 24 kinematic parameters, 10 non-geometric parameters, 6 camera origin errors and 3 parameters for target location relative to the base frame. Some of these parameters are redundant and can be combined, resulting in a vector equation of only 35 unknowns.

6.4 Data Acquisition and Results

A series of views of the same target from many joint angle configurations is used to solve for the 35 unknowns. The choice of target, its position, and the widely varying joint angle robot poses are primary considerations in the data collection process. The target is a small nylon ball bearing illuminated by a laser. It is located above the robot where it can be viewed from different joint angle poses. These poses are automatically generated based on concentrated spheres centered at the target point and which radii are within the vision system calibration range. The poses are chosen to provide a uniform distribution of joint angles for the first five axes. Data collected consists of recorded joint angles from the robot and centroid calculation of the target from the vision system. Mathematical morphology [7] is used for image processing. From this data the 35 parameter corrections are inferred.

7. Linear Perspective Camera Error Model

It was clear from the above work that a good camera calibration is essential to an accurate robot/vision system. Based on the differential error model technique, a linear perspective error model for camera calibration is formulated [12]. Recall that

$$[INT] = \begin{bmatrix} k_1 & 0 & u_0 & 0 \\ 0 & k_2 & v_0 & 0 \\ 0 & 0 & 1 & 0 \end{bmatrix} \quad \text{and} \quad [EXT] = \begin{bmatrix} R & T \\ 0 & 1 \end{bmatrix}$$

where $R = f(r_x, r_y, r_z)$ and $T = (t_x, t_y, t_z)^t$. Express the difference between measured pixel and distored pixel coordinates as

$$\Delta u_d = u_m - u_d \quad \text{and} \quad \Delta v_d = v_m - v_d,$$

where u_d and v_d are cubic functions of $(G, (u_p - u_0)/k_1, (v_p - v_0)/k_2)$ with (u_p, v_p) being the ideal pixel coordinates. The above differences can be approximated by first order corrections as

$$\Delta u_d = \frac{\partial u_d}{\partial k_1} \Delta k_1 + \cdots + \frac{\partial u_d}{\partial t_z} \Delta t_z + \frac{\partial u_d}{\partial r_z} \Delta r_z + \frac{\partial u_d}{\partial G} \Delta G$$

where

$$\frac{\partial u_d}{\partial k_1} = Q_0 * (u_p - u_0)/k_1$$

with $Q_0 = 1 + (G * ((u_p - u_0)/k_1)^2 + ((v_p - v_0)/k_2)^2)$

$$\frac{\partial u_d}{\partial k_2} = \frac{\partial u_d}{\partial v_0} = 0 \quad , \text{ and } \quad \frac{\partial u_d}{\partial u_0} = 1$$

Similar equations can be obtained for Δv_d.

To obtain partials with respect to [EXT], one should consider

$$d[EXT] = [EXT]\, \delta[EXT]$$

where

$$\delta[EXT] = \begin{bmatrix} 0 & -R_z & R_y & T_x \\ R_z & 0 & -R_x & T_y \\ -R_y & R_x & 0 & T_z \\ 0 & 0 & 0 & 0 \end{bmatrix}$$

with $(R_x, R_y, R_z, T_x, T_y, T_z) = f(r_x, r_y, r_z, t_x, t_y, t_z)$. Since initial estimates of all camera parameters are known, one can treat (R,T) as the unknown corrections to [EXT]. This change of coordinates from (r,t) to (R,T) provide an easy representation of d[EXT]. Then

$$\frac{\partial u_d}{\partial S_i} = Q_0 \frac{\partial u_p}{\partial S_i} + \frac{u_p - u_0}{k_1} * 2G \frac{u_p - u_0}{k_1} \frac{\partial u_p}{\partial S_i} + 2G \frac{v_p - v_0}{k_2} \frac{\partial v_p}{\partial S_i}$$

where $S_i \epsilon (R_x, R_y, R_z, T_x, T_y, T_z)$,

$\frac{\partial u_p}{\partial S_i}$ and $\frac{\partial v_p}{\partial S_i}$ are functions of (c_{ij}, u_p, v_p, w),

with c_{ij} being elements of the camera calibration matrix.

An error model of 11 unknowns $(\Delta k_1, \Delta k_2, \Delta u_0, \Delta v_0)$, ΔG, $(\Delta R_x, \Delta R_y, \Delta R_z)$, and $(\Delta T_x, \Delta T_y, \Delta T_z)$, in two equations with the partial derivatives as coefficients has been formulated. These 11 unknowns can be solved by a least squares method using n/2 imaged points to form n equations.

To summarize, the steps to obtain a corrected camera model are

1) compute camera calibration matrices without distortion,
2) form matrices [INT] and [EXT],
3) solve for the 11 unknowns,
4) update [INT] and G by addtion,
5 update [EXT] by using [EXT] = [EXT] (I + δ[EXT]) and orthonormalize,
6) recompute C = [INT] [EXT], and
7) iterate.

The benefits of the perspective error transform method are that

- a single camera accuracy of 1 part in 6750 is achived,
- least squares is the sole optimization technique,
- all parameters are estimated at the same time which allows easy inclusion of abberations due to departure from the ideal pin hole model, and
- no prior camera knowledge is required where other methods required knowledge about u_0, v_0, k_2, or spacings between vertical image elements of the CCD array.

8. Summary

A robot-mounted stereo camera system has been developed for 3D visual guidance of the robot arm. Camera and robot calibration methods have been developed. Local calibration and local volume constraints have been used in the demonstration of vision guided robotic assembly of components into an instrument panel. Local calibration accuracy is about 1.5 mm in a 125 mm cube while the vision system accuracy is about 0.1 mm. Also demonstrated are the two open loop visual control schemes for robotic guidance. To further increase the accuracy of the system, a global robot/vision calibration method has been developed using the stereo cameras mounted on the end-effector. Realizing the importance of the accuracy of the vision system, a linear perspective camera error model has also been developed.

9. References

1. D. H. Ballard and Brown, C. M., Computer Vision, Prentice-Hall, Englewood Cliffs, NJ, 1982.

2. M. Brown and E. Wolf, Principles of Optics, 4th ed., Pergamon Press, Oxford, 1970

3. Duda, R. O. and Hart, P. E., Pattern Recognition and Scene Analysis, Wiley, New York, NY, 1973.

4. L. P. Foulloy and R. B. Kelley, "Improving the Precision of a Robot", Proceedings IEEE International Conference on Robotics and Automation, 1984.

5. S. Ganapathy, "Decomposition of Transformation Matrices for Robot Vision", Proceedings IEEE International Conference on Robotics and Automation, 1984.

6. R. M. Haralick, "Using Perspective Transformations in Scene Analysis", Computer Graphics and Image Processing, 13 (1980).

7. R. M. Haralick, et. al., "Image Analysis Using Mathematical Morphology", Pattern Analysis and Machine Intelligence, Pami-9, No. 4, July 1987.

8. F. G. King, et. al., "Vision Guided Robots for Automated Assembly", Proceedings IEEE International Conference on Robotics and Automation, 1988.

9. D. E. B. Lee, P. Trepagnier, "Guiding Robots with Stereo Vision", Robotics Today, April 1984.

10. Paul, R. P., Robot Manipulators: Mathematics, Programming and Control, MIT Press, Cambridge, Ma, 1981.

11. G. V. Puskorius and L. A. Feldkamp, "Global Calibration of a Robot/Vision System", Proceedings IEEE International Conference on Robotics and Automation, 1987.

12. G. V. Puskorius amd L. A. Feldkamp, "Camera Calibration Methodology Based on a Linear Perspective Transformation Error Model", Proceedings IEEE International Conference on Robotics and Automation, 1988.

13. Y. C. Shiu and S. Ahmad, "Finding the Mounting Position of a Sensor by Solving a Homogeneous Transform Equation of the Form AX = XB", Proceedings IEEE International Conference on Robotics and Automation, 1987.

14. I. E. Sutherland, "Three-Dimensional Data Input by Tablet", Proceedings IEEE 62 (4), April 1974.

15. C. K. Wu, et. al., "Acquiring 3-D Spatial Data of a Real Object", Computer Vision, Graphics and Image Processing, 28 (1984).

16. F. Yamagata, et. al., "Miniature CCD Cameras: A New Technology for Machine Vision", Robotics Age, March 1985.

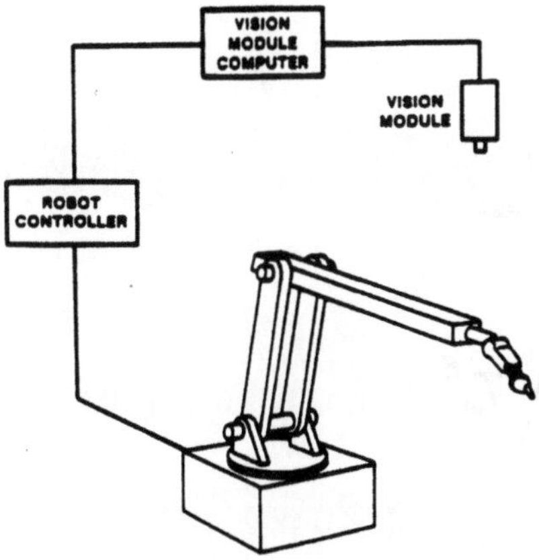

Figure 1. Vision System for Robotic Guidance

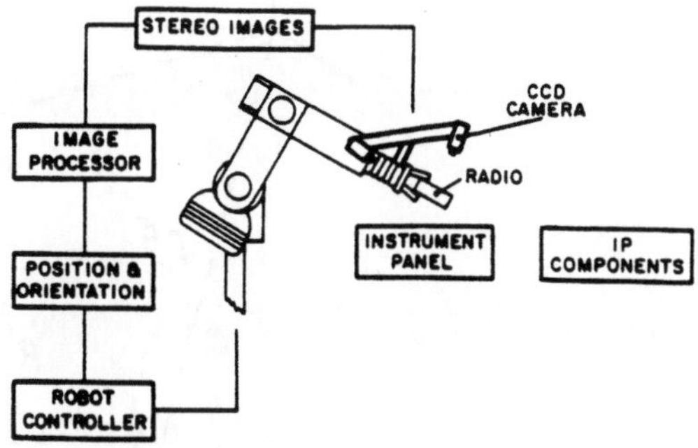

Figure 2. Vision-Guided Robots for Automated Assembly

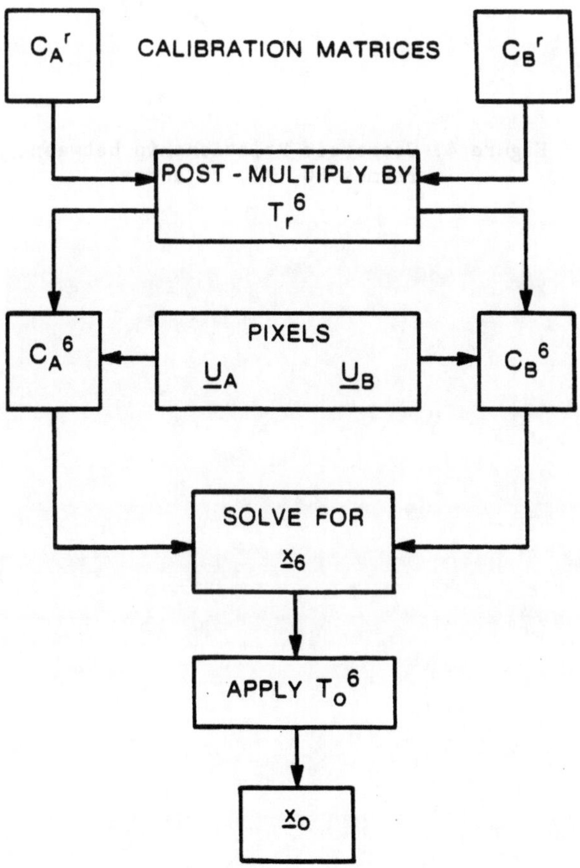

Figure 3. Use of Robot/Vision System

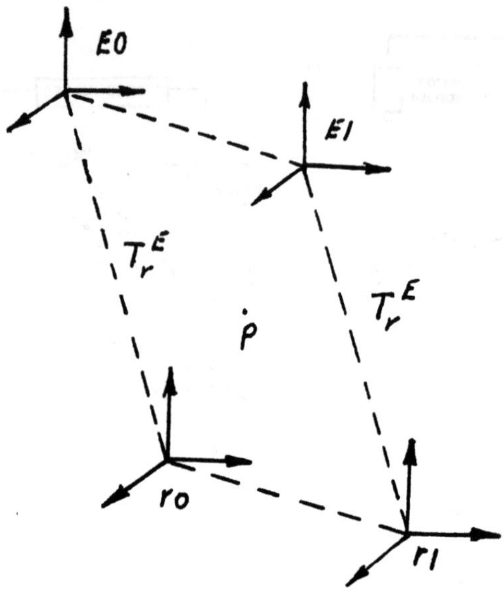

Figure 4. Geometric Relationship between frames E0, E1, r0 and r1

Part III

Neural Networks, Parallel Algorithms and Control Architectures

A Unified Modeling of Neural Networks Architectures*

S. Y. Kung

Princeton University
Department of Electrical Eng.
Princeton, NJ 08540

J. N. Hwang

University of Washington
Department of Electrical Eng., FT-10
Seattle, WA 98195

Abstract

Although neural networks can ultimately be used for many applications, their suitability for a specific application depends on the acquisition/representation, performance vs. training data, response time, classification accuracy, fault tolerance, generality, adaptability, computational efficiency, size and power requirement. In order to deal with such a multiple-spectrum consideration, there is a need of unified examination of the theoretical foundations of neural network modeling. This can lead to more effective simulation and implementation tools. For this purpose, the paper proposes a unified modeling formulation for a wide variety of artificial neural networks (ANNs): single layer feedback networks, competitive learning networks, multilayer feed-forward networks, as well as some probabilistic models. The existing connectionist neural networks are parameterized by nonlinear activation function, weight measure function, weight updating formula, back-propagation, and iteration index (for retrieving phase) and recursion index (for learning phase). Based on the formulation, new models may be derived and one such example is discussed in the paper. The formulation also leads to a basic structure for a universal simulation tool and neurocomputer architecture.

*This research was supported in part by the National Science Foundation under Grant MIP-87-14689, and by the Innovative Science and Technology Office of the Strategic Defense Initiative Organization, administered through the Office of Naval Research under Contract No. N00014-85-K-0469 and N00014-85-K-0599.

1 Characterizations of the Generic Iterative ANN Model

A basic ANN model consists of a large number of neurons, linked to each other with connection weights. Each, say i-th, neural processing unit (PU) has an activation value a_i. This value (either discrete or continuous) is propagated through a network of unidirectional connections to other PUs in the network. Associated with each connection, there is a *synaptic weight* denoted as w_{ij} which indicates the effect the j-th PU has on the i-th PU (see Figure 1(a)). In order to provide some biological fidelity, all of the existing artificial neural networks (ANNs) adopt an information storage/retrieval process which involves *altering* the connectivity pattern of synapses, and/or by *modifying* synaptic weights associated with the connections [3]. From algorithmic point of view, there are two separate phases of ANN processing: *retrieving phase* and *learning phase.*

1.1 Retrieving Phase of the Generic Model

Suppose that the connectivity pattern and synaptic weights of a neural network are known and fixed. In the retrieving phase, responding to the inputs (test patterns), the activation values of all neurons are iteratively updated based on the *system dynamics* until they reach the L-iterations and produce the responding outputs. The system dynamics in the retrieving phase of a generic iterative model for ANNs can be written as:

$$u_i(l+1) = \sum_{j=1}^{N_l} w_{ij}(l+1)a_j(l) \quad (1)$$

$$a_i(l+1) = f_i(u_i(l+1),\ \theta_i(l+1),\ a_i(l)) \quad (2)$$

where $1 \le i \le N_{l+1}$ and $0 \le l \le L-1$. The initialization activation values are often denoted by α_i, i.e., $\{a_i(0) = \alpha_i\}$. The termination activation values are often denoted by β_i, i.e., $\{a_i(L) = \beta_i\}$.

There are two types of inputs are observed: the stimulus inputs $\{a_i(0)\}$ and the external inputs $\{\theta_i(l)\}$. If the stimulus inputs are used to represent the test/training patterns, then the external inputs are often used as non-modifiable thresholding elements (e.g., [13, 1]), or as modifiable parameters to control the bias (e.g., [33, 32, 29]). It is also possible that the external inputs are used to represent the test/training patterns [24, 28], then the stimulus inputs will be purely used as initialization purpose. The system dynamics in Eqs. 1 and 2 may be graphically represented by an L-level feed-forward neural network (with N_l neural units at l-th level) shown in Figure 1(b), where one mathematical iteration is corresponding to one level of the network.

Equation (1) defines the *propagation rule.* Each PU, say i-th neuron at $(l+1)$-th level, receives the weighted inputs from other PUs at l-th level to yield the net input $u_i(l+1)$. Equation (2) defines the *nonlinear activation* function $f_i(l+1)$ which determines the new activation value $a_i(l+1)$ as a function of the net input value $u_i(l+1)$, the external input $\theta_i(l+1)$, and in some models, the previous activation value $a_i(l)$.

Iteration Index l in the Retrieving Phase The iteration index l used in the generic iterative ANN model (see Eqs. 1 and 2) can be used to represent one of the three possible iterations: *time*, *layer*, or *pattern.*

1. If l represents the time iteration, then the network is a single layer feedback network with each neuron being updated synchronously (in parallel) at each level. The resulting

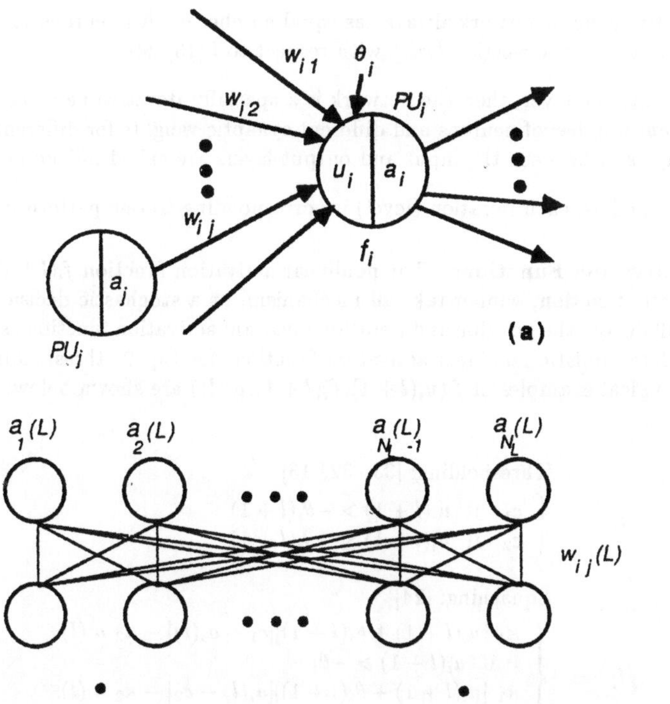

Figure 1: (a) A basic ANN model with two operations: propagation rule, and nonlinear activation, where iteration index is not considered. (b) A generic iterative model (L-iterations) for ANNs, where $w_{ij}(l)$ and N_l may be homogeneous or heterogeneous with respect to l.

time-iterative generic network always has equal number of (N) neurons at each iteration, and constant synaptic weights $\{w_{ij}\}$ with respect to l [13, 24].

2. If l represents the layer, then the network is a spatially iterative network which usually has different number of neurons and different synaptic weights for different levels l. The neuron layers in between the input and output layers are called *hidden layers* [32, 29].

3. In certain models, each iteration (level) is corresponding to one pattern input [28].

Nonlinear Activation Functions The nonlinear activation function $f_i(l+1)$ in Eq. 2 can be a deterministic function, winner-take-all mechanism, or a stochastic decision (for simpler notation, we will denote the position and iteration invariant activation function as f). There are three popular deterministic nonlinear activation functions for Eq. 2: thresholding, squashing, and sigmoid. Typical examples of $f_i(u_i(l+1), \theta_i(l+1), a_i(l))$ are shown below:

Thresholding: [33, 32, 13]

$$f_i = \begin{cases} c_1 & \text{if } u_i(l+1) > -\theta_i(l+1) \\ c_2 & \text{if } u_i(l+1) \leq -\theta_i(l+1) \end{cases}$$

Squashing: [24]

$$f_i = \begin{cases} \kappa_1\,[u_i(l+1)+\theta_i(l+1)][c_1 - a_i(l)] - \kappa_2\, a_i(l) \\ \quad \text{if } u_i(l+1) > -\theta_i \\ \kappa_1\,[u_i(l+1)+\theta_i(l+1)][a_i(l) - c_2] - \kappa_2\, a_i(l) \\ \quad \text{if } u_i(l+1) \leq -\theta_i(l+1) \end{cases}$$

Sigmoid: [29]

$$f_i = \frac{1}{1+e^{-u_i(l+1)-\theta_i(l+1)}}$$

In some applications, winner-take-all type of nonlinear mechanism are adopted [17, 10, 30]:

$$a_i(l+1) = \begin{cases} 1 & \text{if } u_i(l+1) > u_k(l+1)\ \forall k \neq i \\ 0 & \text{if } else \end{cases}$$

Note that the function is more general than Eq. 2, since it depends not only $u_i(l+1)$ but also all other $u_j(l+1), \forall j \neq i$. Nevertheless, this can be easily implemented by lateral inhibitions so that only the neuron receives largest input is activated. The typical stochastic decision rule may be represented as

$$Pr(a_i(l+1)) = f_i(u_i(l+1),\ \theta_i(l+1),\ a_i(l))$$

where $Pr(\cdot)$ represents the probability function [1, 28].

1.2 Learning Phase of the Generic ANN Model

In the learning phase, the synaptic weights for all the connections are also iteratively updated. The weight updating problem, sometimes called credit assignment problem for network in Figure 1(b), is to find the synaptic weights (sometimes also external inputs) which optimize certain predefined measure function E based on a set of input training patterns. The learning phase usually involves two steps: In the first step, the input training patterns are processed by the network based on the retrieving phase equations and generate some actual responses. In the second step, the weights are updated according to the responses generated and the chosen learning rules. Recursive procedures are often adopted. A unified recursive weight updating formulation (learning rule) for the generic ANN model can be expressed as following:

$$w_{ij}^{(m+1)}(l) = \Phi(w_{ij}^{(m)}(l),\ \eta_{ij}(l),\ \frac{\partial E}{\partial w_{ij}^{(m)}(l)}) \tag{3}$$

The new weight value $w_{ij}^{(m+1)}(l)$ at $(m+1)$-th recursion can be determined by the current weight value $w_{ij}^{(m)}(l)$, the updating rate parameter $\eta_{ij}(l)$, and most importantly the gradient $\frac{\partial E}{\partial w_{ij}^{(m)}(l)}$. The updating rate $\eta_{ij}(l)$ is introduced to regulate the rate of change of each weight at each recursion, it can be a global constant or can be a locally-dependent variable. In the following sections, for simplicity, we shall use η to denote $\eta_{ij}(l)$.

Measure Function E as Training Criterion A criterion function E in Eq. 3 has to be selected first, then the weight training may be formulated as a problem of iterative optimization (maximization or minimization) of the function E. In order to provide more flexibility, the measure function can be a global function E, or a local function of l, i.e., $E(l)$. When E is a global function, the training for weights at one level can affect that of other levels, since the optimization is over the weithts of all levels. When E is a local function, a hierarchical network can be established by cascading individually optimized iterations in Eq. 3 [30, 9, 23].

Recursion Index m in the Learning Phase To distinguish from the operations for the retrieving phase, we use a new *recursion index* m for the recursive weight updating formulation (see Eq. 3). The recursion index m may represent either a *pattern* index or a *sweep* index.

1. When the network updates the synaptic weights after the presentation of each training pattern, then m represents the pattern recursion.

2. When the network updates the synaptic weights only after all the P training patterns are presented, then m represents the sweep recursion.

Examples of Updating Formulation The updating formulation function Φ in Eq. 3 may be in an additive form, multiplicative form, or others. The additive formulations lead to the gradient descent (for minimization) or gradient ascent (for maximization) approach:

$$w_{ij}(l) \Longleftarrow w_{ij}(l) + \eta \frac{\partial E}{\partial w_{ij}(l)} \quad \text{or} \tag{4}$$

$$w_{ij}(l) \Longleftarrow w_{ij}(l) - \eta \frac{\partial E}{\partial w_{ij}(l)} \tag{5}$$

where $\pm$ is determined based on either the maximization or minimization formulation. One popular example is the back propagation learning, an iterative gradient descent algorithm designed to minimize the mean squared error between the the desired target values and the actual output values [29].

1.3 Useful Mechanisms for Weight Training

Constant-Sum Constraints In some competitive learning applications the magnitude of $w_{ij}(l)$ are bounded. For example,

$$\sum_{j=1}^{N_{l-1}} w_{ij}^2(l) = c$$

or

$$w_{ij}(l) \geq 0 \quad \text{and} \quad \sum_{j=1}^{N_{l-1}} w_{ij}(l) = c$$

If the additive updating formulation is used, then updating step η should be carefully selected to satisfy the constraints. Note that the convergence is not always guaranteed.

On the other hand, if $\{w_{ij}(l)\}$ are non-negative, i.e., $w_{ij}(l) \geq 0$, and satisfy either of the following constraints,

$$\sum_{i=1}^{N_{l+1}} w_{ij}(l) = 1 \quad \text{or} \quad \sum_{j=1}^{N_l} w_{ij}(l) = 1$$

then the iterative constraint optimization problem leads to a multiplicative or additive/multiplicative formulations [5, 4, 6, 31].

$$w_{ij}(l) \Longleftarrow \eta\, w_{ij}(l) \cdot \frac{\partial E}{\partial w_{ij}(l)} \quad \text{or} \tag{6}$$

$$w_{ij}(l) \Longleftarrow \eta\, [\alpha\, w_{ij}(l) + (1-\alpha)\, w_{ij}(l) \cdot \frac{\partial E}{\partial w_{ij}(l)}] \tag{7}$$

where E should be an arbitrary polynomial of $\{w_{ij}(l)\}$ with positive coefficients.

Note that only proper choice of the updating step η can ensure that the new weights $\{w_{ij}(l)\}$ satisfy the constraints and also lead to convergence. One popular example is the Baum-Welch reestimation learning rule used in the hidden Markov models [22, 28], which iteratively choose the weights to maximize the likelihood of the training patterns.

A possible extension of the constant-sum constraint is a weighted-sum constraint which may be exploited to shape the network so that it would highlight or depress certain properties.

Back-Propagation of Corrective Signals In the multi-level ANN model, the direct gradient updating of the weights in all the levels often incurs enormous computational burden. To alleviate this burden, back-propagation of corrective signals based on chain rule derivation may prove computationally very cost-effective. Note that the gradient term can be decomposed into:

$$\begin{aligned} \frac{\partial E}{\partial w_{ij}(l)} &= \frac{\partial E}{\partial a_i(l)} \frac{\partial a_i(l)}{\partial w_{ij}(l)} \\ &= \delta_i(l) \frac{\partial a_i(l)}{\partial w_{ij}(l)} \end{aligned} \tag{8}$$

where the backward propagated corrective signal $\delta_i(l)$ is defined to be $\frac{\partial E}{\partial a_i(l)}$. The backward propagated corrective signal can be computed directly by approximation if the nonlinear activation function is not differentiable [33, 26]. By adopting appropriate nonlinear differentiable activation functions, the corrective signal $\delta_i(l)$ can be recursively calculated as shown below [29]: For $L-1 \geq l \geq 1$.

$$\begin{aligned}
\delta_i(l) &= \frac{\partial E}{\partial a_i(l)} \\
&= \sum_{j=1}^{N_{l+1}} \frac{\partial E}{\partial a_j(l+1)} \frac{\partial a_j(l+1)}{\partial a_i(l)} \\
&= \sum_{j=1}^{N_{l+1}} \delta_j(l+1)\, r_{ji}(l+1) \qquad (9)
\end{aligned}$$

This defines the basic formulation of the back-propagation of corrective signals of all the levels. Then Eq. 9 may be used to compute the gradients.

Homogeneity Consideration in Homogeneous Iterative Model A special case of the generic iterative network is when the synaptic weights $\{w_{ij}(l)\}$ are constant with respect to l, i.e.,

$$w_{ij}(l) = w_{ij}, \quad \forall\, l \qquad (10)$$

A simple way of computing the gradient is to consider only the last iteration [24, 13],

$$\frac{\partial E}{\partial w_{ij}} = \frac{\partial E}{\partial w_{ij}(L)} \qquad (11)$$

However, in order to derive a more robust estimate of the gradient, some models [29, 28] adopt a more desirable approach of *averaging* the gradients over all the iterations. This may be mathematically derived as below:

$$\begin{aligned}
\frac{\partial E}{\partial w_{ij}} &= \sum_{l=1}^{L} \frac{\partial E}{\partial w_{ij}(l)} \frac{\partial w_{ij}(l)}{\partial w_{ij}} \\
&= \sum_{l=1}^{L} \frac{\partial E}{\partial a_i(l)} \frac{\partial a_i(l)}{\partial w_{ij}(l)} \frac{\partial w_{ij}(l)}{\partial w_{ij}} \\
&= \sum_{l=1}^{L} \delta_i(l)\, \frac{\partial a_i(l)}{\partial w_{ij}(l)} \qquad (12)
\end{aligned}$$

Note that $\frac{\partial w_{ij}(l)}{\partial w_{ij}} = 1$.

1.4 Characterizations of Neural Network Examples

In the proposed generic modeling, neural networks may be characterized by several common factors in the retrieving and learning phases. The factors in the retrieving phase are iteration index l, and nonlinear activation function $f_i(l)$. The factors in the learning phase are recursion index m, measure function E for the training criterion, updating formulation Φ, back-propagated corrective signals $\delta_i(l)$, and homogeneity consideration in certain models (see Table 1). More detailed illustration of such classification method will be discussed in the next section.

Neural Networks	Iteration Index l	Nonlinear Activa. f_i	Recursion Index m	Measure Function E	Updating Formulation $w_{ij}(l) \Longleftarrow \Phi(\cdot)$	Back Prop. $\delta_i(l)$	Comments
Single Layer Adaline	Non-Iterative	Threshold	Pattern	$E = \sum_{i=1}^{N_1} (t_i - u_i)^2$	$w_{ij} + \eta (t_i - u_i)\alpha_j$	No	[33, 32], LMS Learning
Memory & Learning Module	Time	Squash	Pattern	$E = \sum_{i=1}^{N} (\theta_i - u_i(L))^2$	$w_{ij}(L) + \eta (\theta_i - u_i(L))\ \beta_j$	No	[24], Delta-Like Learning
Multilayer Adaline/ Committee Machine	Layer	Threshold	Pattern	$E = \sum_{i=1}^{N_L} \|t_i - \beta_i\|$	$w_{ij}(l) + \eta\ \psi(\frac{\Delta E}{\Delta a_i(l)}) a_j(l-1)$	No	[32, 26], † $\psi(\frac{\Delta E}{\Delta x}) = \begin{cases} 0 & \text{if } \Delta E \geq 0, \\ \frac{\Delta E}{\Delta x} & \text{if } \Delta E < 0 \end{cases}$
Back-Propagation Network	Layer	Sigmoid	Pattern	$E = \sum_{i=1}^{N_L} (t_i - \beta_i)^2$	$w_{ij}(l) - \eta\delta_i(l)\ f_i'(u_i(l))a_j(l-1)$	Yes	[29, 12], $\delta_i(L) = -(t_i - \beta_i)$
Recurrent Back-Propagation	Time	Sigmoid	Pattern	$E = \sum_{i=1}^{N} (t_i - \beta_i)^2$	$w_{ij} - \eta \sum_{l=1}^{L} \delta_i(l)\ f'(u_i(l))a_j(l-1)$	Yes	[29, 12], $\Delta w_{ij} = \sum_{l=1}^{L} \Delta w_{ij}(l)$
Parallel Hopfield Associative Network	Time	Threshold	Sweep	$E = -\frac{1}{2} \sum_{i=1}^{N} \sum_{j=1}^{N} w_{ij}\beta_i\beta_j - \sum_{i=1}^{N} \theta_i\beta_i$	$w_{ij} + \eta (2\beta_i - 1) \cdot (2\beta_j - 1)$	No	[13, 14], ‡ No Adaptive Learning
Self-Organized Perceptual Network	Layer	Linear	Sweep	$E(l) = \sum_i [-\frac{1}{2} \sum_{j=1}^{N_{l-1}} \sum_{k=1}^{N_{l-1}} Q_{jk}(l) w_{ij}(l) w_{ik}(l) - \frac{c_5}{2} (\sum_{j=1}^{N_{l-1}} w_{ij}(l))^2 - c_6 \sum_{j=1}^{N_{l-1}} w_{ij}(l)]$	$w_{ij}(l) + \eta [\sum_{k=1}^{N_{l-1}} Q_{jk}(l) w_{ik}(l) + c_5 \sum_{j=1}^{N_{l-1}} w_{ij}(l) + c_6]$	No	[23], $Q_{jk}(l) =< (a_j^{(m)}(l-1) - \bar{a}_j(l-1)) \cdot (a_k^{(m)}(l-1) - \bar{a}_k(l-1)) >$
Self-Organized Feature Map	Non-Iterative	Winner-Take-All	Pattern	$E = \frac{1}{2} \sum_{i=1}^{N} \beta_i (\alpha_j - w_{ij})^2$	$w_{ij} + \eta\ \beta_i (\alpha_j - w_{ij})$	No	[17, 18], Shrinking Neighborhood
Adaptive Resonance Theory	Non-Iterative	Winner-Take-All	Pattern	$E = \frac{1}{2} \sum_{i=1}^{N} \beta_i (\alpha_j - w_{ij})^2$	$w_{ij} + \eta\ \beta_i (\alpha_j - w_{ij})$	No	[10], Vigilance Test
Rumelhart Feature Detector	Layer	Winner-Take-All	Pattern	$E(l) = \frac{1}{2} \sum_{i=1}^{N} a_i(l) (a_j(l-1) - w_{ij}(l))^2$	$w_{ij}(l) + \eta\ a_i(l) (a_j(l-1) - w_{ij}(l))$	No	[30], Leaky Learning
Boltzmann Machine	Time	Stochastic	Sweep	$E = \sum_{\forall a_v} Pr^+(a_v)\ ln\ \frac{Pr^+(a_v)}{Pr^-(a_v)}$	$w_{ij} + \eta (p_{ij}^+ - p_{ij}^-)$	No	[1], $p_{ij}^{\pm} = \sum_{\forall a} Pr^{\pm}(a)\ a_i a_j$
Hidden Markov Model	Pattern	Stochastic	Sweep	$E = Pr(O\|\lambda) = \sum_{i=1}^{N} \beta_i$	$\eta\ w_{ij} \sum_{l=1}^{L} \delta_i(l)\ f_i(\theta(l)) a_j(l-1)$	Yes	[22, 28], $w_{ij} \geq 0$, $\sum_{i=1}^{N} w_{ij} = 1$

Table 1: Neural network examples which are classified according to the factors considered in the retrieving and learning phase. † Due to the difficulty (for sake of nonlinear thresholding elements) of calculating $\delta_i(l)$ quantitatively, approximations were made. ‡ Instead of adaptive learning, a batch processing is adopted to determine the weights.

1.5 Architectural Aspects of ANNs

This generic formulation also leads to a basic structure for a parallel neurocomputer architecture and/or a universal simulation tool. Summarized below are several key considerations for parallel processing and array architecture implementations of neural networks:

- Convergence issues of synchronous (parallel) updating of system dynamics in the retrieving phase.
- The architectural design should ensure that the processing in both the retrieving phase and the learning phase share the same array configuration, storage, and processing hardware. This will not only speed up real-time learning but also avoid the difficulty in the reloading of synaptic weights for retrieval.
- The digital design must identify a proper digital arithmetic technique to efficiently compute the necessary operations required in both phases.
- The array architecture design should maximize the strength of VLSI in terms of intensive and pipelined computing and yet circumvents its main limitation on communication. It is desirable to find a local interconnectivity of systolic solution to the implementation of the global interconnectivity of neural networks.
- VLSI arrays can be systematically derived from the dependency structure of the neural network algorithms [25, 19].
- A digital design must offer a greater flexibility, so a general-purpose programmable array architecture is derived for implementing a wide variety of neural network algorithms in both the retrieving and the learning phases.

Based on Table 1, it may be further derived that operations in both the retrieving and learning phases of the generic iterative ANN models can be formulated as *consecutive matrix-vector multiplication*, *consecutive vector-matrix multiplication*, or *outer-product updating* problems [21]. In terms of the array structure, all these formulations lead to a same universal ring systolic array architectures. In terms of the functional operations at each processor, all these formulations calls for a MAC (multiply and accumulation) processor and a nonlinear operator. The choice of arithmetic processing unit can be determined only after an extensive simulation and numerical analysis. Preliminarily speaking, for a *time efficient* design, a parallel array multiplier (with piecewise linear approximation of sigmoid functions) may still be a favored option. For an *area efficient* design, a Cordic processor (which can implement both MAC and nonlinear sigmoid function) might be a good alternative [2, 21, 14].

2 Unification of Existing Connectionist Neural Networks

This section illustrates how the generic iterative ANN models may be used as a unifying model for the existing connectionist neural networks. The case studies include:

- An example of convergence study of parallel updating in Hopfield associative neural networks.
- Unifying several competitive learning networks under the generic iterative ANN formulation.
- A unified viewpoint of the multilayer feed-forward neural networks, especially the link between multilayer perceptrons and hidden Markov models.

2.1 Parallel Hopfield Associative Neural Network

A Hopfield associative neural network is a single layer (time iterative) feedback network, which consists of N binary-valued neurons linked to each other with symmetric weights $\{w_{ij} = w_{ji}\}$. In the Hopfield model, thresholding elements are added to linear associators to perform iterative feedback auto-association tasks. Moreover, a notion of energy function is adopted to prove that the feedback system exhibits a number of locally stable points (attractors) in the state space, which provides a basic mechanism for signal retrieval and error correction from partial or noisy missing information [13].

Retrieving Phase The system dynamics in the retrieving phase of a Hopfield associative network is

$$\begin{aligned} u_i(l+1) &= \sum_{j=1}^{N} w_{ij} a_j(l) \\ a_i(l+1) &= \begin{cases} 1 & \text{if } u_i(l+1) > -\theta_i \\ 0 & \text{if } u_i(l+1) < -\theta_i \\ a_i(l) & \text{if } u_i(l+1) = -\theta_i \end{cases} \end{aligned} \tag{13}$$

The original Hopfield associative network requires each neuron to be updated asynchronously (sequentially) to guarantee the convergence of the system dynamics. The dynamic evolution of the system state can be regarded as an energy minimization that continues until a stable state (local energy minimum) is reached. To demonstrate the convergence, a notion of Liapunov energy function is often instrumental.

In the following, we will show that synchronous (parallel) updating of all the neurons at each iteration is also possible for non-negative definite symmetric weight matrix $\{w_{ij}\}$. To demonstrate the convergence of the parallel updating iterations, it is useful to adopt the following Liapunov energy function $E(l)$ (after the l-th updating):

$$E(l) = -\frac{1}{2}\sum_{i=1}^{N}\sum_{j=1}^{N} w_{ij} a_i(l) a_j(l) - \sum_{i=1}^{N} \theta_i a_i(l)$$

If all the neurons are updated in parallel at each time iteration, then the energy change between any two iterations is

$$\begin{aligned} \triangle E(l+1) &= E(l+1) - E(l) \\ &= -\frac{1}{2}\sum_{i=1}^{N}\sum_{j=1}^{N} w_{ij} a_i(l+1) a_j(l+1) - \sum_{i=1}^{N} \theta_i a_i(l+1) \\ &\quad +\frac{1}{2}\sum_{i=1}^{N}\sum_{j=1}^{N} w_{ij} a_i(l) a_j(l) + \sum_{i=1}^{N} \theta_i a_i(l) \\ &= -\sum_{i=1}^{N}(a_i(l+1) - a_i(l)) \, [\sum_{j=1}^{N} w_{ij} a_j(l) + \theta_i] \\ &\quad -\frac{1}{2}\sum_{i=1}^{N}(a_i(l+1) - a_i(l))[\sum_{j=1}^{N} w_{ij}(a_j(l+1) - a_j(l))] \\ &= -\triangle \mathbf{a}^T(l+1)\,[\mathbf{u}(l+1) + \theta] \end{aligned}$$

$$
\begin{aligned}
& -\frac{1}{2} \triangle \mathbf{a}^T(l+1)\mathbf{W} \triangle \mathbf{a}(l+1) \\
= & \triangle E_1(l+1) + \triangle E_2(l+1)
\end{aligned} \tag{14}
$$

Since the nondecreasing thresholding functions are assumed, the signs of $\{\triangle a_i(l+1)\}$ and $\{u_i(l+1)+\theta_i\}$ are the same (see Eq. 13), and $\triangle E_1(l+1) \leq 0$. In order to also guarantee $\triangle E_2(l+1) \leq 0$, the weight matrix should be a non-negative definite matrix, this can be ensured by setting weight matrix to be

$$
w_{ij} = \sum_{p=1}^{P}(2\beta_i^{(p)}-1)(2\beta_j^{(p)}-1), \quad \mathbf{W} = \sum_{p=1}^{P}[2\mathbf{v}^{(p)}-\mathbf{1}]\,[2\mathbf{v}^{(p)}-\mathbf{1}]^T \tag{15}
$$

where P binary reference patterns, represented by $\{\mathbf{v}^{(p)} = [\beta_1^{(p)}, \beta_2^{(p)},, \cdots, \beta_N^{(p)}], p = 1, 2, \ldots, P\}$, are stored in and to be retrieved from the associative network. Note that the $\mathbf{W}$ matrix is formed by an outer product without diagonal nullification (as opposed to original Hopfield network [13]), so it is a non-negative definite matrix.

Learning Phase The weight determination process based on Eq. 15 can be interpreted by the recursive weight updating formulation given in Eq. 3. The measurement function E to be optimized bears the same form as the energy function in the retrieving phase, i.e.,

$$
\begin{aligned}
E & = -\frac{1}{2}\sum_{i=1}^{N}\sum_{j=1}^{N} w_{ij}a_i(L)a_j(L) - \sum_{i=1}^{N}\theta_i a_i(L) \\
& = -\frac{1}{2}\sum_{i=1}^{N}\sum_{j=1}^{N} w_{ij}\beta_i\beta_j - \sum_{i=1}^{N}\theta_i\beta_i
\end{aligned}
$$

If we use the additive updating updating formulation (cf Eq. 5), then

$$
w_{ij} \Longleftarrow w_{ij} - \eta\frac{\partial E}{\partial w_{ij}} = w_{ij} + \frac{\eta}{2}\beta_i\beta_j \tag{16}
$$

where the the recursion index is corresponding to each training pattern used. This leads to the simplest form of Hebbian learning rule [11], and the stored pattern $\{\beta_i\}$ are the desired outputs (desired auto-associative retrieval information).

Instead of going through the iterative Hebbian weight updating learning procedures, Hopfield used the ensemble average of $\{w_{ij}\}$ over P (training) pattern recursions to determine the fixed connection weights with approximate *zero statistical mean* [13], this leads to the weight determination formulation given in Eq. 15.

2.2 Competitive Learning Networks

Competitive learning is an unsupervised procedure that classify a set of input patterns into a number of disjoint clusters in such a way that the input patterns within each cluster are all similar to one another [12]. Most competitive learning networks are single layer feed-forward networks using winner-take-all mechanism. It is possible to provide a unified perspective of several competitive learning networks based on the generic iterative ANN formulation (see Table 1). The main idea is to identify a common measurement function E used in the learning phase of all the competitive learning networks, i.e.,

$$
E = \frac{1}{2}\sum_{i=1}^{N}\beta_i(\alpha_j - w_{ij})^2 \tag{17}
$$

Retrieving Phase Without loss of generality, the system dynamics between the inputs $\{\alpha_i\}$ and the outputs $\{\beta_i\}$ in the retrieving phase of a competitive learning network is given:

$$\begin{aligned} u_i &= \sum_{j=1}^{N_0} w_{ij}\alpha_j \\ \beta_i &= \begin{cases} 1 & \text{if } u_i > u_k \ \forall k \\ 0 & \text{if } else \end{cases} \end{aligned} \tag{18}$$

where $1 \leq i \leq N_1$. The *winner-take-all* competition mechanism (e.g., lateral inhibitions) is used in the output layer so that only the neuron receiving largest input is activated.

Learning Phase The learning phase in the competitive learning network can also be interpreted by the recursive weight updating formulation (see Eq. 3 and Eq. 17).

$$\begin{aligned} w_{ij} &\Leftarrow w_{ij} - \eta \frac{\partial E}{\partial w_{ij}} \\ &= w_{ij} + \eta \, \beta_i \, (\alpha_j - w_{ij}) \end{aligned} \tag{19}$$

where one recursion is for one pattern. According to Eqs. 18, 19 implies that only the weights associated with the winning neuron are updated and all the other weights remain unchanged. This is a special feature of the competitive learning networks.

We hasten to note that the above formulation can only describe the basic feature commonly shared by the competitive learning networks. In actuality, each individual model has almost unexceptionally adopted certain special mechanism – see the "comments" in Table 1. Some examples are highlighted below.

- Kohonen's self-organized feature map [17, 18] introduced a neighborhood (whose size slowly decreases with each iteration) of a winning neuron in a two dimensional neuron layer. Weights associated with the winner and the neurons in the neighborhood of the winner are all modified. This has purpose of making the neurons more responsive to the current input pattern.

- The adaptive resonance theory (ART) by Carpenter and Grossberg [10, 7, 8] introduced a *vigilance test* to adaptively create new neuron units for the incoming input patterns which are quite different from the memorized patterns. This test requires additional modifiable feedback weights connecting from output layer to the input layer.

- Rumelhart's competitive learning algorithm [30] introduced a *leaky learning model* to prevent the possibility of totally unlearned neurons. This is done by performing training in Eq. 19 over all the weights in the network. However, the weights associated with the winner get much larger η values.

- Neocognitron can be formed as a hierarchical learning system by cascading many single layer competitive learning networks[9]. The learning for each layer is largely based on single layer analysis. It progresses stage by stage from the input layer to the output layer.

2.3 Multilayer Feed-Forward Neural Networks

Multilayer feed-forward neural networks are spatially iterative neural networks, which have several homogeneous or heterogeneous layers of *hidden neuron units* between the input and output neuron layers (see Figure 1). The weight updating for the hidden layers adopts the mechanism of back-propagated corrective signal from the output layer (see Section 1.3).

This section provides a unified viewpoint of two popular multilayer feed-forward neural networks: multilayer perceptrons [29] and hidden Markov models [28].

2.3.1 Multilayer Perceptrons

Retrieving Phase The system dynamics in the retrieving phase of an L-layer perceptron can be described by the following spatially iterative equations (where the iteration index l denotes the *layer* iterations):

$$\begin{aligned} u_i(l+1) &= \sum_{j=1}^{N_l} w_{ij}(l+1)a_j(l) \\ a_i(l+1) &= f_i(u_i(l+1)+\theta_i(l+1)) \end{aligned} \tag{20}$$

where $1 \leq i \leq N_{l+1}$, $0 \leq l \leq L-1$, and $f_i(l+1)$ is non-decreasing and differentiable. For simplicity, the modifiable external inputs $\{\theta_i(l+1)\}$ are often treated as special synaptic weights $\{w_{i,0}(l+1)\}$ which have clamped inputs $a_0(l)=1$.

Learning Phase The learning phase of an L-layer multilayer perceptron follows the recursive *additive* weight updating formulation (see Eq. 3), i.e., gradient descent approach. Given a pair of input/target training patterns, $\{\alpha_i^{(p)},\ i=1,\cdots,N_0\}$, $\{t_j^{(p)},\ j=1,\cdots,N_L\}$, and $p=1,\cdots,P$, our goal is to iteratively choose a set of $\{w_{ij}(l),\ \forall l\}$ for all layers so that the squared error function E can be minimized;

$$E = \frac{1}{2}\sum_{i=1}^{N_L}(t_i - a_i(L))^2 = \frac{1}{2}\sum_{i=1}^{N_L}(t_i-\beta_i)^2 \tag{21}$$

To be more specific, the iterative gradient descent formulation (with pattern recursion) for the multilayer perceptron can be written as:

$$\begin{aligned} w_{ij}(l) &\Longleftarrow w_{ij}(l) - \eta\,\frac{\partial E}{\partial w_{ij}(l)} \\ &= w_{ij}(l) - \eta\,\frac{\partial E}{\partial a_i(l)}\frac{\partial a_i(l)}{\partial w_{ij}(l)} \\ &= w_{ij}(l) - \eta\,\delta_i(l)\frac{\partial a_i(l)}{\partial w_{ij}(l)} \\ &= w_{ij}(l) - \eta\,\delta_i(l)\frac{\partial a_i(l)}{\partial u_i(l)}\frac{\partial u_i(l)}{\partial w_{ij}(l)} \\ &= w_{ij}(l) - \eta\,\delta_i(l) f_i'(u_i(l))a_j(l-1) \end{aligned} \tag{22}$$

where $1 \leq l \leq L$, and $f_i'(x)$ is the derivative of $f_i(x)$ with respect to x. The corrective signal $\delta_i(l)$ can be recursively calculated as shown below (see Eq. 9): For $L-1 \geq l \geq 1$.

$$\begin{aligned}\delta_i(l) &= \sum_{j=1}^{N_{l+1}} \delta_j(l+1)\,\frac{\partial a_j(l+1)}{\partial a_i(l)} \\ &= \sum_{j=1}^{N_{l+1}} \delta_j(l+1)\frac{\partial a_j(l+1)}{\partial u_j(l+1)}\frac{\partial u_j(l+1)}{\partial a_i(l)} \\ &= \sum_{j=1}^{N_{l+1}} \delta_j(l+1)f_j'(u_j(l+1))\; w_{ji}(l+1) \qquad (23)\end{aligned}$$

Note that the backward initialization error signal on the top layer $\delta_i(L)$ can be computed directly without recursive formulation, i.e., $\delta_i(L) = -(t_i - \beta_i)$.

2.3.2 Recurrent Back-Propagation Networks

A special case of the multilayer perceptrons is the recurrent back-propagation network, where the synaptic weights $w_{ij}(l) = w_{ij},\ \forall\ l$ (see Eq. 10). The recurrent back-propagation network can be considered as a homogeneous multilayer perceptron with the iteration index l interpretated as the *time* iteration. (In essence, it is a single layer feedback network which is synchronously updated over a fixed (L) iterations) [29, 12].

All the derivations from Eqs. 20, 21, and 22, are applicable to the recurrent back-propagation network. In addition, the averaging mechanism (see Eq. 12) introduced in Section 1.3 is adopted. Therefore

$$\begin{aligned}w_{ij} &\Longleftarrow w_{ij} - \eta \sum_{l=1}^{L} \frac{\partial E}{\partial w_{ij}(l)} \\ &= w_{ij} - \eta \sum_{l=1}^{L} \delta_i(l) f_i'(u_i(l)) a_j(l-1) \qquad (24)\end{aligned}$$

Note that the computation of $\delta_i(l)$ follows the back propagation rule of multilayer perceptron (see Eq. 23).

2.3.3 Hidden Markov Models

A hidden Markov model (HMM) is a doubly stochastic process with an underlying stochastic process that is not observable (i.e., hidden), but can only be observed through another set of stochastic process that produces the sequence of observed symbols [28]. From the retrieving phase point of view, HMMs described by a trellis structure can be regarded as a homogeneous multilayer perceptrons with a squashing type of nonlinear activation function. From the learning phase point of view, the homogeneity and constant-sum constraints can be applied to the trellis structure to derive the Baum-Welch reestimation formulation in HMMs.

The basic components of an HMM can be represented by:

1. There are N possible states $\{q_i,\ i = 1,\ 2,\ \ldots,\ N\}$ in an HMM with transition probability $\{w_{ij}\}$:

$$w_{ij} = Pr(\ q_i \text{ at } l \mid q_j \text{ at } l-1), \quad l = 1,\ 2,\ \cdots,\ L$$

2. The initial state probability π_i:

$$\pi_i = Pr(\ q_i \text{ at } l = 0)$$

3. There are possible occurrence patterns $\{v_k\}$ that can be observed at i-th state with probability f_i:

$$f_i(v_k) = Pr(\ v_k \mid q_i\ )$$

Given an input (test or training) pattern sequence $\mathbf{O} = (\theta(0),\ \theta(1),\ \cdots\ \theta(L))$, and a pre-specified HMM, $\lambda = \{w_{ij}, f_i, \pi_i\}$, the retrieving phase of an HMM is to compute the occurrence probability $Pr(\mathbf{O}|\lambda)$, which allows us to choose one among several models that best matches the observations. The learning phase of an HMM is to find a new set of $\lambda = \{w_{ij}, f_i, \pi_i\}$, so that the occurrence probability $Pr(\mathbf{O}|\lambda)$ can be maximized.

Retrieving Phases It is observed that the computation of the occurrence probability in the retrieving phase of an HMM can be greatly facilitated by the trellis structure representation of the algorithm, which leads to the connectionist network structure [16]. The system dynamics in the retrieving phase of an HMM can thus be written:

$$u_i(l+1) = \sum_{j=1}^{N} w_{ij}\ a_j(l) \tag{25}$$

$$a_i(l+1) = f_i(\theta(l+1))\ u_i(l+1) \tag{26}$$

where $0 \le l \le L-1$ and $1 \le i \le N$. The activation value, $a_i(l) = Pr(\theta(0), \cdots, \theta(l),\ q_i \text{ at } l \mid \lambda)$, denotes the forward likelihood. The initial forward likelihood $a_i(0) = f_i(\theta(0))\pi_i,\ 1 \le i \le N$. The occurrence probability can then be calculated [28]:

$$Pr(\mathbf{O}|\lambda) = \sum_{i=1}^{N} a_i(L) = \sum_{i=1}^{N} \beta_i$$

This again leads to the L-level feed-forward network, with l index specifying the pattern iteration.

Learning Phase The *Baum-Welch reestimation* learning algorithm is adopted in the learning phase of an HMM to adjust the model parameters $\{w_{ij},\ f_i,\ \pi_i)$ to maximize the occurrence probability of the input training pattern given the model. Given the training pattern sequence $\mathbf{O}$, our goal is to iteratively choose a homogeneous set of $\{w_{ij}\}$, so that the likelihood $Pr(\mathbf{O}|\lambda)$ can be maximized [22, 28].

$$E = \sum_{i=1}^{N} a_i(L) = \sum_{i=1}^{N} \beta_i \tag{27}$$

Due to the constraints imposed on the weights, i.e., $w_{ij} \ge 0$ and $\sum_{i=1}^{N} w_{ij} = 1$, the multiplicative gradient ascent updating proposed for the iterative constrained optimization tasks [5, 6] can be adopted. As a matter of fact, most of the mathematical derivation follows that of homogeneous multilayer perceptron except the sweep recursion is adopted [15], i.e.,

$$\begin{aligned}
w_{ij} &\Longleftarrow \eta\, w_{ij} \sum_{l=1}^{L} \frac{\partial E}{\partial w_{ij}(l)} \\
&= \eta\, w_{ij} \sum_{l=1}^{L} \delta_i(l)\, \frac{\partial a_i(l)}{\partial w_{ij}(l)} \\
&= \eta\, w_{ij} \sum_{l=1}^{L} \delta_i(l)\, \frac{\partial a_i(l)}{\partial u_i(l)}\, \frac{\partial u_i(l)}{\partial w_{ij}(l)} \\
&= \eta\, w_{ij} \sum_{l=1}^{L} \delta_i(l)\, f_i(\theta(l))\, a_j(l-1) \qquad (28)
\end{aligned}$$

Again the back-propagated corrective (or called backward likelihood) signal $\delta_i(l)$ can be recursively computed:

$$\begin{aligned}
\delta_i(l) &= \sum_{j=1}^{N} \frac{\partial E}{\partial a_j(l+1)}\, \frac{\partial a_j(l+1)}{\partial a_i(l)} \\
&= \sum_{j=1}^{N} \delta_j(l+1)\, f_j(\theta(l+1))\, w_{ji} \qquad (29)
\end{aligned}$$

From Eq. 27, it is obvious that $\delta_i(L) = \frac{\partial E}{\partial a_i(L)} = 1$.

Unlike the back-propagation learning algorithm where η is not explicitly determinable [20], the updating step η in the Baum-Welch reestimation algorithm is constrained by the standard probability property,

$$\begin{aligned}
\sum_{i=1}^{N} w_{ij} &= 1 = \eta \sum_{i=1}^{N} w_{ij} \sum_{l=1}^{L} \delta_i(l)\, f_i(\theta(l))\, a_j(l-1) \\
&= \eta \sum_{i=1}^{N} \sum_{l=1}^{L} w_{ij}\, \delta_i(l)\, f_i(\theta(l))\, a_j(l-1) \\
&= \eta \sum_{l=1}^{L} [\sum_{i=1}^{N} w_{ij}\, \delta_i(l)\, f_i(\theta(l))]\, a_j(l-1) \\
&= \eta \sum_{l=1}^{L} \delta_j(l-1)\, a_j(l-1) \\
&= \eta \sum_{l=0}^{L-1} \delta_j(l)\, a_j(l) \qquad (30)
\end{aligned}$$

where $j = 1, 2, \cdots, N$, and it is obvious that

$$\eta = \eta_{ij} = \frac{1}{\sum_{l=0}^{L-1} \delta_j(l)\, a_j(l)}$$

Similarly, the updating for $\{f_i(v_k)\}$ and $\{\pi_i\}$ can also be performed, more specifically,

$$
\begin{aligned}
f_i(v_k) &\Longleftarrow \eta\ f_i(v_k) \sum_{l=1}^{L} \frac{\partial E}{\partial a_i(l)} \frac{\partial a_i(l)}{\partial f_i(v_k)} \\
&= \eta\ f_i(v_k) \sum_{l=1,\ \theta(l)=v_k}^{L} \delta_i(l)\ u_i(l) \\
&= \eta \sum_{l=1,\ \theta(l)=v_k}^{L} \delta_i(l)\ f_i(\theta(l))\ u_i(l) \\
&= \eta \sum_{l=1,\ \theta(l)=v_k}^{L} \delta_i(l)\ a_i(l)
\end{aligned} \tag{31}
$$

3 Future Extension: Generating New Neural Networks

Based on the better understanding of the generic ANN formulation, new models may be derived. Some multiplicative learning model may prove to be suitable for application where a constant-sum-constraint is imposed. In the following a potential multiplicative recurrent back-propagation network is proposed to serve this purpose.

Retrieving Phase The system dynamics in the retrieving phase of an L-iteration (time iterative) multiplicative recurrent back-propagation network can be written:

$$
\begin{aligned}
u_i(l+1) &= \sum_{j=1}^{N} w_{ij}\ a_j(l) \\
a_i(l+1) &= f_i(u_i(l+1)+\theta_i)
\end{aligned} \tag{32}
$$

where $0 \le l \le L-1$ and $1 \le i \le N$. Note that, for sake of convergence, the f_i should be chosen to be an arbitrary polynomial function of $u_i(l+1)$ with positive coefficients [4, 31]. This model is specifically useful for applications where constraints are imposed on the inputs/targets and weights:

$$
\begin{aligned}
a_i(0) \ge 0, \quad \forall\, i \quad \text{and} \quad \sum_{i=1}^{N} a_i(0) = c \\
w_{ij} \ge 0, \quad \forall\, i,j \quad \text{and} \quad \sum_{j=0}^{N} w_{ij} = 1
\end{aligned}
$$

Learning Phase The learning phase of the multiplicative recurrent back-propagation network follows the same derivation of the recurrent back-propagation algorithm, except that the weights are updated in the multiplicative formulation (additive formulation is also allowed [27]). The goal is to maximize the measurement function E, which should be again a polynomial of $\{a_i(L)\}$ with positive coefficients. One possible choice is

$$
E = \sum_{i=1}^{N} t_i\ a_i(L) = \sum_{i=1}^{N} t_i\ \beta_i \tag{33}
$$

where $t_i \geq 0$, $\forall i$, and $\sum_{i=1}^{N} t_i = c$. This specifies the correlation between the target and the actual outputs. The weight updating follows that of recurrent back-propagation networks:

$$\begin{aligned} w_{ij} &\Longleftarrow \eta\, w_{ij} \sum_{l=1}^{L} \frac{\partial E}{\partial w_{ij}(l)} \\ &= \eta\, w_{ij} \sum_{l=1}^{L} \delta_i(l) \frac{\partial a_i(l)}{\partial u_i(l)} \frac{\partial u_i(l)}{\partial w_{ij}(l)} \\ &= \eta\, w_{ij} \sum_{l=1}^{L} \delta_i(l)\, f_i'(u_i(l))\, a_j(l-1) \end{aligned} \tag{34}$$

where the back-propagated corrective signal $\delta_i(l)$ can again be recursively computed (see Eq. 9).

4 Conclusion

Neural networks offer an attractive new computational tool for many applications in vision, speech, signal processing, and robotics. Their real potential lies in the ability to learn and self-adapt. To fully realization such potential, there is a need of reexamination of the theoretical foundations of existing neural networks. For certain applications, novel neural networks will be necessary. To this end, a unified formulations of iterative neural networks is proposed. The formulation will allow us to better understand several critical issues in neural networks, such as convergence, stability, and connectivity issues. It can also help further our understanding of the relationship between neural networks and conventional approaches. The advanced mathematical theories and new computer tools based on the unified formulation should in many ways benefit the development of the simulation and implementation tools for neural networks research.

References

[1] D. H. Ackley, G. E. Hinton, and T. J. Sejnowski. A learning algorithm for Boltzmann machines. *Cognitive Science*, 9:147–169, 1985.

[2] H.M. Ahmed. Alternative arithmetic unit architectures for VLSI digital signal processors. In *VLSI and Modern Signal Processing*, chapter 16, pages 277–306. Prentice Hall, Inc., Englewood Cliffs, NJ, 1985.

[3] J. A. Anderson. *Neurocomputing - Paper Collections.* MIT Press, 1988.

[4] L. E. Baum and G. R. Sell. Growth transformations for functions on manifolds. *Pacific Journal of Mathematics*, 27(2):211–227, 1968.

[5] L.E. Baum and J. A. Eagon. An inequality with applications to statistical estimation for probabilistic function of Markov processes and a model for ecology. *Amer. Math. Soc. Bull.*, 73:360–363, May 1967.

[6] L.E. Baum, T. Petrie, G. Soules, and N. Weiss. A maximization technique occurring in the statistical analysis of probabilistic functions of Markov chains. *Ann. Math. Statistic.*, 41:164–171, 1970.

[7] G. A. Carpenter and S. Grossberg. ART2: Self-organization of stable category recognition codes for analog input patterns. In *Proc. IEEE ICNN'87, San Diego*, pages II 727– II 736, 1987.

[8] Gail A. Carpenter and Stephen Grossberg. ART2: Self-organized of stable category recognition codes for analog input patterns. *Applied Optics*, 26(23):4919–4930, December 1987.

[9] K. Fukushima. Neocognitron: A self-organizing neural network model for a mechanism of pattern recognition unaffected by shift in position. *Biological Cybernetics*, 36:193–202, April 1980.

[10] S. Grossberg. Adaptive pattern classification and universal recoding: Part 1. parallel development and coding of neural feature detectors. *Biological Cybernetics*, 23:121–134, 1976.

[11] D. O. Hebb. *The Organization of Behavior.* Wiley Inc., New York, 1949.

[12] G. E. Hinton. Connectionist learning procedure. Technical Report CMU-CS-87-115, Carnegie Mellon University, September 1987.

[13] J. J. Hopfield. Neural network and physical systems with emergent collective computational abilities. In *Proc. Natl'. Acad. Sci. USA*, volume 79, pages 2554–2558, 1982.

[14] J. N. Hwang. *Algorithms/Applications/Architectures of Artificial Neural Nets.* PhD thesis, Dept. of Electrical Engineering, University of Southern California, December 1988.

[15] J. N. Hwang and S. Y. Kung. A unifying viewpoint between multilayer perceptrons and hidden Markov models. In *IEEE Int'l Symposium on Circuits and Systems, ISCAS'89, Portland*, pages 770–773, May 1989.

[16] B. H. Juang. On the hidden Markov model and dynamic time warping for speech recogintion – a unified view. *AT&T Bell Laboratories Technical Journal*, 63(7):1213–1243, September 1984.

[17] T. Kohonen. Self-organized formation of topologically correct feature map. *Biological Cybernetics*, 43:59–69, 1982.

[18] T. Kohonen. *Self-Organization and Associative Memory, Series in Information Science, Vol. 8.* Springer-Verlag, New York, 1984.

[19] S. Y. Kung. *VLSI Array Processors.* Prentice Hall Inc., N.J., 1988.

[20] S. Y. Kung and J. N. Hwang. An algebraic projection analysis for optimal hidden units size and learning rate in back-propagation learning. In *IEEE, Int'l Conf. on Neural Networks, ICNN'88, San Diego* , pages Vol.1: 363–370, July 1988. (Also accepted for publication in *Neural Networks.*)

[21] S.Y. Kung and J. N. Hwang. A unified systolic architecture for artificial neural networks. *Journal of Parallel and Distributed Computing, Special Issue on Neural Networks*, 6(2):358–387, April 1989.

[22] S. E. Levinson, L. R. Rabiner, and M. M. Sondhi. An introduction to the application of the theory of probabilistic functions of a Markov process to automatic speech recognition. *The Bell System Technical Journal*, 62:1035–1074, April 1983.

[23] R. Linsker. Self-organization in a perceptual network. *IEEE Computer Magazine*, 21:105–117, March 1988.

[24] J. L. McClelland and D. E. Rumelhart. Distributed memory and the representation of general and specific information. *Journal of Experimental Psychology: General*, 114:158–188, 1985.

[25] C. Mead and L. Conway. *Introduction to VLSI Systems.* Addison-Wesley, 1980.

[26] N. J. Nilsson. *Learning Machines.* McGraw-Hill Book Company, 1965.

[27] L. R. Rabiner. A tutorial on hidden Markov models ans selected applications in speech recognition. *Proceedings IEEE*, 77(2):257–286, February 1989.

[28] L. R. Rabiner and B. H. Juang. An introduction to hidden Markov models. *IEEE ASSP Magazine*, 3(1):4–16, January 1986.

[29] D. E. Rumelhart, G. E. Hinton, and R. J. Williams. Learning internal representations by error propagation. *Parallel Distributed Processing (PDP): Exploration in the Microstructure of Cognition (Vol. 1)*, chapter 8, pages 318–362. MIT Press, Cambridge, Massachusetts, 1986.

[30] D. E. Rumelhart and D. Zipser. Feature discovery by competitive learning. *Cognitive Science*, 9:75–112, 1985.

[31] P. F. Stebe. Invariant functions of an iterative process for maximization of a polynomial. *Pacific Journal of Mathematics*, 43(3):765–783, 1972.

[32] B. Widrow and R. Winter. Neural nets for adaptive filtering and adaptive pattern recognition. *IEEE Computer Magazine*, 21:25–39, March 1988.

[33] G. Widrow and M. E. Hoff. Adaptive switching circuit. *IRE Western Electronic Show and Convention: Convention Record*, pages 96–104, 1960.

PRACTICAL NEURAL COMPUTING FOR ROBOTS: PROSPECTS FOR REAL-TIME OPERATION

I. Aleksander

Department of Electrical Engineering
Imperial College of Science, Technology and Medicine
Exhibition Road, London SW7 2BZ, U.K.

1. Why the Recent Interest in Neural Computing?

The sudden growth of interest in neural computing is a remarkable phenomenon that will be seen by future historians of computer science as marking the 1980s in much the same way as research into artificial intelligence (AI) has been the trademark of the 1970s. There is one major difference, however: in contrast with AI which was largely an outlet for a minority of computer scientists, neural computing unites a very broad community: physicists, statisticians, parallel processing experts, optical technologists, neurophysiologists and experimental biologists. The focus of this new paradigm is rather simple. It rests on the recognition by this diverse community that the brain 'computes' in a very different way from the conventional computer.

This is quite contrary to the focus of the AI paradigm, which is based on the premise that an understanding of what the brain does represents a true understanding only if it can be explicitly expressed as a set of rules that, in turn, can be run on a computer which consequently performs artifically intelligent tasks. Those who contribute to neural computing believe that the brain, given sensors and a body, builds up its own hidden rules through what is usually called 'experience'. When a person activates his muscles in complex sequences driven by signals from his eyes, from sensory receptors in his muscles and even from his ears when performing an every-day act such as getting on a bus, or when he notices a 'polite chill' in a colleague's voice, these are manifestations of large numbers of implicit rules at work in the brain in a simultaneous and coordinated fashion. In neural computing it is believed that the cellular structures within which such rules can grow and be executed are the focus of important study in contrast to the AI concern of trying to extract rules from a human being in order to run them on a computer.

Neural computing therefore is concerned with a class of machines that compute by absorbing experience, a class which in that sense includes

NATO ASI Series, Vol. F 66
Sensor-Based Robots: Algorithms and Architectures
Edited by C. S. George Lee

the brain, but may include other machines with similar properties. It is important to stress that neural computing scientists are not latter-day Frankensteins in the business of making brains. They are however, united in trying to understand computing structures that are brain-like in the sense that they acquire knowledge through experience rather than pre-programming. So, neural computing is not necessarily about the details of mimicking the neurons of the brain and their interconnections, but is more about the nature of the broad class of machines which behave in brain-like ways, and through this, add both to our armoury of knowledge in computing and to our ability to apply such knowledge through the design of novel machinery.

Perhaps from all this it may be possible to ferret out a definition of neural computing:

'Neural computing is the study of cellular networks that have a natural propensity for storing experiential knowledge. Such systems bear a resemblance to the brain in the sense that knowledge is acquired through training rather than programming and is retained due to changes in node functions. The knowledge takes the form of stable states or cycles of states in the operation of the net. A central property of such nets is to recall these states or cycles in response to the presentation of cues.

2. Relevance to Sensor-based Robots

Every aspect of neural computing points to its usefulness in the marshalling of complex sensory information into symbols that can be subjected to conventional computing which, in turn, generates robot actions. The concept of training by example is natural to robot operation in determining trajectories. With neural computing the training can be extended to include this complex sensory information and distinguish subtleties in such data.

The rest of this paper will first summarise some general aspects of the neural computing paradigm which carry the penalty of slowness when performed on conventional computers. The methods developed in the author's laboratory are then described. These are better placed for implementation and real-time performance since they are based on conventional logic design methodology. A system that has gone into industrial use (the WISARD) will be described. Current work which makes use of probabilistic parameters will be presented at the end of the paper.

3. Some History

There is an undoubted degree of hype associated with this field. Phrases such as 'the dawn of a new era' are used by conference organisers and the press talks of 'new computers that are built like the brain and really think for themselves'. But there is nothing new about neural computing: it is as fundamental as the more conventional or 'algorithmic' mode. Norbert Wiener in his 1947 book 'Cybernetics' wrote:

'Mr. Pitts had the good furtune to come under Dr. McCulloch's influence (in 1943) and the two began working quite early on problems concerning the union of nerve fibres by synapses into systems with given overall properties..... They added elements suggested by the ideas of Turing in 1936: the consideration of nets containing cycles...'

So some of the discussions that echo in the auditoria of contemporary conferences were begun more than ten years before the invention of the computer that we know and love. The McCulloch and Pitts model of the neuron is still the basis for more neural node models (1943), and Turing's concern about nets and cycles is the very stuff of neural computing. Indeed the 1960s were most productive in this area. Well known is the work of Rosenblatt of Cornell Unversity on 'perceptrons' (Rosenblatt, 1962) and the destruction of the credibility of this work by Minsky and Papert of MIT in 1969 which led to a halt to such work in the USA.

But in Europe, neural net researchers were not as prone to the winds of change that blew from the direction of MIT as their colleagues in the USA. Eduardo Caianiello in Italy and Teuvo Kohonen in Finland continued to develop an understanding of neural computers to great depth and elegance. I too, largely through a fascination with how well and fast the brain performs tasks of pattern recognition with components much slower than those found in computers, continued designing machines based on neuron models that I first defined in 1965. These are characterised by the fact that they are easily implemented in electronics and understood through formal logic. This has led to the commercialisation of practical systems, and is pointing to new high-performance systems for the future (Aleksander 1984).

Despite the historical appeal of these approaches, there is no doubt that the work of the 'Parallel Distributed Processing' (PDP) group in the USA has been fundamental in nailing down both the language and the targets of the current paradigm (Rumelhart and McClelland, 1986). But what is the rapidly expanding band of workers in neural computing hoping to achieve?

4. Four Promises

There appear to be be four major reasons for developing neural computing methods, the first of which is a rebuttal of the Minsky and Papert criticism. Although this is not the place to debate the technical issues, it is helpful to note that the criticism was founded on a demonstration that there are simple pattern recognition tasks that neural nets appeared not to be able to accomplish. It is now clear that this conclusion was mistaken because it considered only a restricted class of neural system. In fact, the first promise of neural computing it that it is computationally complete. This means that, give an appropriate neural structure, and appropriate training, there are no computational tasks that are not available to neural nets. This does not mean that a neural net is as efficient at performing certain tasks as a conventional computer. For example, in order to perform multiplications, the net may have to learn multiplication tables in the way that a human being does, but these can be easily performed by a fast arithmetic unit in a conventional computer. Also there are tasks for which the neural net not only outperforms the conventional computer but is the only way of performing the task.

This leads to the second promise: functional use of experiential knowledge. It is here that the neural net can perform functions that are beyond the capability of rule-based, conventional systems. Typical are the Achilles' heels of Artificial Intelligence: speech, language and scene understanding. The problem with conventional approaches to these tasks is either that rules are hard to find, or the number of such rules explodes alarmingly even for simple problems. Imagine having to distinguish betweeen the faces of two people. What information should be extracted? What should be measured in this information? How can we be sure that what we meaure will distinguish between the faces? Although a fair amount of study may provide the answers to some of these questions and when compiled into a program may actually differentiate between the faces in question, there is no guarantee that the same measures can be applied to another pair of faces. In contrast, 20 seconds of exposure to a neurally based system such as the WISARD (Aleksander et al. 1984 and as described below) will allow the net to select among a vast number of rules (node functions) in a very short time, to provide the best discrimination between the images in question. The third promise is performance: rapid solutions to problems which in conventional computers would take a long time. For example it has been possible to solve the 'travelling salesman' problem* in many fewer steps than by conventional (exhaustive) algorithms.

---

*The travelling salesman problem is the finding of the shortest route between geographically scattered points. This is traditionally difficult for conventional machines because it relies on trying out an astronomically large parallel and then allowing these to interact finds solutions very rapidly.

But there is a snag to the exploitation of this performance: neural systems have actually to be built or run on general purpose parallel machines. It is worth pointing out that machines such as the Connection Machine (Hillis 1986) are not neural systems. They are general purpose parallel systems that require programs just as much as any conventional machine. But the program could be the structure of the neural net, that is, an emulation, which due to the parallelism of the host machine, exploits the speed with which the neural system is capable of solving some problems. Indeed, several 'neural computers' that are appearing on the market are emulations of this kind. A useful function that they perform is to provide a tutorial vehicle that gives their users experience in the way such systems work. The first serious neural computer that is capable of solving real-life problems in real time is still to be built. There are many opportunities open for the design of the neural node (e.g. by optical means, conventional memory chips or special Very Large Scale Integrated systems).

The fourth and final promise of neural computing is the provision of an insight into the computational characteristics of the brain. This is very much the stated aim of the authors of the PDP books. In fact, it is becoming apparent that the nature of the research that is done in neural computing will differ depending on whether one is concerned with the understanding of principles and the design of machines on the one hand, or with brain modelling on the other. In the first one of these, general structures are investigated, while in the second, certain structure characteristics may be ruled out of court should they not conform with what is known of the brain, even if such structures may be computationally highly competent. Of course some work faces both ways, being concerned both with (as an example) the creation of novel machinery and with providing a deeper analysis of what may be happening in the brain when it is 'understanding' language.

5. The WISARD

The WISARD (Aleksander et al. 1984) was probably the first machine based on neural principles to arrive on the market. It is largely an image processing machine which is shown examples of sets of patterns together with their classification. An example is the recognition of safe and unsafe bayonet lamp connectors as seen through a television camera. The net is trained to respond in one sector for safe ones and another sector for unsafe ones.

It then builds internal rules in its 64,000 neurons which cause the net to respond appropriately to previously unseen lamps. The neurons used are of the LOGIC type first proposed by ourselves in 1965 (Aleksander 1965). A conventional Random-Access Memory is used as the neural node. In Fig. 1 the RAM is compared to the classical McCulloch and Pitts model of the neuron. Figure 2 shows how the WISARD is constructed out of these devices and is compared to the historical Perceptron. Here is how it works.

The WISARD (Wilkie Stonham and Aleksander's Recognition Device) is the simplest possible single-layer network of logical neurons. The image to be recognised is presented as K bits (typically these may be organised into a 512x 512 bit matrix in the frame store of the WISARD). A group of K/n logical neurons, where n is the number of inputs, is connected randomly to the image bits. Such a group is called a 'discriminator' (S) and one discriminator for each desired recognition class is generally used.

The basic mode of learning and remembering may be formally stated as follows. Say that at K there is an image T (1) of unit area. Starting with all RAM locations set to zero the system is taught to recognise T (1) by setting all the current RAM locations 'addressed ' by T (1) to 1. Say that the discriminator is similarly trained on a second image, T (2). Given some unknown image T (U) which overlaps bit by bit over an area A (1) with T (1) and A (2) with T (2), it may be easily shown that the proportion of RAMs responding with a 1 (the response of the discriminator) is:

$$r\,(U) = (A\,(1)\,)^n + (A\,(2)\,)^n - (A\,(1,2)\,)^n$$

If yes, A (1,2) is the overlap between T(U), T(1) and T(2) which forms the last term. This needs to be subtracted so that it is not counted twice. Similar equations may be generated for larger training sets. They are just bigger equations but similar in character to the above. This character is largely defined by the following properties. If T(U) is close to any of the training patterns, r(U) will be high. If the training patterns are similar to one another there will be good interpolation in r(U), while if there are not, r(U) will only be high in the vicinity of the training patterns. The key factor is that this response is given after only one pass through the system and therefore is equivalent to a parallel search. WISARD uses several discriminators (one for each class) as mentioned, so the decision can go to the the highest response, and its difference from the next highest may be used as a measure of confidence. So, at its extremes of operation the system can both discriminate between similar patterns and classify together dissimilar patterns. It is the parameter n which is crucial in determining this performance. The actual structure of the machines is based on a patented way of using fast, state-of-the-art silicon memory.

The patent has been purchased by a UK company which engineered the system into a product which is currently used in a variety of quality control and security applications. Curiously, this system has never been marketed as a neural computer. Plans are in progress to capitalise on this technique in building a general purpose high-performance neural computer which will not only recognise patterns, but perfoRm the generality of experiential knowledge-based tasks discussed earlier. This is based on probabilistic nodes and pyramids as described below.

6. A Logical Calculus Based on Probabilistic Pyramids

In this area of work we provide a framework of three nested levels of components: a LEARNING ATOM, a PYRAMID which is made up of learning atoms and a NEURAL NET which is made up of pyramids. ('Higher-level' systems can be made up of neural nets, but we leave that aside.) A NEURON is best approximated by a pyramidal structure of atoms.

The learning atom has what seems to be an irreducible set of four attributes. The first is an output which either fires (denoted by 1), does not fire (0) or does not 'know' whether to fire or not (d). In the last case, the output becomes 0 or 1 with equal probability, the decision whether to fire or not being taken at discrete time intervals*. It is the addition of this probabilistic state that is one of the factors that distinguishes this approach from that used in the WISARD.

So 1, 0 and d are the INTERNAL STATES of the learning atom. The second attribute is the set of n inputs, say i(1), ...i(n). The third attribute is a variable function which uniquely associates each possible pattern of Os and 1s over the n inputs with an internal state value (0, 1 or d) of the atom. The function is said to be variable because of the fourth attribute of the learning atom: its learning property. The nature of this (which also seems to be irreducible) is that the atom must receive information (say through the value of some variable T) as to whether, at a considered time interval and only for the current input pattern, its response is correct (T=1) or incorrect (T=-1). When nothing is to be done T=0. If T=1 and the state associated with the current input is 0 or 1 nothing happens, as the function is acting correctly. But if T=1 and the state associated with the input is d, the d is changed to whatever the current (arbitrarily generated) output many be. If T=-1 it is assumed that the atom has consistently, over may time intervals, produced the wrong output. If the state is 0 or 1, it is changed to d. If the state is d, nothing happens as the atom is not responsible for whatever leads to the conclusion that there is a consistent error.

---

*Much of our work uses several states with graded probabilites of generation a logical 1. This has been omitted from this paper for the sake of clarity, but without loss of generality.

It is noted that this definition of a learning atom is a rough approximation to the 'squashing function' of Rumelhart e al. (1986, Chapter 8) with only three probability states (0,1 and 0.5) and with no interdependence on the way in which the firing probabilities are set for individual input patterns. It can be shown that pyramids of such atoms yield closer approximations to the squashing function.

A pyramid (as shown in Fig. 3) has two main parameters, N the number of inputs for each node (the case of pyramids with different atoms is not considered here), and D the depth (or number of layers). There is a dependent variable, the 'width' of the base of the pyramid W. The pyramid learns to determine the probability of a 1 at its output atom in response to a pattern of 0s and 1s at the W inputs. The correctness or error of the behaviour is detected at the output of the top node only and it is from this that action flows to the other 'hidden' units. Pyramids require the training of hidden units which is one of the most challenging aspects of artificial neural net studies. Hinton and Sejnowski (see Rumelhart and McClelland 1986, Chapter 7) have suggested that in Boltzmann nets that are symmetrically connected (i.e. if node A transmits to node B via a given symaptic weight, then B transmits to A via the same weight) hidden units may be treated in the same way as visible units.

Weights may be optimally adjusted by measuring the probability with which the nodes at the ends of a weight fire simultaneously, (a) when the net is allowed to run freely and (b) when the training information is 'clamped' on visible (non-hidden) parts of the net. The weights are adjusted to minimise the difference between these two measurements. There is a serious lack of biological credence in this model: first, neurons are not bidirectionally connected and second, supervisory elements which retain knowledge of firing probabilities under the two stated conditions (a) and (b) are not known to exist.

Recently much attention has been paid to the 'error-back-propagation' method of dealing with hidden units due to Hinton, Rumelhart and Williams (Rumelhart and McClelland 1986, Chapter 7). Here, note of an error occurring at an output unit is propagated backwards, unit by unit, multiplied by the weights found along these propagation paths. Such signals are used to adjust the weights to reduce the error. This too is biologically unreal as there are no known ways for such signals to propagate backwords along axons.

In our logical calculus, we only assume that all the atoms in a pyramid should receive information on whether the output is erroneous or correct without defining special paths for the propagation of such information. This provides a less demanding framework for biological enquiry than either of the two models quoted above. We call this PYRAMIDAL

LEARNING. It goes as follows. Every training instance consists of applying data to all of the W inputs of the pyramid, and comparing the output to a desired output. Initially all the atoms of the pyramid are assumed to be in the d state for all the possible input patterns. This means that for any given input pattern the output atom generates 0s and 1s with equal probability. As the desired input can only be 0 or 1, T=1 is generated by the error detector (Fig. 3) as soon as the correct output is achieved. The learning property, discussed above, is such that all the atoms associate their current output with their current input. This ensures that if the data of the training instance were to be presented again, and the system were no longer in the training mode, then all the atoms would enter their learnt state, generating the correct output as the overall output of the pyramid. Small changes from training inputs reduce the probability of firing in the trained way by an amount that may easily be calculated.

At the early stages of learning the degree of arbitrariness offered by the d states is high. This diminishes as these states are committed to 0 and 1. Clearly, this form of training ensures that the states of the learning atoms are always correct with respect to the past sequence of training instances. Errors can only occur with respect to as yet unseen instances. These are characterised by the pyramid's consistent inability to generate the desired output. On detecting such an error the error detector floods the pyramid with the T=1 sign returning the currently addressed atom states responsible for the error to the d state. We state a law without proof.

7. The Law of Pyramidal Learning

Pyramidal learning converges, provided that the desired fuction can be achieved by the pyramid.

This is not the place for providing rigorous proofs of our findings; suffice it to say that the above law is based on the fact that, at most, only one memory state per atom reverts to the d state whenever an error is detected, preserving other states that are still correct with respect to the prior training sequence. The learning is therefore conservative of successes, leading to convergence, albeit less rapidly as the desired function is approached than at the beginning of the process.

8. The Future: Neural Robotics?

The considerable exaggeration that surrounds much current work on neural computing is by no means constructive. However, it is of some consolation that it is self-defeating as well. Many laboratories new to neural computing are discovering that it is not fruitful to cobble together any simulation of a neural net, and then hope that it will compute the first thought-of task. This quickly diverts the thrust towards a need to understand what can and cannot be expected of a particular net, and the way the parameters of a net are optimised. It is the aim of many dedicated scientists to contribute to such understanding, which is the best way of fighting the exaggeration.

In robotics, it may be important to revisit the 'blocksworld' notion as framed by Winograd (1973). This methodology relied first, on a highly stylised and simplified environment for the robot so that symbols such as 'box' and 'red' and 'pyramid' could be extracted from some digitized image. Second, it used a logic-based extraction of 'understanding' from simple sentences such as 'put the pyramid on the red box'. The methodology could not extend beyond its assumed simplicity. Neural computing provides an alternative approach to both of these problems. There is promise of being able to extract features from much more complex scenes (e.g. 'road', 'firtree', 'picket fence') and of rapid extraction of meaning from more elaborate sentences.

So what is likely to be the ultimate neural computing architecture of the future? This is an area on which researchers may differ, mainly because of their dedication to the understanding of specific approaches. But one thing does seem to be evident. Neural computing of the future is not likely to be a replacement of conventional computing and AI programs, but is likely to form a complementary technology. It would border on the frivolous to create, with difficulty, neural computations that can be performed with ease through conventional methods. The key issue however is that the two methods must be made to exist under the same roof (or in the same metal box). So the ultimate challenge for experts in computer architecture is to exploit the two technologies within the box, and present a single, flexible interface to the user. If they succeed, we may well witness a quantum step forward in the ease with which humans will be able to interact with machines and hence a step forward in the usefulness of the machines themselves.

References

1. Wiener, N.: Cybernetics, MIT Press, Cambridge MA, 1947

2. McCulloch, W.S. and Pitts, W.: A Logical calculus of the ideas imminent in nervous activity'. Bull, Math. Biophys., Vol 5, pp 113-115, 1943.

3. Rosenblatt, F.: The Principles of Neurodynamics, Spartan Books, New York, 1962.

4. Minsky, M. and Papert, S.: Perceptrons: an introduction to computational geometry. MIT Press, Boston, 1969.

5. Aleksander, I., Thomas, W,V., and Bowden P.A. WISARD, a radical step forward in image recognition. Sensor Review, vol 4, no. 3, pp. 120-124.

6. Rumelhart, D.E. and McClelland J.L. (eds.) : Parallel Distributed Processing, Vol 1 & 2 MIT Press, Cambridge, MA, 1986.

7. Hillis, W. D.: 'The connection machine, MIT Press, Cambridge, MA, 1986.

8. Aleksander, I.: Fused adaptive circuit which learns by example Electronics Letters, August 1965.

9. Winograd, T.: 'A procedural model of language understanding', Schank, R.C. and Colby, K.M. (eds.) In computer models of thought and language, . pp. 152-186, Freeman, San Francisco, 1973.

THE MCCULLOCH AND PITTS MODEL 1943

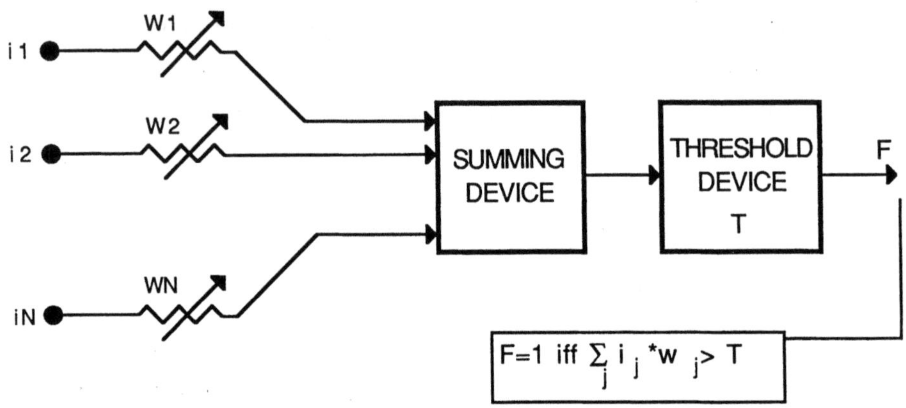

Training: (Widrow/Hoff 1960) Must know desired output 0 or 1 for a particular input pattern of 0s and 1s.
Then measure Σ and distribute the duty to remove the error among the active weights (by some degree).

THE RAM MODEL 1965

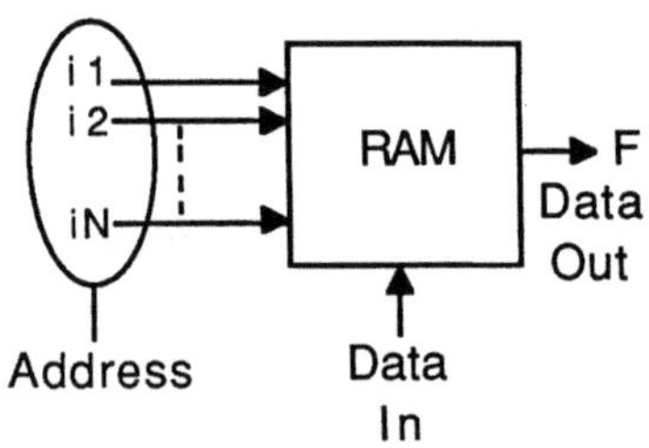

$$F = M_1 \cdot \bar{i}_1 \cdot \bar{i}_2 \cdot \bar{i}_3 \cdots\cdots \cdot \bar{i}_N$$
$$+ M_2 \cdot \bar{i}_1 \cdot \bar{i}_2 \cdot \bar{i}_3 \cdots\cdots \cdot i_N$$
$$\vdots \qquad \vdots$$
$$+ M_{2^N} \cdot i_1 \cdot i_2 \cdot i_3 \cdots\cdots \cdot i_N$$
$$M_j = 0 \text{ or } 1$$

Training: Must know desired F for particular input.
Apply desired value to Data In

Figure 1: Logic Neural Nets: the RAM-Neuron Analogy

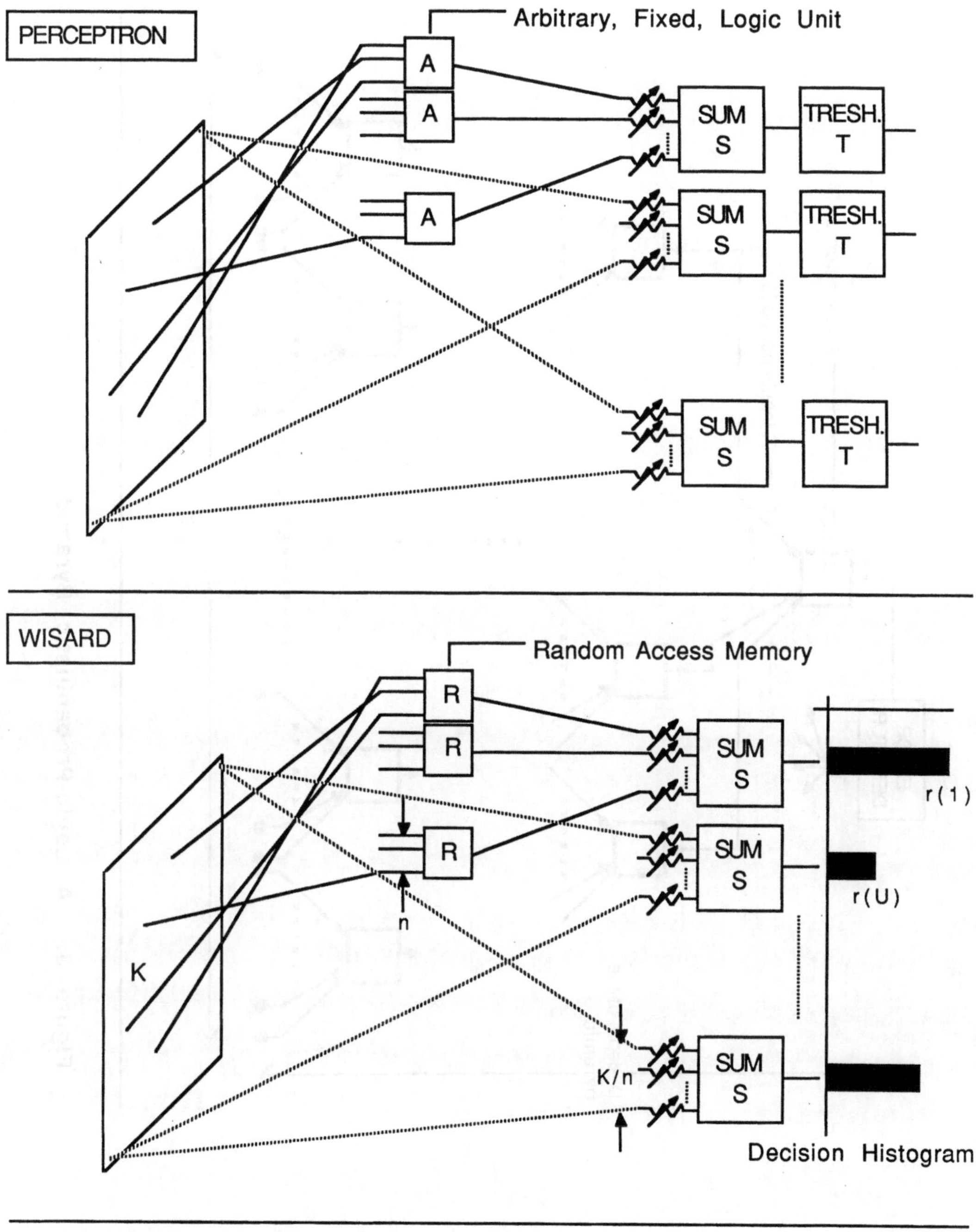

Figure 2: The WISARD-PERCEPTRON Analogy

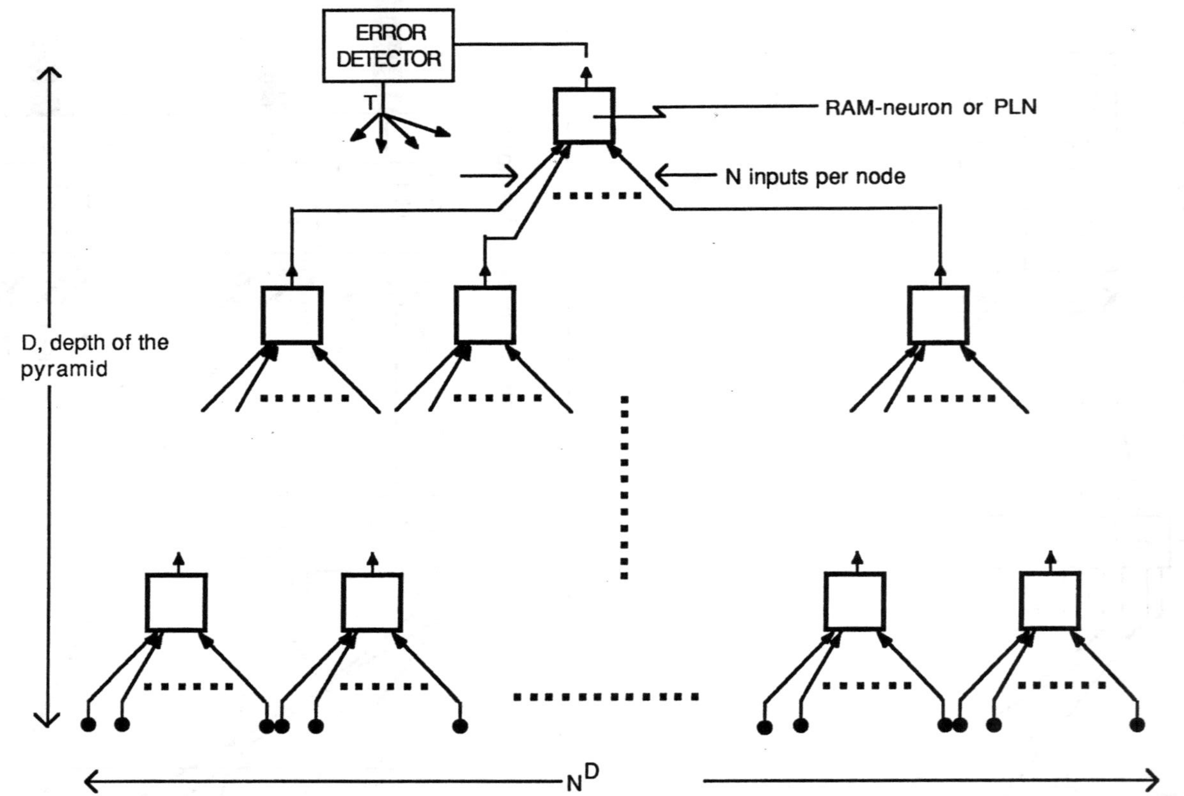

Figure 3: A Logic-Probabilistic Pyramid.

Self-Organizing Neuromorphic Architecture for Manipulator Inverse Kinematics

Jacob Barhen   Sandeep Gulati

Jet Propulsion Laboratory
California Institute of Technology
4800 Oak Grove Drive
Pasadena, CA 91109

ABSTRACT

We describe an efficient neuromorphic formulation to accurately solve the inverse kinematics problem for redundant manipulators, thereby enabling development of enhanced anthropomorphic capability and dexterity. Our approach involves a dynamical learning procedure based on a novel formalism in neural network theory: the concept of "terminal" attractors, that are shown to correspond to solutions of the nonlinear neural dynamics with infinite local stability. Topographically mapped terminal attractors are then used to define a neural network whose synaptic elements can rapidly encapture the inverse kinematics transformations using *a priori* generated examples and, subsequently generalize to compute the joint-space coordinates required to achieve arbitrary end-effector configurations. Unlike prior neuromorphic implementations, this technique can also systematically exploit redundancy to optimize kinematic criteria, e.g. torque optimization, manipulability etc. and is scalable to configurations of practical interest. Simulations on 3-DOF and 7-DOF redundant manipulators, are used to validate our theoretical framework and illustrate its computational efficacy.

1. INTRODUCTION

The successful deployment of industrial teleoperated and reprogrammable robots is leading to a rapidly increasing interest in applying this technology to more exacting scientific applications in unstructured and hazardous environments, such as space missions, maintenance activities in nuclear plants, undersea operations etc. In these envisioned applications, significantly enhanced capability, dexterity and reliability, is essential to achieve real-time operational responses in a semi-autonomous decision environment characterized by severe constraints on size, weight and power consumption. Despite a tremendous spurt in research activity and growing literature on the subject, provision of the above attributes entails a level of paradigmatic complexity far exceeding that what can be provided by the existing modeling strategies. Traditional computing paradigms have typically focussed on problems that

NATO ASI Series, Vol. F 66
Sensor-Based Robots: Algorithms and Architectures
Edited by C. S. George Lee

are clearly defined and deterministic, and can best be handled by computers employing rigorous, precise logic, algorithms or production rules. But, the anthropomorphic capability and perception necessitated by the unstructured applications to be performed by the next generation intelligent machines entails providing for situations which may have received no prior treatment or thought. These problems are in general ill-posed, ill-conditioned and plagued with incomplete information and uncertainty and often must satisfy large numbers of competing constraints. These problems typically involve acquisition and processing of large amounts of sensory data. It is however observed that living systems handle analogous problems of sensor-motor coordination and vision with remarkable ease, and reveal a spontaneous emergent ability that enables them to adapt their structure and function. Consequently, the latter class entails a level of computational complexity that necessitates recourse to alternate paradigms which are inherently amenable to emulating characteristics of concurrent processing.

Artificial neural networks are defined as massively parallel adaptive dynamical systems modeled on the general features of biological networks, that are intended to interact with the objects of the real world and its statistical characteristics in the same way the biological systems do. In contrast to the existing notions on *imperative* and *symbolic* computing, the potential advantages of neuronal processing arise as a result of their ability to perform concurrent, asynchronous and distributed information processing, in a dynamic self-organizing manner typical of living systems. These individual neurons having simple properties, and interacting according to relatively simple rules can accomplish collectively complex functions such as generalization, error correction, pattern classification, learning etc. However their paradigmatic strength for potential applications, which require solving intractable computational problems and adaptive modeling, arises from their emergent ability to achieve functional synthesis, i.e., extract invariances and establish relationships between multiple continuous-valued inputs and outputs, based on a presentation of a large number of examples. Once the underlying invariances have been encaptured in the synaptic interconnections, the networks can generalize to solve arbitrary problem instances. In addition, the operational versions of these trained networks can be dynamically "regularized" or adapted to overcome additional constraints imposed by the environment or the particular application. Thus, neural networks provide an adequate basis for developing a rudimentary learning capability towards the design of autonomous robots that can self-organize and adapt to changes in structure and function.

Also, integral to the realization of any application envisioned for intelligent robots is the ability to dexterously and adaptively manipulate in a nonstationary task workspace. There are two aspects to the provision of this capability. First, given the initial and final end-effector task coordinates, simultaneously generate, in real-time a Cartesian-space trajectory that can achieve a goal (the path planning problem), and a set of joint space trajectories which cause the end-effector to follow the desired trajectory (inverse kinematics problem) while satisfying additional constraints. Secondly, provide adaptive mechanisms for responding to any unforseen changes in the workspace or the manipulator geometry. In addition, some applications may require online strategies for optimizing trajectories with respect to certain kinematic constraints, e.g., obstacle avoidance, servo-motor torque minimization, joint availability etc. Currently, there is no analytical formulation that can satisfactorily address the problem in real-time.

In this paper we have chosen to address the simplest problem which coalesces two of the issues fundamental to the development of autonomous intelligent robots, namely, enabling a rudimentary learning capability and improving dexterous manipulation by redundancy. In particular, we demonstrate a powerful new neural learning paradigm for solving a large class of "inverse problems", e.g. manipulator inverse kinematics, commonly encountered during the design of real-time, adaptive systems operating in redundant environments. The organization of the remaining paper is as follows. In section 2 we briefly review some of the existing strategies for solving the inverse kinematics of redundant manipulators in order to motivate a departure from the traditional jacobian-manipulation based strategies. We also review proposed neuromorphic approaches to this transcedental function approximation problem, and discuss some of the currently available implementations based on backward error-propagation type algorithms. In section 3 we specify the neural network architecture, and derive corresponding learning equations in terms of new algorithms for constrained differential optimization which strictly enforce the Lyapunov stability criteria. In particular, we introduce the notion of "terminal attractors" based on non-Lipschitzian dynamics, and describe their implications towards neural modeling. Section 4 presents the results of our investigation with 3-DOF and 7-DOF redundant manipulators. The last section presents the conclusions of this paper.

2. MANIPULATOR INVERSE KINEMATICS

A forward kinematics operator Φ is defined as a nonlinear differentiable function which uniquely relates a set of N_Q joint variables, $\bar{q}$, to a set of N_X task space-coordinates, $\bar{x}$: $\bar{x} = \Phi(\bar{q})$, assumed by the manipulator. However, the primary practical interest in manipulator kinematics is the inverse problem,

$$\bar{q} = \Phi^{-1}(\bar{x}) \tag{2.1}$$

i.e. determine one or more sets of joint configurations which take the end-effector into a desired task position and orientation in the operational workspace. Though the kinematics equations relating unknown joint-coordinates to specified end-effector coordinates are nonlinear, closed form analytical solutions can be found for a number of non-redundant manipulators with special architecture. In theory, complete positioning capability can be achieved in Cartesian space using only six degrees of freedom. However, most manipulators have degenerate configurations or kinematic
singularities, near which small displacement of the end-effector require physically unrealizable joint speeds, thereby leading to severe inaccuracy in the resultant motion. Since these singularities effectively lead to a loss of usable workspace and capability, there is a strong incentive to design redundant robots with additional degrees of freedom. Thus, a robot manipulator is kinematically redundant if the number of its degrees of freedom is greater than the dimension of the end-effector task space. In contradistinction to other engineering contexts, where redundancy per se, implies *fault-tolerance*, i.e component duplication allowing for continued system functionality in the event of an element failure

or *superfluity* i.e, an unneeded excess capacity, redundancy in robotics is determined relative to the task [5]. For example, a 6-DOF manipulator could be redundant with respect to tasks with symmetry about one axis, while an arm with 3 or more joints is redundant for achieving any end-effector position in a two-dimensional space. The major objective motivating introduction of redundancy in robot design and control is to use the additional degrees of freedom to improve performance in complex and unstructured environments. It helps overcome kinematic, mechanical and other design limitations of non-redundant manipulators, and simultaneously satisfy additional constraints, such as obstacle avoidance [15], minimization of actuator torques [13], singularity avoidance [1], providing greater dexterity [6], minimization of kinetic energy, improvement of some measure of manipulability, etc.

However, incorporation of redundancy injects additional complexity into the problem. For redundant manipulators, the kinematic equations relating the specified end-effector task coordinates to the unknown joint angles may not have a unique solution, and in general the problem is both ill-posed (see Fig. 1) and ill-conditioned. Often an infinite number of joint-configurations can be obtained to satisfy a given end-effector configuration. However, it can be shown [5] that the infinity of solutions can be mapped into a finite set of manifolds. Because of this infinity of solutions, many redundant manipulator investigators have chosen to focus on the instantaneous or differential kinematics, which uses a jacobian-matrix to relate end-effector velocities to the joint velocities. The jacobian is defined as

$$\dot{\bar{x}} = \mathbf{J}(\bar{q})\,\dot{\bar{q}} \tag{2.2}$$

For redundant robots the manipulator Jacobian is not uniquely invertible, and pseudo-inverse techniques can be used to select a solution from the infinity of possible solutions in the null space of $J(\bar{q})$. Eq. (2.2) is often referred to as the inverse kinematics solution, although (2.1) is the true inverse kinematics problem.

Given a desired end-effector velocity, $\dot{\bar{x}}$, the joint velocities can simply be determined by:

$$\dot{\bar{q}} = \mathbf{J}^{\dagger}(\bar{q})\dot{\bar{x}} \tag{2.3}$$

where $\mathbf{J}^{\dagger}(q)$ is the pseudo-inverse, or a weighted pseudo-inverse, of the manipulator Jacobian matrix. This redundancy resolution solution minimizes a weighted quadratic norm of instantaneous joint velocities. The end-effector velocities are typically generated by a path planning algorithm, and the joint velocities computed by (2.3) are used as the reference input to a joint-space control system.

This solution can be modified by adding a null space component to the joint velocities [17]:

$$\dot{\bar{q}} = \mathbf{J}^{\dagger}(\bar{q})\dot{\bar{x}} + (I - \mathbf{J}^{\dagger}(\bar{q})\mathbf{J}(\bar{q}))\bar{z} \tag{2.4}$$

where $\bar{z}$ is an arbitrary vector. The term $(I - \mathbf{J}^{\dagger}(\bar{q})\mathbf{J}(\bar{q}))$ projects this arbitrary vector into the null space of the manipulator. Physically, any motion in the null space is an instantaneous internal motion of the manipulator which causes no motion of the end-effector. Many redundancy resolution criteria can be developed as potential functions, and $\bar{z}$ might be the gradient of the resolution potential function, i.e., , $\bar{z} = \alpha \nabla \Psi(\bar{q})$, where α is a weighting factor. Then for a given end-effector configuration, the gradient of this function is used to control joint velocity in the redundant directions, in a manner that forces the manipulator to seek an optimal configuration. However, the pseudo-inverse resolution techniques are generally not cyclic [1,15], i.e. these techniques do not generate closed joint-space trajectories corresponding to closed end-effector trajectories, thereby posing a serious limitation for practical implementations. Other researchers have used the null space of the jacobian, which corresponds to the self-motion of the robot, to optimize various performance criteria. For example, Liegeois [17] has developed a gradient projection scheme that utilizes the null space of the Jacobian to optimize a joint-position dependent, scalar performance criterion.

Recently focus has been on *redundancy resolution* techniques based on global or local resolution of redundancy. The primary objective is to determine the motion of the joints to simultaneously achieve end-effector trajectory control while optimizing an additional kinematic constraint. Hollerbach and Suh [13] have suggested that extra degrees of freedom be used to minimize the magnitude of applied torque during motion, thereby resolving redundancy at the acceleration level. Whitney [25] has resolved redundancy at the velocity level by minimizing the kinetic energy of the manipulator. Yoshikawa [26] proposed a powerful geometric technique that uses kinematic redundancy to increase the end-effector manipulability. In a similar vein, Chiu has exploited redundancy induced posture variation to maximize the coincidence of optimal directions of the manipulator with those of the task geometry. Chang [6] developed an extended Jacobian technique to optimize joint rotations for dexterous manipulation. Klein et al [15] on the other hand have focused on improving obstacle avoidance. Dubey et al [8] have used the gradient projection algorithm to improve the efficiency, mechanical advantage and flexibility of the manipulator. Nonetheless, existing methods are in general very expensive computationally, and are unable to find global redundancy resolution optima with respect to multiple criteria in real-time. Also the manipulators can have more than one distinct internal motion for a given end-effector location but the instantaneous methods only optimize over one internal motion, and therefore can miss the true optimum which lies on another internal motion [5].

In the absence of closed form solutions, off-line iterative approximation techniques based on "local-methods" have been used to solve the inverse transformation problem. In this context, Goldenberg et al [9] have proposed an "augmented task method" that uses a modified Newton-Raphson method to simultaneously obtain all the joint variables. They partition the augmented Jacobian matrix into an invertible non-redundant component and a redundant component to obtain approximate bounds on the magnitude of the joint angles. A nonlinear constrained optimization is then performed to determine the angular displacements for the redundant joints by satisfying some auxiliary criteria. The resulting values are used to compute the Newton-Raphson correction that minimizes an error-residual between desired and current end-effector coordinates. Despite its versatility, this techniques suffers from algorithmic singularities, since it fails to ensure the non-singularity of the Jacobian-matrix partition prior to start of each iteration. Also, for a large number of degrees of freedom, the nonlinear optimization algorithm during each iteration induces a significant computational complexity.

In a significantly different approach Burdick [5] conducts a topological and geometrical analysis of the kinematics of redundant manipulators. Formulating inverse kinematics as a global manifold mapping problem, he uses the singularities of the forward kinematics to partition the configuration space manifold into disjoint regions. The topological characteristics of these regions and their forward mapping are then used to rigorously analyze kinematic properties, such as bounds nature and number of singularities that must be encountered along an arbitrary cyclic path and bounds on the number of inverse kinematic solutions. Currently formal procedures are being developed for translating this qualitative insight to quantitative algorithms that could aid the design and control of redundant manipulators.

In contrast to the algebraic and iterative strategies mentioned above, neuromorphic approaches to the inverse kinematics problem entail systems composed of many simple processors ("neurons"), fully or sparsely interconnected, whose function is determined by the topology and strength of the interconnections. The synaptic elements of such neural systems must capture the transcendental kinematic transformations using *a priori* generated examples enabling subsequent generalization to other points in the workspace. Thus, the inverse transformation equations do not need to be explicitly programmed or derived. Once they have been learned, the network's inherent self-organizing abilities enable it to adapt to changes in the environment, e.g. planning joint trajectories in the presence of obstacles, or to any unforseen changes in the mechanical structure of the manipulator, with little effort [14]. Within a neuromorphic framework, a solution of the inverse kinematic involves two phases, a training phase and a recall phase. The training phase involves encoding the inverse mapping in the network's synaptic weight space, through repeated presentations of a finite set of *a priori* generated examples, linking cartesian space end-effector coordinates to the corresponding joint angles. Once the network has acquired the nonlinear mapping imbedded within the training set, it can be used to rapidly recall, or generalize the joint configuration corresponding to any arbitrary cartesian-space orientation within it's workspace of training, thereby eliminating the intensive computational overheads that plague the existing iterative techniques. Also, once the training cycle is completed, the time required to obtain a solution depends in a weak fashion on the number of degrees of freedom.

In the past, Josin [14], Guez et al [10] and Tawel et al [24] have applied this generic neuromorphic paradigm to the inverse kinematics problem for a 3-DOF redundant manipulator. In particular, they train a heteroassociative, multi-layered feed-forward neural network using the backpropagation algorithm [22]. The following principle is commonly used during the training process. When the system produces a wrong output on presentation of an I/O pair, the learning update rule simply changes each weight in the direction which makes the size of the error decrease as rapidly as possible. The components of this steepest descent direction in weight space are evaluated by using the chain rule to compute the partial derivatives of an error function with respect to each weight. The implementation of this weight change requires recursively propagating an error signal backward through the network, changing weights that had a large effect on the output more than those that did not. This process is repeated until the residual error between the network and target output, over all patterns, falls below a minimum acceptable tolerance.

Despite its conceptual simplicity, there are a number of non-trivial issues, both from the kinematics perspective and from the computational cost perspective that have hitherto limited the efficacy of such neuromorphic solutions to the inverse kinematics problem for redundant motion control. The major limitations, as discerned from the existing implementations, include an unacceptably large number of training iterations ($O(10^6)$ even for generalizing over small manifolds, see Tawel et al [24]). Also the interpolated angular coordinates have relatively poor precision as compared to their algebraic or iterative counterparts. Besides, the backpropagation algorithm fails to efficiently scale-up to configurations with large number of degrees of freedom. For example manipulators with seven or more degrees of freedom could not be satisfactorily trained using the standard back-propagation algorithm even after several million iterations. Furthermore the back-propagation algorithm *per se* does not provide any intrinsic mechanism to simultaneously exploit redundancy to increase the task workspace (design constraints) and satisfy additional requirements inherent to operations in an unstructured environment such as obstacle avoidance in real-time. Since the latter flexibility is quintessential to the purpose of redundant manipulators, there is a strong incentive to develop an alternative neural network paradigm that can be applied to the inverse kinematics problem. In this context we introduce a neuromorphic formalism for learning nonlinear mappings, that obviates many of these limitations, and can provide efficient and accurate solutions to a large class of inverse problems.

3 NEURODYNAMICS MODEL

3.1 NEURAL NETWORK SPECIFICATION

Consider a fully connected neural network with N graded-response neurons, implementing a functional mapping from the N_X-dimensional end-effector cartesian space to the N_Q-dimensional joint space of the redundant manipulator. As shown in Fig. 2, the network is topographically partitioned into three mutually exclusive regions comprising of a set of input neurons, S_X, that receive the end-effector task coordinates, an output set S_Q, which provides the angular coordinates required to achieve the desired end-effector motion and a set of "hidden" neurons, S_H, whose sensitizations partially encode the input / output mapping being learnt. The network is presented with K training pairs of cartesian- and joint-space variables, $\{ \bar{x}^k, \bar{q}^k \mid k = 1,\ldots,K\}$ obtained from the forward kinematics relations (see Paul [19]).

Our goal is to determine the synaptic interconnection strengths that can correctly encapture the inverse kinematics mapping, Φ^{-1}, imbedded within the training samples. Our approach is based upon the minimization of a constrained objective ("energy") function given by the following expression

$$E = \frac{1}{2K} \sum_{k=1}^{K} \left(\frac{1}{N_X} \sum_{l \in S_X} [\, u_l^k - x_l^k \,]^2 + \frac{1}{N_Q} \sum_{l \in S_Q} [\, u_l^k - q_l^k \,]^2 \right) + \sum_r \lambda_r \, g_r(\cdot) \quad (3.1.1)$$

where u_l^k denotes the l-th neuron's activity when processing the k-th training sample, $g_r(\cdot)$ reflect network design considerations related to specific applications e.g., manipulability, and λ_r denotes the Lagrangian multiplier corresponding to the r-th application or design requirement. The proposed objective function includes contributions from two sources. Firstly, it enforces the convergence of every neuronal state in S_X and S_Q to attractors corresponding to the presented end-effector task coordinates and joint coordinates respectively, for every sample pair in the training set. Secondly, it exploits some of the structural redundancy to optimally satisfy auxiliary design criteria e.g., motion-time of joints, operational ranges, manipulability, torque optimization, etc. We now proceed with the formal derivation of the learning equations (time evolution of the synaptic weights) by minimizing the energy function given in eqn. (3.1.1).

In the past, several neuromorphic algorithms have been proposed for constrained minimization of non-convex energy functions. For details the reader may refer to Hopfield and Tank [12], Barhen et al [2] and Platt and Barr [21]. In order to motivate and distinguish our optimization approach from the existing techniques, we first briefly examine some of the features which limited the general applicability of previous approaches. Hopfield and Tank's method for the Traveling Salesman problem [12] involved the minimization of an energy function of the type,

$$E = f(\bar{u}) + \sum_r W_r \left[g_r(\bar{u}) \right]^2 \tag{3.1.2}$$

A first difficulty with this model is that the specific constraint strengths, W_r, were determined heuristically, i.e., by "*anecdotal exploration*". Furthermore, the adopted penalty function construction was known to easily lead to constraint violation. Also, as the dimensionality of constraints increases the constraint strengths get harder to set [21]. A recently proposed alternative, i.e., Platt and Barr's Basic Differential Multiplier Method, [21], alleviates some of these limitations by modifying the objective function to

$$E = f(\bar{u}) + \sum_r \lambda_r \, g_r(\bar{u}) \tag{3.1.3}$$

where λ_r denote the *Lagrange multipliers* corresponding to the constraints $g_r(\bar{u}) = 0$. A straightforward (but naive) application of Lyapunov's stability requirements, (i.e. $\dot{E} < 0$), would result in the following equations of motion:

$$\dot{u}_i = -\frac{\partial E}{\partial u_i} = -\frac{\partial f}{\partial u_i} - \sum_r \lambda_r \frac{\partial g_r(\bar{u})}{\partial u_i} \tag{3.1.4}$$

and

$$\dot{\lambda}_r = -\frac{\partial E}{\partial \lambda_r} = -g_r(\bar{u}) \tag{3.1.5}$$

However, for some pathological cases the above algorithm could result in $\lambda_r \rightarrow 0$, i.e., the constraints might no longer be satisfied. Hence, Platt and Barr suggested the following heuristic change :

$$\dot{\lambda}_r \;=\; +\, g_r(\bar{u}). \tag{3.1.6}$$

Notice that their proof of correctness upon inclusion of the above heuristic is based on assumptions which are extremely restrictive in nature. Specifically, the necessary condition to achieve stability requires establishing equivalence to a damped mass system, which in itself is a nontrivial mathematical exercise. In contrast (see below) , the methodology we propose, guarantees rapid convergence for arbitrary problem situations.

Lyapunov's stability criteria require an energy function to be monotonically decreasing in time. Since in our model the internal dynamical parameters of interest are the synaptic interconnection strengths T_{nm} and the Lagrange multipliers λ_r, this implies that

$$\dot{E} \;=\; \sum_n \sum_m \frac{\partial E}{\partial T_{nm}}\, \dot{T}_{nm} \;+\; \sum_r \frac{\partial E}{\partial \lambda_r}\, \dot{\lambda}_r \;<\; 0 \tag{3.1.7}$$

One can choose

$$\tau_T\, \dot{T}_{nm} \;=\; -\frac{\partial E}{\partial T_{nm}} \tag{3.1.8}$$

where τ_T is an arbitrary but positive time-scale parameter. Then substituting in Eqs. (3.1.7) we have

$$\sum_r \frac{\partial E}{\partial \lambda_r}\, \dot{\lambda}_r \;<\; \tau_T\, \dot{T} \oplus \dot{T}. \tag{3.1.9}$$

In the above expression $\oplus$ denotes tensor contraction , i.e.,

$$\dot{T} \oplus \dot{T} \;\equiv\; \sum_i \sum_j \dot{T}_{ij}\, \dot{T}_{ij}$$

This will be true *a fortiori* if for some $\theta > 0$,

$$\sum_r \dot{\lambda}_r \frac{\partial E}{\partial \lambda_r} + \theta \;<\; \tau_T\, \dot{T} \oplus \dot{T}.$$

The equations of motion for the Lagrange multipliers λ_r must now be constructed in such a way that Eq. (3.1.9) is strictly satisfied. Noting that the analytic expression for the energy function results in $\frac{\partial E}{\partial \lambda_r} \;=\; g_r(\cdot)$, we adopt the following model:

$$\dot{\lambda}_r \;=\; \tau_T\, \frac{\dot{T} \oplus \dot{T} \;-\; \theta}{\bar{g} \oplus \bar{g} \;+\; \theta}\, g_r(\cdot) \tag{3.1.10}$$

where $\bar{g} \,\oplus\, \bar{g} \;\equiv\; \sum_r g_r(\cdot)\, g_r(\cdot)$, and θ is an arbitrary positive constant. It is easy to see that $\dot{E} \;<\; 0$ is then strictly satisfied.

Now we can proceed with the formal derivation of the learning equations. On differentiating (3.1.1) with respect to T_{nm} we get

$$\frac{\partial E}{\partial T_{nm}} = \frac{1}{K}\sum_k \left\{ \frac{1}{N_X}\sum_{l\in S_X} [\,u_l^k - x_l^k\,]\, \frac{\partial u_l^k}{\partial T_{nm}} \right.$$

$$\left. + \frac{1}{N_Q}\sum_{l\in S_Q} [\,u_l^k - q_l^k + N_Q \sum_r \frac{\partial g_r}{\partial u_l^k}\,]\, \frac{\partial u_l^k}{\partial T_{nm}} \right\} \qquad (3.1.11)$$

If we define,

$$\hat{I}_l^k = \begin{cases} \frac{1}{KN_Q}[\,u_l^k - q_l^k + N_Q\sum_r \frac{\partial g_r}{\partial u_l^k}] & \text{if } l \in S_Q \\ 0 & \text{if } l \in S_H \\ \frac{1}{KN_X}[\,u_l^k - x_l^k\,] & \text{if } l \in S_X \end{cases} \qquad (3.1.12)$$

we can rewrite (3.1.11) as

$$\frac{\partial E}{\partial T_{nm}} = \sum_l \sum_k \hat{I}_l^k \frac{\partial u_l^k}{\partial T_{nm}} \qquad (3.1.13)$$

where the index l is defined over the entire set of neurons. Equations [3.1.8, 3.1.12 and 3.1.10] constitute a dissipative nonlinear dynamical system, the flow of which generally converges to a manifold of lower dimensionality. Our initial effort described in this article focuses on convergence to point attractors, i.e., state-space vector locations where information of interest is stored. Of crucial importance is to know how stable those attractors are and how fast they can be reached. In this vein, we first briefly review a novel concept in nonlinear dynamical systems theory, the *terminal attractor* , and its properties, that subsequently will enable us to formalize neural network algorithms for learning the inverse kinematics mapping.

Hopfield and others [12,16] have shown that artificial neural networks store memory states or patterns in terms of the fixed points of the network dynamics, such that initial configurations of neurons in some neighborhood or *basin of attraction* of that memory state will be attracted to it. But the static attractors considered so far in nonlinear dynamical system formulations in general, and in neural network models in particular, have represented regular solutions of the differential equations of motion as shown in figure 3(a). The theoretical relaxation time of the system to these "regular attractors" can theoretically be infinite, and they suffer from convergence to spurious states and local minima. The concept of *terminal attractors* in dynamical systems, was initially introduced by Zak [27], to obviate some of the above limitations, thereby significantly improving the performance characteristics of associative memory neural network models.

The existence of terminal attractors was established by Zak using the following argument. At equilibrium, the fixed points, $\bar{p}$, of an N-dimensional, dissipative dynamical system

$$\dot{u}_i + f_i(u_1, u_2, ,\cdots, u_N) = 0 \quad i = 1, 2, \cdots, N \qquad (3.1.14)$$

are defined as its constant solutions $\bar{u}^{\infty}(\bar{p})$. If the real parts of the eigenvalues, η_{μ} of the matrix $M_{ij} = \left[\frac{\partial f_i}{\partial u_j}(\bar{p})\right]$ are all negative, i.e., Re $\{\eta_{\mu}\} < 0$ then these points are globally asymptotically stable [4]. Such points are called static attractors since each motion along the phase curve that gets close enough to $\bar{p}$, i.e., enters a so called basin of attraction, approaches the corresponding constant value as a limit as $t \to \infty$. An equilibrium point represents a repeller if at least one of the eigenvalues of the matrix **M** has a positive real part. Usually, nonlinear neural network deal only with systems which satisfy the Lipschitz conditions, i.e., $\mid \partial f_i \;/\; \partial u_j \mid < \infty$ This condition guarantees the existence of a unique solution for each of the initial phase space configurations. That is why a transient solution cannot intersect the corresponding constant solution to which it tends, and therefore, the theoretical time of approaching the attractors is always infinite. Fig. 3(a) shows the temporal evolution to such an attractor.

In contrast, Zak's [27] notion of terminal attractors is based upon the violation the of Lipschitz conditions. As a result of this violation the fixed point becomes a singular solution which envelops the family of regular solutions, while each regular solution approaches the terminal attractor in finite time, as displayed in figure 3(b). To formally exhibit a terminal attractor which is approached by transients in finite time, consider the simplest one-dimensional example:

$$\dot{u} = -u^{1/3} \tag{3.1.15}$$

This equation has an equilibrium point at $u = 0$ at which the Lipschitz uniqueness condition is violated, since

$$\frac{d\dot{u}}{du} = -\frac{1}{3}u^{-2/3} \longrightarrow -\infty \;\; at \;\; u \longrightarrow 0 \tag{3.1.16}$$

Since here the Re $\{\eta\} \longrightarrow -\infty < 0$ this point is an attractor with "infinite" local stability. As a consequence the dynamical system is bestowed with "infinite attraction power", enabling rapid clamping of neuronal potentials to the fixed points; in our case this implies immediate relaxation to the desired attractor coordinates, x_l and q_l. Also the relaxation time for the solution corresponding to initial conditions $u = u_0$ to this attractor is finite. It is given by

$$t_0 = -\int_{u_0}^{x \to 0} \frac{du}{u^{1/3}} = \frac{3}{2}u_0^{2/3} < \infty \tag{3.1.17}$$

i.e., this attractor becomes terminal. As shown in Fig. 3(b), it represents a singular solution which is intersected by all the attracted transients. In particular, static terminal attractors occur for $k = (2n+1)^{-1}$ and $n \geq 1$, while for $k = 2n+1$ all attractors are regular. It has been shown (Zak [27]) that that incorporation of terminal attractor dynamics leads to the elimination of all spurious states. This property is critical to providing an accurate generalization ability, since it ensures that no interpolation is performed over false attractors. For details on implication of terminal attractor dynamics for neural learning algorithms see [3,27]. In our proposed neuromorphic framework, terminal attractor dynamics then provides a mechanism that can implicitly exploit the time-bounded terminality of phase trajectories and the locally infinite stability, thereby enabling an efficient and accurate solution to the manipulator inverse kinematics.

3.2 Adaptive Conservation Equations

To capture the emergent invariants of the inverse kinematics relationship we consider a fully connected neural network, defined by the following system of coupled differential equations

$$\tau_u \dot{u}_l^k + u_l^k = \varphi_\gamma \, [\sum_{l'} T_{ll'} u_{l'}^k \,] - I_l^k \tag{3.2.1}$$

Here u_l represents the mean soma potential of the lth neuron (u_l^k is the neuron's activity when processing the kth training sample), $T_{ll'}$ denotes the synaptic coupling from the l'-th to the l-th neuron, and I_l^k captures the input/output contribution in a terminal attractor formalism. Though I_l^k influences the degree of stability of the system and the convergence to fixed points in finite time, it does not further affect the location of existing static attractors. In eqn. (3.2.1), $\varphi_\gamma(\cdot)$ denotes the sigmoidal neural response function with gain γ; typically,

$$\varphi_\gamma(z) = \tanh(\gamma \cdot z).$$

In topographic maps, N_T neurons are generally used to compute a single value of interest in terms of spatially-coded response strengths. Here we use the simplest possible model (where $N_T = 1$), but encode the information through terminal attractors. Thus, the topographic map is given by

$$I_l^k = \begin{cases} (\, u_l^k - x_l^k \,)^{1/3} & \text{if } l \in S_X \\ 0 & \text{if } l \in S_H \\ (\, u_l^k - q_l^k \,)^{1/3} & \text{if } l \in S_Q \end{cases} \tag{3.2.2}$$

where x_l^k and q_l^k are the attractor coordinates provided by the training sample, to be denoted for brevity as a_l^k. Our basic operating assumption for the dynamical system defined by (3.2.1) is that at equilibrium, for l = 1,..,N :

$$\dot{u}_n \longrightarrow 0 \quad \text{and} \quad u_n \longrightarrow a_n$$

This yields the fixed point equations :

$$a_n = \varphi_\gamma \left[\sum_m T_{nm} \, a_m \right] \tag{3.2.3}$$

In associative memory applications, these equations can in principle be used to determine the synaptic coupling matrix T, resulting in each memory pattern being stored as a fixed point. The key issue is that some of these fixed points may actually be repellers. The terminal attractors are thus used to guarantee that each fixed point becomes an attractor, i.e., spurious states are suppressed. Here however, we are in the process of learning a mapping between two spaces and as indicated in Fig. 2, attractor coordinates have been defined for only two of the three topographic regions of the network, i.e., the input set S_X, and the output set S_Q. Consequently, the fixed point equation $\bar{a} = \varphi(T\bar{a})$ may not necessarily be defined, since for $| S_H | > 0$, $\{ a_n \mid n \in S_H\}$ are not defined, and cannot be used for directly computing T.

This necessitates the development of an alternative strategy, whereby "virtual" attractor coordinates are first determined for the hidden units. These coordinates are virtual since they correspond to a current estimate $\hat{T}$ of the synaptic connectivity matrix. This is achieved by considering the fixed point equations as ***adaptive conservation equations*** which use the extra degrees of freedom made available by the hidden neurons in S_H. Let $\{ \hat{u}_j = a_j \mid j \in S_H \}$ denote the ***virtual attractors*** to which the unknowns, $\{ u_j \mid j \in S_H \}$ are expected to converge to. Then at equilibrium, Eqs. (3.2.3) yield

$$\begin{aligned}
\varphi^{-1}(x_i) &= \sum_{i' \in S_X} \hat{T}_{ii'} x_{i'} + \sum_{j' \in S_H} \hat{T}_{ij'} \hat{u}_{j'} + \sum_{l' \in S_Q} \hat{T}_{il'} q_{l'} && \forall i \in S_X \\
\varphi^{-1}(\hat{u}_j) &= \sum_{i' \in S_X} \hat{T}_{ji'} x_{i'} + \sum_{j' \in S_H} \hat{T}_{jj'} \hat{u}_{j'} + \sum_{l' \in S_Q} \hat{T}_{jl'} q_{l'} && \forall j \in S_H \\
\varphi^{-1}(q_l) &= \sum_{i' \in S_X} \hat{T}_{li'} x_{i'} + \sum_{j' \in S_H} \hat{T}_{lj'} \hat{u}_{j'} + \sum_{l' \in S_Q} \hat{T}_{ll'} q_{l'} && \forall l \in S_Q \qquad (3.2.4)
\end{aligned}$$

where $\hat{T}_{jl}$ denotes the current estimate of synaptic coupling from lth neuron to the jth neuron, and $\hat{u}_j$ represents a virtual attractor whose value is isomorphic to the current level of knowledge in the network. Now define,

$$\begin{aligned}
\psi_i &= \varphi^{-1}(x_i) - \sum_{i'} \hat{T}_{ii'} x_{i'} - \sum_{l'} \hat{T}_{il'} q_{l'} && \forall\, i \in S_X \\
\psi_j &= \sum_{i'} \hat{T}_{ji'} x_{i'} + \sum_{l'} \hat{T}_{jl'} q_{l'} && \forall\, j \in S_H \\
\psi_l &= \varphi^{-1}(x_l) - \sum_{i'} \hat{T}_{li'} x_{i'} - \sum_{l'} \hat{T}_{ll'} q_{l'} && \forall\, l \in S_Q. \qquad (3.2.5)
\end{aligned}$$

Then consistency with the terminal attractor dynamics assumptions requires that $\{ \hat{u}_j \mid j \in S_H \}$ be simultaneous solutions to the following "conservation" equations

$$\begin{aligned}
\sum_{j' \in S_H} \hat{T}_{ij'} \hat{u}_{j'} &= \psi_i && \forall\, i \in S_X \\
\varphi^{-1}(\hat{u}_j) - \sum_{j' \in S_H} \hat{T}_{jj'} \hat{u}_{j'} &= \psi_j && \forall\, j \in S_H \\
\sum_{j' \in S_H} \hat{T}_{lj'} \hat{u}_{j'} &= \psi_l && \forall\, l \in S_Q \qquad (3.2.6)
\end{aligned}$$

The above system of equations for $\hat{u}$ is generally overdetermined. A number of standard algorithms exist to obtain a good approximate solution to such a system. In our implementation we use an iterative approach (e.g. conjugate gradient descent) to minimize the function

$$\hat{E} = \frac{1}{2N_X} \sum_i \Big(\psi_i - \sum_{j'} \hat{T}_{ij'} \hat{u}_{j'} \Big)^2 + \frac{1}{2N_H} \sum_j \Big(\hat{u}_j - \varphi[\sum_{j'} \hat{T}_{jj'} \hat{u}_{j'} + \psi_j] \Big)^2 + \frac{1}{2N_Q} \sum_l \Big(\psi_l - \sum_{j'} \hat{T}_{lj'} \hat{u}_{j'} \Big)^2 \tag{3.2.7}$$

We can now return to the computation of $\partial u_l^k \; / \; \partial T_{nm}$ in Eq. (3.1.12). Let us define

$$z_l^k = \sum_{l'} T_{ll'} \, u_{l'} \tag{3.2.8}$$

and denote

$$\varphi'_{lk} = \frac{\partial \varphi(\cdot)}{\partial z_l^k}. \tag{3.2.9}$$

Then at equilibrium, as $\dot{u}_l^k \longrightarrow 0$ and $I_l^k \longrightarrow 0$, we have

$$\frac{\partial u_l^k}{\partial T_{nm}} = \varphi'_{lk} \left[\sum_{l'} \frac{\partial T_{ll'}}{\partial T_{nm}} u_{l'}^k + \sum_{l'} T_{ll'} \frac{\partial u_{l'}^k}{\partial T_{nm}} \right] \tag{3.2.10}$$

which can be rewritten as

$$\sum_{l'} [\; \delta_{ll'} - \varphi'_{lk} \, T_{ll'} \;] \frac{\partial u_{l'}^k}{\partial T_{nm}} = \varphi'_{lk} \delta_{ln} u_m^k \tag{3.2.11}$$

In the above expression δ_{ij} denotes the Kronecker symbol. We now define, following Pineda [20], a weighted coupling matrix

$$A_{ll'}^k = \delta_{ll'} - \varphi'_{lk} \, T_{ll'}. \tag{3.2.12}$$

Then, substituting (3.2.12) in (3.2.11), and premultiplying both sides with $[A^{-1}]_{nl}^k$ and summing over l yields

$$\sum_l [A^{-1}]_{nl}^k \sum_{l'} A_{ll'}^k \frac{\partial u_{l'}^k}{\partial T_{nm}} = \sum_l [A^{-1}]_{nl}^k \, \varphi'_{lk} \, \delta_{ln} \, u_m^k. \tag{3.2.13}$$

Carrying out the algebra, and relabeling the dummy indices results in :

$$\frac{\partial u_l^k}{\partial T_{nm}} = [A^{-1}]_{ln}^k \, \varphi'_{nk} \, u_m^k. \tag{3.2.14}$$

The above expression can now be substituted in Eq. (3.1.12); the learning equation thus takes the form

$$\tau_T \dot{T}_{nm} = -\sum_l \sum_k \hat{I}_l^k \, [\, A^{-1} \,]_{ln}^k \, \varphi'_{nk} \, u_m^k \tag{3.2.15}$$

where the indices l and k run over the complete sets of neurons and training samples.

3.3 ADJOINT DYNAMICS

A computation of the synaptic interconnection matrix as suggested by Eq. (3.2.15) would involve a matrix inversion. Since direct matrix inversion is typically nonlocal, we adopt the relaxation procedure suggested by Pineda [20] to compute the synaptic updates defined by (3.2.15). Consider the following change of variable

$$v_n^k = \sum_l [\, A^{-1} \,]_{ln}^k \, \hat{I}_l^k \, \varphi'_{nk} \tag{3.3.1}$$

Then substituting (3.3.1) in (3.2.15) we have

$$\begin{aligned} \sum_n A_{np}^k \frac{v_n^k}{\varphi'_{nk}} &= \sum_l \hat{I}_l^k \sum_n [A^{-1}]_{ln}^k \, A_{pn}^k \\ &= \sum_l \hat{I}_l^k \, \delta_{lp} \\ &= \hat{I}_p^k \end{aligned} \tag{3.3.2}$$

One can also use the explicit form of A_{np}^k from (3.2.12) and by substitution in (3.2.15), we obtain

$$\begin{aligned} \sum_n A_{np}^k \frac{v_n^k}{\varphi'_{nk}} &= \sum_n \delta_{np} \frac{v_n^k}{\varphi'_{nk}} - \sum_n \varphi'_{nk} \, T_{np} \frac{v_n^k}{\varphi'_{nk}} \\ &= \frac{v_p^k}{\varphi'_{pk}} - \sum_n T_{np} \, v_n^k \end{aligned} \tag{3.3.3}$$

Regrouping the previous equations (3.3.2) and (3.3.3), and relabeling the dummy indices yields

$$v_n^k = \varphi'_{nk} \cdot [\, \sum_p T_{pn} v_p^k + \hat{I}_n^k \,]. \tag{3.3.6}$$

We see that v_n^k represents a fixed point solution of an "adjoint" neural network having the following coupled dynamics

$$\tau_v \dot{v}_n^k \ + \ v_n^k \ = \ \varphi'_{nk} \cdot [\sum_p T_{pn} \, v_p^k \ + \ \hat{I}_n^k \,] \tag{3.3.7}$$

Recall that $\hat{I}_l^k$ was defined in Eq. (3.2.2). By comparing Eqs. (3.2.15, 3.3.1 and 3.3.7) we see that the resulting neural learning equations couple the terminal attractor dynamics for u_m^k with the adjoint dynamics for v_n^k, i.e.,

$$\tau_T \dot{T}_{nm} \ = \ - \sum_k v_n^k u_m^k \tag{3.3.8}$$

The complete algorithm is summarized below.

SUMMARY OF LEARNING ALGORITHM

[0] Initialize $\hat{T}$, $\hat{\lambda}$
  [1] Learn $\hat{T}$: **iterate IT** = 1, .., NIT
    [1.0] **Loop** over training samples, **k** = 1, .., K
      [2] Initialize $\bar{u}^k$, $\bar{v}^k$
        [2.1] Estimate virtual attractors, $\{\hat{u}_j^k \mid j \in S_H\}$ from conservation equations (3.2.6)
        [2.2] Evolve $\bar{u}^k$ for inverse mapping $\bar{x}^k \implies \bar{q}^k$ using terminal attractor dynamics (3.2.1)-(3.2.2)
        [2.3] Compute v^k using the adjoint network (3.3.7)
        [2.4] Store outer product $u^k \wedge v^k$ increment
        [2.5] **Enddo** {**k**}
    [1.2] Update $\hat{T}$ using Eq. (3.3.8)
    [1.3] Update $\hat{\lambda}$ using Eq. (3.1.10)
    [1.4] Check for convergence:
        If yes **then** exit **else** go to [1]
    [1.5] **Enddo** {**IT**}
  [2] **exit**

4. SIMULATION AND RESULTS

The neural learning framework developed in the preceding section was applied to a planar 3-degree of freedom redundant manipulator and to a spatial 7-degree of freedom human-arm like manipulator. Though either of the manipulators encompasses configurations which exhibit sufficiently the problematic complexity presented by the inverse kinematics mapping, we experimented with both examples for a variety of reasons. The 3-DOF planar manipulator was primarily used to ascertain the algorithmic correctness of our terminal-attractor-based neural learning algorithm. It provided benchmarks for comparison with the existing backpropagation based neural network solutions [10,14,24], in terms of number of training iterations and training samples needed to stabilize the network, estimates on the number of synaptic elements or neurons required to successfully capture the inverse mapping and the accuracy of recalled or "generalized" joint configurations corresponding to the input end-effector coordinates. Though backpropagation-based neural networks can be applied to planar redundant manipulators with 3-DOF, they failed to scaleup to cases involving seven or more joints. Their practical applicability would thus appear to be severely limited for kinematic control of most industrial robot manipulators. We also empirically analyzed the robustness and computational speed of our artificial neural system on a 7-DOF redundant manipulator to illustrate its efficacy as a viable, real-time alternative to the existing quantitative techniques [9,8,15,18,25]. To provide a context for the subsequent analysis we precede our discussion on the simulation results with a brief description of each candidate manipulator.

The continuous-time, dynamical training procedure was simulated using parameters corresponding to a constrained configuration of the six-jointed PUMA 560 industrial robot. By suppressing the motion of the shoulder, elbow and the wrist joints, the PUMA robot arm was restricted to motion in a vertical plane only. Figure 4(a) illustrates the worst-case normalized behavior of the state variables and the synaptic elements during the learning phase, as the neural network acquires the inverse mapping. Figure 4(b) shows the convergence of the output neurons to the presented attractors during the learning phase for a particular sample. When learning has stabilized over all the training samples the network switches to its operational mode. Figure 5(a) displays the normalized convergence behavior of the neuronal activity, $\bar{u}$ as the network generalizes the joint angles in response to arbitrary cartesian inputs. Figure 5(b) illustrates the convergence of interpolated joint angles. Notice the rapid rate of convergence in computing the joint configuration as juxtaposed to conventional techniques.

In addition, the dynamical training algorithm was applied to learn the inverse kinematics transformations for Hemami's simplified 7-DOF manipulator formulation [11] of the human-arm. The following joint motions are available :the joint θ_1 provides back and forth motion about the shoulder, θ_2 provides effector elevation in the vertical plane, θ_3 enables rotation around the upper-arm axis, while θ_4 provides the elbow motion, θ_5 is around the forearm axis and θ_6 and θ_7 lead to the pitch and yaw motions of the wrist, respectively. Details on the geometric parameters, namely link length, twist angle, joint limits and offsets may be found in [11]. Forward kinematics equations (see Paul [19]), were used to generate training samples of end-effector and joint-space coordinates over the workspace volume of the robot. Figure 6(a) displays the normalized, worst-case temporal behavior of state variables $\bar{u}$, i.e., $\max_{\forall\, n} \mid (u_n^{t+1} - u_n^t)/u_n^t \mid$, adjoint variables $\bar{v}$ and the synaptic elements, $\hat{T}$, during the learning phase. Figure 6(b) shows the variation in activity at the output neurons during the learning cycle corresponding to one of the training pairs being presented to the network. Note that the system learns the inverse mapping in a few hundred iterations only, as compared to the several million iterations required by the gradient descent-based backpropagation algorithm. The computational efficacy of our neuromorphic learning algorithm may be estimated from the rate of variation in network activity during the operational phase of the network as shown in figures 7 (a) & (b). As depicted in Fig. 7 (a), once the network has acquired the inverse transformations, it may be used to recall or generate the joint angles needed to achieve any arbitrary end-effector coordinate, within the workspace of the manipulator, in very few dynamical iterations. The detailed results of this study will be reported elsewhere.

5. CONCLUSIONS

In this paper we have attempted to address a complex and important problem in robotics research, which enables the enhancement of manipulative capability and reliability. Our novel learning paradigm for neural network models, based on the terminal attractor concept, is shown to be computationally competitive with iterative methods currently used in robotics to solve the inverse kinematics of redundant manipulators. The neuromorphic framework is expected to facilitate the development of robust real-time algorithms for computing joint configurations to achieve arbitrary end-effector trajectories. In addition, this strategy does not appear to suffer from non-cyclicity of motion, as encountered in the pseudo-inverse resolution techniques [13,15,6] or the algorithmic singularities common to augmented task approaches [9]. Furthermore, unlike the feed forward, backpropagation neural learning approaches, the adaptive dynamical system formulation presented here, provides the flexibility for incorporating arbitrary combinations of kinematic optimization criteria, without imposing high computational overheads. Two options are available for including the redundancy resolution criteria in the algorithm to resolve the nonuniqueness of joint configurations that may satisfy a given end-effector configuration. The constraints may either be included *a priori* , i.e., while generating the training samples themselves, thereby forcing the network to learn only limited aspects of inverse kinematics mapping with a bias towards a particular criterion; or they could be selectively applied in real-time to an operational version of the network (trained to encode the emergent invariants of the inverse kinematic mapping), to regularize the solutions (i.e. provide unique best answers). In addition, it was found that this strategy scales-up to configurations of practical interest, where conventional neural learning techniques, e.g., back propagation appear to fail.

Despite the emphasis on real-time performance, the dexterous nature of applications envisaged for the next generation robots impose uncompromising demands on the resultant end-effector trajectory. Consequently, this entails the generation of intermediate joint angles with a high degree of precision, currently achievable only through off-line programming techniques (e.g acceptable error tolerances are less than 0.05%). In this context, our future directions for research include development of true neural topographic map techniques, enabling the much higher resolution needed to achieve the desired precision in interpolated joint angles.

ACKNOWLEDGEMENTS

The research described in this paper was performed at the Jet Propulsion Laboratory, California Institute of Technology, under contract with the National Aeronautics and Space Administration. The authors wish to acknowledge fruitful discussions with Michail Zak, Ken Kreutz and Joel Burdick.

6. REFERENCES

[1] J. Baillieul, "Kinematic Programming Alternatives for Redundant Manipulators ", in *Proc. IEEE Int'l Conf. on Robotics and Automation* , San Francisco, April 1986, pp. 1698-1704.

[2] J. Barhen, N. Toomarian and V. Protopopescu, "Optimization of the Computational Load of a Hypercube Supercomputer Onboard a Mobile Robot, " *Applied Optics* , Vol. 26, No. 23, Dec. 1987, pp. 5007-5014.

[3] J. Barhen, S. Gulati and M. Zak, "Neural Learning of Constrained Nonlinear Transformations", *IEEE Computer*, Vol. 22, June 1989, pp. 67-76.

[4] E. Beltrami, *Mathematics for Dynamic Modeling* , Academic Press Inc., New York, 1987.

[5] J.W. Burdick IV, " Kinematic Analysis and Design of Redundant Robot Manipulators," *Ph. D Thesis*, Dept. of Mech. Engg., Stanford Univ., 1988.

[6] P.H. Chang, " A Closed Form Solution for the Control of Manipulators with Kinematic Redundancy ", in *Proc. IEEE Int'l Conf. on Robotics and Automation* , San Francisco, CA, 1985, pp. 9-14.

[7] P.J. Denning, "Blindness in Designing Intelligent Systems," *American Scientist* , Vol. 76, March 1988, pp. 118-120.

[8] R.V. Dubey, J.A. Euler and S.M. Babcock," An Efficient Gradient Projection Optimization Scheme for a 7-DOF Redundant Robot with a Spherical Wrist ", in *Proc. IEEE Intern. Conf. on Robotics and Automation*, Philadelphia, April 1988, Vol. I, pp. 28-36.

[9] A.A. Goldenberg, "A Complete Generalized Solution to the Inverse Kinematics of Robots," *IEEE Jour. of Robotics and Automation* , Vol. RA-1, No. 1, March 1985, pp. 14-20.

[10] A. Guez, "Solution to the Inverse Kinematics Problem in Robotics by Neural Networks," in *Proc. of 2nd Int'l Conf. on Neural Networks* , San Diego, 1988, Vol. 2, pp. 617-624.

[11] A. Hemami, "On a Human-Arm Like Mechanical Manipulator", *Robotica*, Vol. 5, 1987, pp. 23-28.

[12] J.J. Hopfield and D.N. Tank, "Neural Computations of Decisions in Optimization Problems ", *Biological Cybernetics* , Vol. 52, 1985, pp. 141-153.

[13] J.M. Hollerbach and K.C. Suh, " Redundancy Resolution of Manipulators through Torque Optimization," in *Proc. of IEEE Int'l Conf. on Robotics and Automation* , St. Louis, 1985, pp. 1016-1021.

[14] G. Josin, " Integrating Neural Networks with Robots ", *AI Expert* , Vol. 3, No. 8, Aug. 1988, pp. 50-60.

[15] C.A. Klein and C.H. Huang, " Review of Pseudo-Inverse Control for use with Kinematically Redundant Manipulators ", *IEEE Trans. System, Man and Cybernetics* , Vol. SMC-13, No. 2, 1983, pp. 245-250.

[16] J.D. Keeler, " Basins of Attraction of Neural Network Models ", *Proc. of AIP Conference* , Vol. 151, Neural Networks for Computing, Snowbird, UT, 1986, pp. 259-265.

[17] A. Liegeois, "Automatic Supervisory Control of the Configuration and Behavior of Multibody Mechanisms", *IEEE Trans. System, Man and Cybernetics*, Vol. SMC-7, No. 12, 1977, pp. 868-871.

[18] A.A. Maciejewski and C.A. Klein, "Obstacle Avoidance for Kinematically Redundant Manipulators in Dynamically Varying Environments", *Int'l Journal on Robotics Research* , Vol. 4, No. 3 , 1985, pp. 109-117.

[19] R.P. Paul, *Robot Manipulators : Mathematics, Programming and Control* , Cambridge, Mass., MIT Press, 1981.

[20] F.J. Pineda, "Generalization of Back-Propagation to Recurrent Neural Networks," *Physical Review Letters* , Vol. 59, No. 19, Nov. 1987, pp. 2229-232.

[21] J.C. Platt and A.H. Barr, "Constrained Differential Optimization," in *Proc. of IEEE Conf. on Natural Information Processing Systems* , Denver, 1987.

[22] D.E. Rumelhart, J.L. McClelland and the PDP Research Group, *Parallel Distributed Processing* , Vol. I, Cambridge, Mass.: MIT/Bradford Book, 1986.

[23] H. Seraji, " Configuration Control of Redundant Manipulators ", presented at the *NATO Advanced Research Workshop on Robots with Redundancy* , Salo, Italy, June 1988.

[24] R. Tawel, S. Eberhardt and A.P. Thakoor, "Neural Networks For Robotic Control," in *Proc. Conf. on Neural Networks for Computing* , Snowbird, Utah, 1988.

[25] D.E. Whitney, " Resolved Motion Rate Control of Manipulators and Human Prosthesis," *IEEE Trans. on Man-Machine Systems* , MMS-10, 1969, pp. 47-53.

[26] T. Yoshikawa, " Analysis and Control of Robot Manipulators with Redundancy ", *Robotics Research First Int'l Symp.* , ed. M. Brady and R. Paul, Cambridge, Massachusetts: MIT Press, 1984, pp. 735-748.

[27] M. Zak, " Terminal Attractors for Addressable Memory in Neural Networks," *Physics Letters A*, Vol. 133, No. 1,2, Oct. 1988, pp. 18-22.

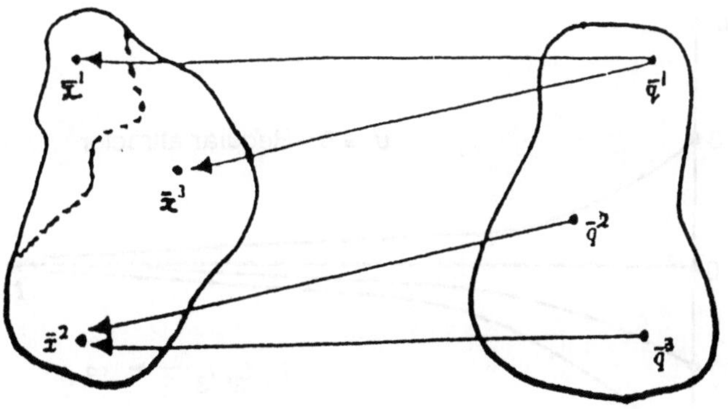

Figure 1. **Ill-posedness of inverse problem $\bar{q} = \Phi^{-1}(\bar{x})$: no solution (from $\bar{x}^1$); non-unique solution (from $\bar{x}^2$)**

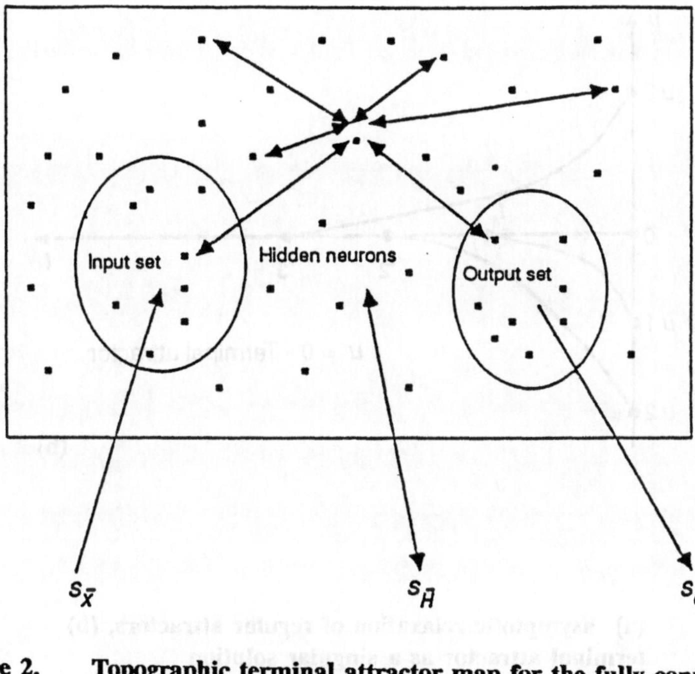

Figure 2. **Topographic terminal attractor map for the fully connected neural network model.**

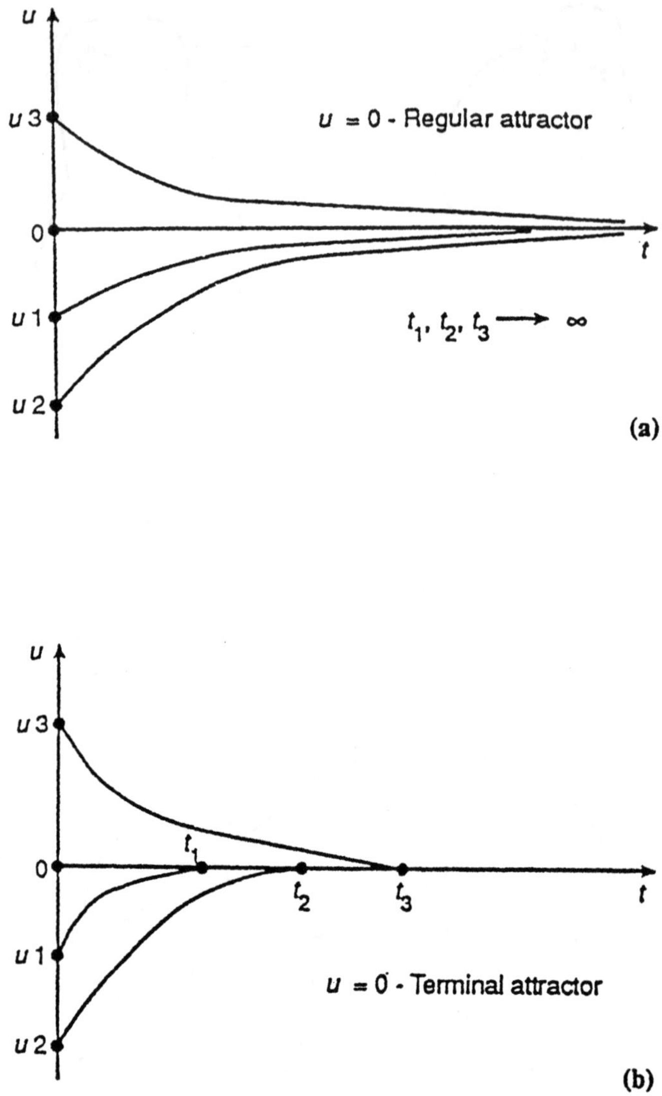

Figure 3. **(a) asymptotic relaxation of regular attractors, (b) terminal attractor as a singular solution**

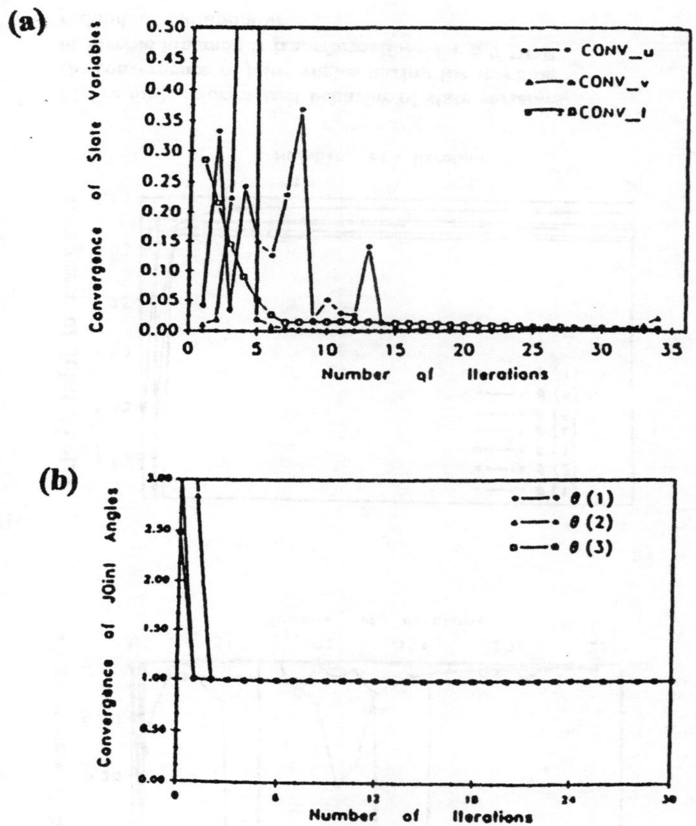

Figure 4(a). Convergence of state variables, (b) convergence of joint angles, during the training phase for a planar 3-DOF redundant manipulator

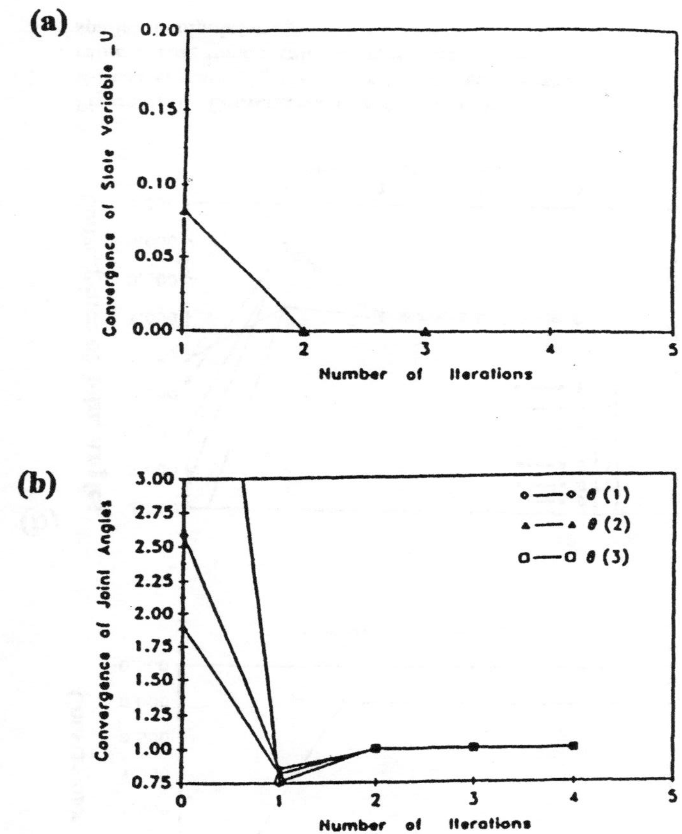

Figure 5(a). Network activity (b) normalized behavior of output neurons, $S_{\_Q}$, during generation of joint angles in the operational phase for the planar 3-DOF configuration

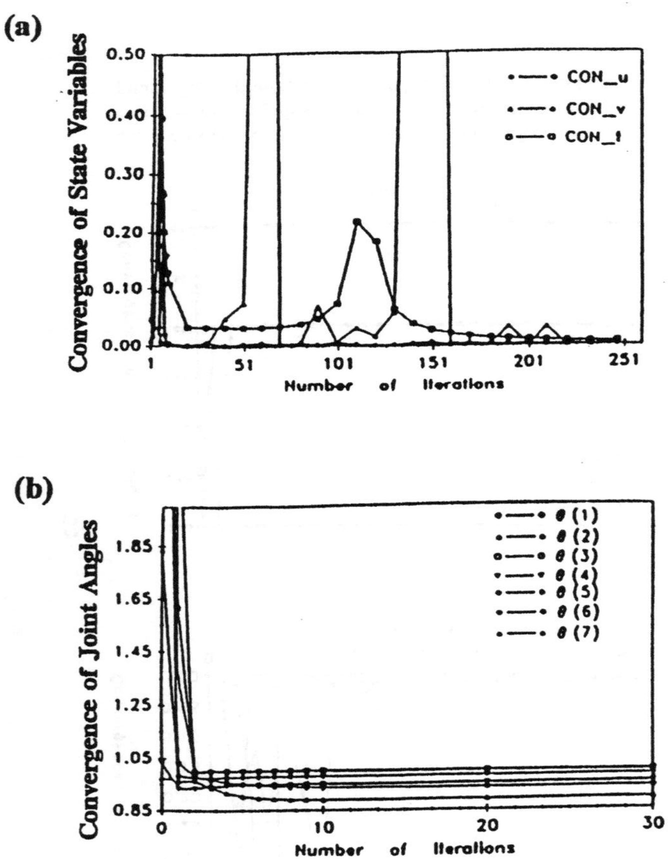

Figure 6(a). Normalized behavior of state variables, (b) convergence of joint angles during the learning of inverse kinematics transformations for a 7-DOF redundant manipulator

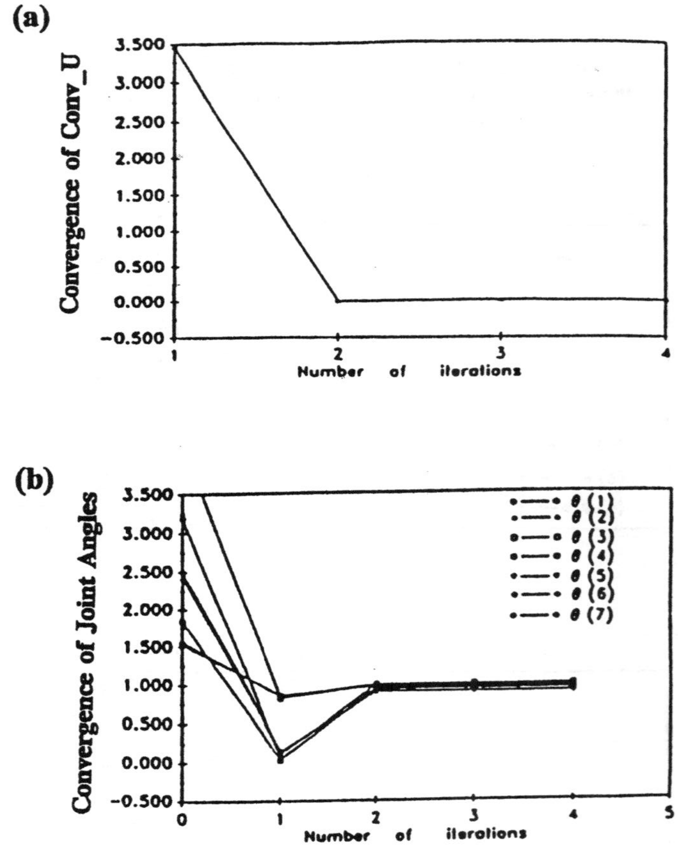

Figure 7(a). Convergence of network activity (b) convergence of joint angles to their generalized value during their recall phase, for the 7-DOF spatial manipulator

ROBOTICS VECTOR PROCESSOR ARCHITECTURE FOR REAL-TIME CONTROL

David E. Orin
Department of Electrical Engineering
The Ohio State University
2015 Neil Avenue
Columbus, OH 43210 USA

P. Sadayappan[‡], Y.L.C. Ling[§], and K.W. Olson[¶]

Abstract

This paper proposes a restructurable architecture based on a VLSI Robotics Vector Processor (RVP) chip. It is specially tailored to exploit parallelism in the low-level matrix/vector operations characteristic of the kinematics and dynamics computations required for real-time control. The RVP is comprised of three tightly synchronized 32-bit floating-point processors to provide adequate computational power. Besides adder and multiplier units in each processor, the RVP contains a triple register-file, dual shift network and dual high-speed input/output channels to satisfy the storage and data movement demands of the computations targeted. Efficiently synchronized multiple-RVP configurations, that may be viewed as Variable-Very-Long-Instruction-Word (V²LIW) architectures, can be constructed and adapted to match the computational requirements of specific robotics computations. The use of the RVP is illustrated through a detailed example of the Jacobian computation, demonstrating good speedup over conventional microprocessors even with a single RVP. The RVP has been developed to be implementable on a single VLSI chip using a 1.2 μm CMOS technology, so that a single-board multiple-RVP system may be targeted for use on a mobile robot.

[‡]Department of Computer & Information Science, The Ohio State University, Columbus, OH 43210

[§]High Tech Center, Boeing Electronics Co., Bellevue, WA 98808.

[¶]Dept. of Electrical Engineering, The Ohio State University, Columbus, OH 43210.

NATO ASI Series, Vol. F 66
Sensor-Based Robots: Algorithms and Architectures
Edited by C. S. George Lee

1. INTRODUCTION

Many robotics control algorithms are not implementable in real time on state-of-the-art microprocessors due to their severe computational needs. Use of hardware parallelism and specialization of hardware are two attractive approaches to improve the performance of such computationally intensive algorithms. In fact, hardware accelerated performance enhancement may be absolutely essential for real-time dynamic control of multiple closed-chain systems such as multi-manipulators and multi-legged vehicles to be feasible, due to their demanding computational requirements. This paper presents a design for a VLSI Robotics Vector Processor tailored to exploit the low-level parallelism in the matrix/vector operations that are characteristic of the kinematics and dynamics computations required in robotics control.

The improvement of performance through use of parallelism and through hardware specialization has been considered by others. Multiprocessors have been used to speed up robotics control computations for the Utah/MIT hand [1], for robotic manipulator control at Stanford [2] and for control of a hexapod walking machine at the Ohio State University [3]. Multiprocessors have also been considered extensively for speedup of the Inverse Dynamics computation for robot manipulators [4-6]. Their use has mostly been targeted at the exploitation of coarse-grained parallelism, as opposed to fine-grained parallelism, due to the significant overhead incurred when attempting to utilize parallelism at the lowest level with general-purpose machines.

Parallel algorithms have been devised for specific robotics control algorithms, for example, Inverse Dynamics [7,8], the Jacobian [9] and the Inertia Matrix [10]. These endeavors have focussed primarily at transforming conventional sequential algorithms for these computations into a parallel form that exhibits greater concurrency. The design of these algorithms has usually been coupled with the derivation of an appropriate processor interconnection/data synchronization scheme to effectively exploit the structural regularity typically exhibited by the algorithms.

The decreasing cost of hardware and the tremendous potential of application-specific VLSI integrated circuits has prompted considerable interest in specialized architectural designs at the circuit board/VLSI chip level [11-15]. A board level design of a robotic processor is outlined in [11] and is targeted at the low-level parallelism available in the matrix/vector operations of Inverse Kinematics and Inverse Dynamics, to give an order of magnitude speedup over state-of-the-art systems. A specialized VLSI chip for solution of Direct Kinematics is described in [12], using fixed-point arithmetic for the computations. A floating-point VLSI robotics processor was developed in [13] and was used as a processing element in systolic architectures for the Jacobian and Inverse Dynamics computations. Specialized architectures have been directed to speedup of Inverse Kinematics through use of a robust numerical algorithm and off-the-shelf floating-point unit in [14] and with a closed-form solution and VLSI CORDIC processors in [15].

In [16], a layered architectural framework was proposed to effectively exploit parallelism at multiple levels in robotics control computations. The multiprocessor architecture was built around a Robotics Vector Processor (RVP) chip. In this paper, the architecture of the RVP will be described in detail. Its arithmetic units, interconnection structure, and instruction set are specialized so as to efficiently compute the three-dimensional spatial quantities required. It also has an appropriate input/output interface to permit interconnection of tightly synchronized sets of processors, resulting in an architecture similar to that of Very-Long-Instruction-Word (VLIW) architectures [21]. The inter-RVP communication mechanism is made flexible so that multiple-RVP systems are restructurable into different configurations to suit the requirements of specific robotics computations.

Even though high-performance floating-point adders/multipliers with an approximate 10 Mflops performance are now commercially available, and may appear to provide the computational power required, the realized performance when incorporated into a general-purpose workstation (e.g. Sun 3) is considerably less than the potential peak performance. The exploitation of the high "raw" performance of these ALU's for robot kinematics and dynamics computations would require a carefully tailored architecture of the system built using them, addressing the critical issues of data movement and execution control. The need for a specialized architectural design is especially pronounced in the context of a high-performance robotics multiprocessor, such as the one proposed in [16]. Therefore, the emphasis in this paper is the architectural design of a system with adequate computational power, along with the appropriate input/output and control structures that are essential to avoid loss of computational power in practical application.

The paper is organized as follows. Section 2 describes the nature of potential parallelism in robot kinematics and dynamics computations and motivates the architectural approach pursued here. In Section 3, the architecture of the Robotics Vector Processor and its base instruction set are described. Section 4 discusses how fine-grained parallelism is exploited through the efficient implementation of matrix/vector operations, such as the vector cross product and matrix-vector multiply, using the base instructions of the processor. The issue of overlapped execution of matrix/vector operations is treated and the implementation of an example robotics computation, the Jacobian, is detailed. Section 5 elaborates on the use of multiple tightly coupled RVP's to effectively implement a restructurable Variable-Very-Long-Instruction-Word-Processor

(V^2LIWP) for achieving higher performance through further exploitation of parallelism in robotics computations, again using the Jacobian as an example. Section 6 concludes the paper with a summary of the work.

2. PARALLELISM IN ROBOTICS COMPUTATIONS

Robot kinematics and dynamics computations have considerable potential for parallel execution. Though the potential parallelism is easily identified, its actual effective exploitation is nontrivial. Besides use of sufficient number of arithmetic units to satisfy the computational demands, appropriate tailoring of the hardware for input/output and control is essential. In this section, we identify different levels at which parallelism is available in robotics computations. We then motivate an architectural approach that aims at being efficient in exploiting maximal parallelism while simultaneously maintaining flexibility to be restructurable so as to adapt to the requirements of a wide variety of robotics computations.

At the highest level, a robotics control scheme can be represented in terms of generic computational blocks, such as for the Jacobian, Inverse Dynamics etc., with potential **inter-block level** parallelism in the execution of two or more of these blocks [17]. Each generic computational block can be expressed in terms of a number of identical or similar subcomputations, often one for each of the links of the robotics system being modeled. Parallelism is thus available at the **intra-block level.** Such a subcomputation can be further broken down and expressed in terms of matrix/vector operations on 3x1 vectors and 3x3 matrices [18]. At this **matrix/vector operation level,** there typically is potential for parallelism in the execution of independent matrix/vector operations comprising the higher level subcomputation. Finally, there is obvious potential for parallel execution at the **primitive operation level,** of operations that comprise a matrix/vector operation. For example, multiplication of two 3x3 matrices requires 27 multiplications and 18 additions, where all the multiply operations are completely independent and potentially executable in parallel.

Thus the very same computational algorithm can be viewed in terms of tasks, or operations at different levels of granularity. This can be captured using a task graph [5,9,15], where vertices of such a task graph represent computations and directed edges between tasks denote data flow and implied precedence in scheduling. In utilizing any general-purpose multiprocessor to implement such a computation, the appropriate choice of granularity of the task graph will typically be dictated by a compromise between the degree of parallelism desired and the execution overheads that result from task scheduling and interprocessor communication/synchronization.

The finer the granularity of the chosen tasks, the greater is the degree of potential parallelism, making load-balancing easier and processor-utilization higher. However, the greater degree of parallelism is often obtained at the cost of higher task scheduling overhead and the need for increased data movement and more frequent interprocessor synchronization. The degree of parallelism that is effectively exploitable using any currently available asynchronous general-purpose multiprocessor is thus limited [19].

An approach that overcomes the above problems of asynchronous multiprocessors is the use of systolic architectures [20]. Since systolic architectures are synchronous systems, task scheduling and interprocessor synchronization are implicit and involve no overhead. Further, all data movement is explicitly factored into the design, so that only controlled neighbor-to-neighbor communication is required. Thus fine-grained parallelism is exploitable, and the design is typically scalable with the number of degrees of freedom of the robotics system modeled. However, there are drawbacks to the systolic approach. Different generic computational blocks may require different systolic architectures for their solution, and systolic solutions are not appropriate or available for all robot kinematics and dynamics computations. Further, systolic architectures generally pipeline the computations they implement and significant improvements to the pipeline initiation rate (throughput) of the computation are achieved, but the latency of execution may not be decreased very much [9,10,15].

Most robot kinematics and dynamics computations are conditional-free, i.e. the sequence of operations at the lowest levels is not a function of the input data from the sensors. Thus, a one-time analysis of the computation is sufficient for the purpose of partitioning the computation amongst the processors of a multiprocessor. Further, if the multiprocessor system is tightly synchronized, where all processors execute under lock-step control, the conditional-free nature of the robotics computation makes it feasible to use implicit interprocessor synchronization. Thus, of the three sources of execution overhead in exploiting fine-grained parallelism in robotics computations with an asynchronous multiprocessor, namely dynamic task scheduling, interprocessor synchronization and data movement, the first two can be easily eliminated by using a tightly coupled multiprocessor system. However, viewing the total computation as a task graph at the primitive operation level can pose a problem with respect to the data movement required. Though an analysis of the available parallelism in a task graph at the primitive operation level is relatively straightforward, and heuristics for effective load-balanced scheduling of task graphs on parallel processor systems have been proposed [4,5], the data-movement costs can become comparable or even dominate the computational cost. Minimizing the data movement costs requires a design of the interprocessor communication structure that matches the structure of data communication required by the algorithm, but there is little recognizable structure in the data movement pattern of a robotics computation represented at the primitive operation level. Thus, despite the greater potential parallelism in viewing the computation at the primitive operation level, the difficulty in effectively exploiting it in practice motivated the use of a two-level approach pursued in this work.

In summary, the regular, local communication and tight synchronization of the systolic approach make it highly efficient but relatively inflexible whereas the attributes of asynchronous multiprocessors that make them flexible also cause them to suffer relatively high execution overhead. The architectural approach pursued here attempts to simultaneously achieve efficiencies comparable to the systolic approach as well as flexibility closer to asynchronous multiprocessors. This is attained by exploiting the conditional-free nature of robotics computations and adopting a two-level architectural viewpoint. A small number of tightly synchronized arithmetic units coupled together by a specially tailored data section constitute the basic building block, a single chip Robotics Vector Processor (RVP). The specialized design of the RVP facilitates efficient exploitation of fine-grained parallelism in executing computations at the level of matrix/vector operations. Higher level computations are expressed as task graphs with matrix/vector operations as the constituent tasks. Such task graphs can then be scheduled onto multiple-RVP configurations. Viewed thus as task graphs at the matrix/vector operation level, the amount of inter-task data movement required is relatively small compared to the amount of intra-task data movement and hence not critical in determining overall data communication efficiency. The inter-RVP connection mechanism is therefore made very flexible so as to make these multiple-RVP configurations restructurable to suit the requirements of specific computations. The architectural details of the basic building block, the Robotics Vector Processor, are presented next.

3. THE ROBOTICS VECTOR PROCESSOR (RVP)

3.1 Architecture of the RVP

The Robotics Vector Processor (RVP) is a single-chip processor tailored to the efficient execution of 3x1 vector / 3x3 matrix operations. Fig. 1 shows the block level architecture of the RVP. It contains three floating-point processors (FPP's) for performing arithmetic operations. The three FPP's operate in a lock-step SIMD (Single-Instruction-Multiple-Data) mode, thus controlled by a single controller. The controller initiates execution of an externally supplied vector instruction every clock cycle. Each FPP has an associated register-file so that intermediate quantities in a vector computation can be retained on-chip and avoid large off-chip data movement costs. A dual shift/broadcast network (SBN) is used to facilitate communication between the FPP's, and is especially useful in implementing vector operations such as vector cross product and matrix-vector multiply. A dual input/output channel exists for communication of data values in and out of the RVP. It permits flexible interconnection of a number of RVP's, for example in a ring configuration, among others.

Fig. 2 shows an FPP in greater detail. An FPP is comprised of a floating-point adder (FPA), a floating-point multiplier (FPM) and a register-file (RF). A dual-bus interconnects the register-file and the floating-point units of an FPP. The RF contains 56 32-bit words and is interfaced to the other FPP units through a buffer. It is dual-port readable, but only single-port writeable, due to the design constraints in VLSI. Thus any two registers in the file may be read simultaneously using the two buses (BusA and BusB). During a write, however, only one register can be written into at a time, using either one of the two buses. Besides the registers in the RF, each FPP also has eight other special-registers, explained below, for a total of 64 registers. The size of the register-file was determined by a chip-area vs. computational need trade-off. The chosen size is adequate for many robot kinematics and dynamics computations, as exemplified by the Jacobian computation discussed in detail later in the paper.

A similar trade-off between chip-area and execution time was made in the design of the FPA and the FPM. The FPA uses a Manchester carry chain and carry lookahead and takes three cycles (prenormalization, add/subtract, followed by postnormalization) for an operation.The FPM takes 5 clock cycles for a multiplication, and is designed as a two-stage pipeline with a 4-cycle first stage (four 6-bit sets of carry save multiplication for the mantissa) and a 1-cycle second stage (carry propagate addition and exponent adjustment). Both floating-point units operate on 32-bit operands and use clamping for overflow/underflow, thus obviating the need to use interrupts and traps.

Most robot kinematics and dynamics computations may be expressed in terms of 3x1 vector operations on spatial quantities. A 3x1 vector is stored in the RVP by distributing the three components among the three FPP's. The RVP directly implements simple 3x1 vector operations such as vector add, vector multiply and vector shift (permute). More complex vector operations such as vector dot product, vector cross product, matrix-vector multiply and matrix-matrix multiply are implemented as sequences of the basic vector operations. A simple vector operation such as a vector add is executable in parallel without inter-FPP data transfer, by keeping the corresponding components together in the same FPP. The result produced by the FPA may be written back into any register of the RF, or may be retained in a special output latch and used in a succeeding vector operation. The FPA's output latch is thus treated as a special-register (R6)

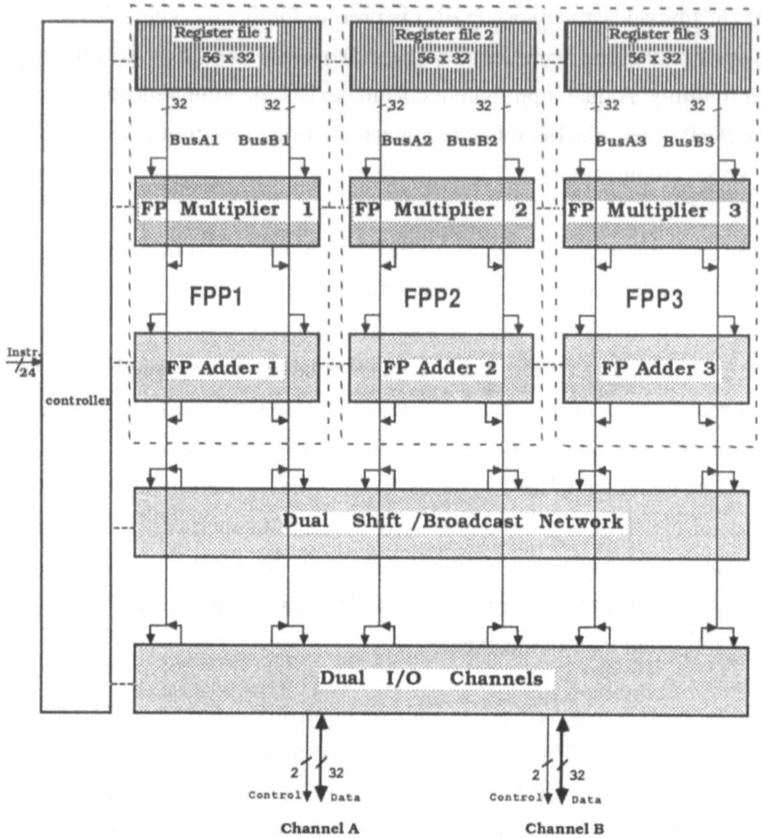

Figure 1. Robotics Vector Processor Block Diagram

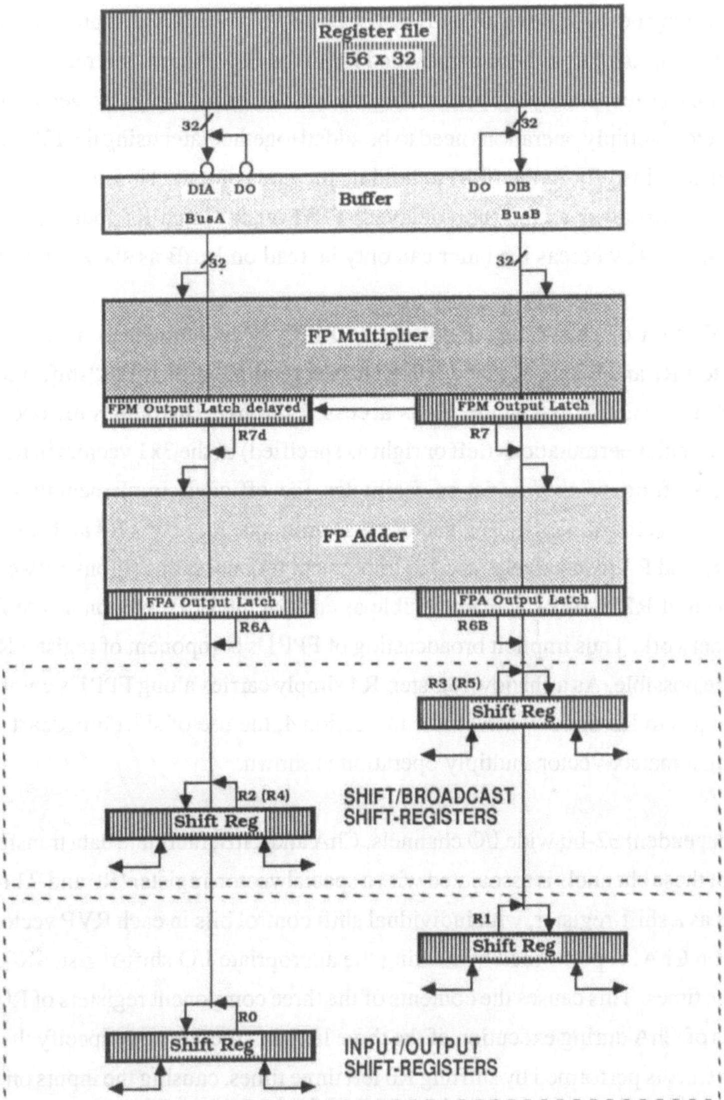

Figure 2. Floating Point Processor Block Diagram

whose contents may be input along either BusA or BusB, to be used as an input operand to a vector operation. Likewise, an output latch is provided for in the FPM, that is referenced as register R7. Since in many cases (especially for performing matrix-vector multiply) the vector results of two successive vector multiply operations need to be added together later using the FPA, an additional output latch is provided for in the FPM to hold its previous output. This delayed output register R7d has the same address as the (non-delayed) FPM output latch R7, but the former is only readable along BusA, whereas the latter can only be read on BusB as shown in Fig. 2.

Communication of data values between the FPP's is accomplished using the dual SBN. Special registers R2 and R3 are used to perform data communication in the "shift" mode. Register R2 is loaded/read using BusA while R3 is accessed through BusB. A shift operation causes independent circular permutations (left or right as specified) of the 3x1 vectors in R2 and R3. The simultaneous shifting of two registers facilitates the efficient implementation of the data movement required for implementing vector cross products. Registers R4 and R5 are "shadow" registers of R2 and R3 respectively, used to implement broadcast operations between the FPP's. The component of R2 in FPP1 is made visible as each of the FPP components of R4 through a multiplexor network. Thus implicit broadcasting of FPP1's component of register R2 to all three FPP's is made possible. As a shadow register, R4 simply carries along FPP1's component of R2, and thus changes to R2 directly affect R4. In Section 4, the use of shift/broadcast operations in implementing a matrix-vector multiply operation is shown.

Two independent 32-bit wide I/O channels, ChA and ChB, facilitate data transfer to/from the RVP. Each of these channels is associated with a special vector-register (R0 and R1 respectively), that operates as a shift-register, with individual shift control bits in each RVP vector instruction. Output, say on ChA, is performed by loading the appropriate I/O shift-register R0, and shifting R0 right three times. This causes the contents of the three component registers of R0 to appear on the data lines of ChA during execution of the three later instructions that specify the right shift of R0. Input on ChA is performed by shifting R0 left three times, causing the inputs on the data lines of ChA to get shifted into the three components of vector-register R0. Thus, two 3x1 vectors can be input/output in three clock cycles using the two I/O channels. The dual I/O channels are especially important in using several RVP's together in multiple-RVP configurations, as explained in Section 5.

3.2 Instruction Set of the RVP

The RVP is an SIMD (Single-Instruction-Multiple-Data) machine, with identical operations being performed in each of the three FPP's at any cycle. Each instruction takes one or more clock cycles to completely execute, but the controller (Fig. 1) performs an instruction initiation every cycle. An appropriate number of no-op instructions may need to be inserted into the instruction sequence to ensure proper synchronization between instructions and to avoid conflicts in their use of the functional units of the RVP.

The RVP has only seven instructions (plus a no-op) and uses a format as shown in Fig. 3. The instruction set is designed to be simple but allow flexible control of all RVP components - the FPA's, FPM's, SBN, I/O Channels, register-files and dual-buses. The single FPP resource that is shared by the various component units is the dual-bus. The instructions may be classified on the basis of their use of this common resource. The bus is used to either i) transfer data from any special-register to the register-file, or ii) move data from a register (special-register/register-file) to one of the functional units. A single instruction Write File (WF) is used for the former (to move data into the register-file from outside it), while individual instructions (OUT, FPM, FPA, FPS, LSR) are available for the latter, serving in the case of the FPM/FPA functional units to also initiate the respective operation after data movement into the unit.

The shift-registers associated with the I/O channels (ChA and ChB) have individual shift control bits (IOSa/IOSb) in each instruction, so that data in and out of the RVP chip can be controlled on each channel, independent of other data movement/functional unit control. This is especially important for synchronization in multiple-RVP configurations. Each I/O channel is also provided with two external control bits (IOCa/IOCb) that are simply gated to the external control lines of the channel (Fig. 1), for the purpose of controlling external I/O devices connected to the channel. Finally, a separate instruction, Shift (SH) is provided to control the shift-registers of the SBN. While individual shift control bits in each instruction for these shift-registers could also have been used, similar to that of the I/O shift-registers, this was not done so as to reduce the number of bits in an instruction, since the overall expected use of this function did not justify it.

Since the number of addressable vector-registers is 64, each explicit register operand uses a 6-bit address. A 4-bit opcode is used (even though 3 bits will suffice) to make the instruction size a whole number of bytes. Each of the RVP instructions is elaborated on below:

WF (Write register File): One of the seven special vector-registers (R0-R7) serves as the source and one of the vector-registers (R8-R63) in the register-file is the destination, so that data from one of the functional units may be stored in the register-file. Thus, when the source is R0/R1, input data received from ChA/ChB is moved into the register-file; when the source is R2/R3(R4/R5) data from the primary (shadow) registers of the SBN are stored; and if the source is R6(R7), the output result from an FPA(FPM) operation is

	00 XX	01 XX	11 XX	10 XX
00	NOP	WF	FPA	OUT
01	-	-	FPS	-
11	SH	-	FPM	-
10	-	-	LSR	-

a) Instruction Opcodes

	23 22 21 20	19 … 14	13 … 8	7 6	5 4	3 2	1 0
	Op Code	Register Addressing		I/O Control			
OUT	1 0 0 0	Rsa	Rsb	IOSa	IOCa	IOSb	IOCb
FPA,FPS, FPM,LSR	1 1 x x	Rsa	Rsb	IOSa	IOCa	IOSb	IOCb
WF	0 1 0 0	Rs	Rd	IOSa	IOCa	IOSb	IOCb
SH	0 0 1 1	0 0 0 0 SBNa	0 0 0 0 SBNb	IOSa	IOCa	IOSb	IOCb

SBNa/SBNb

01 : Right Circular shift (RC)
10 : Left Circular shift (LC)
00 : No Shift (NS)

IOSa/IOSb

01: Right Shift
10: Left Shift
00: No Shift

b) Instruction Format

Operands / Operation	Rs		Rd	
	Rsa	Rsb	Rda	Rdb
WF	0-7		8-63	
FPA/FPS	8-63;0,2,4,6,7d	8-63;1,3,5,6,7	FPAa	FPAb
FPM	8-63;0,2,4,6,7d	8-63;1,3,5,6,7	FPMa	FPMb
OUT	8-63,0,2,4,6,7d	8-63,1,3,5,6,7	0,none	1,none
LSR	8-63;0,2,4,6,7d	8-63;1,3,5,6,7	2,none	3,none

c) Instruction Operands

Figure 3. Robotics Vector Processor (RVP) Instruction Set

stored back into the register-file. Only one of the data buses is used during a WF and the operation takes one clock cycle to complete.

OUT (OUTput data): This is an I/O instruction, used to transfer the contents of any of the vector-registers to the shift-registers R0/R1 associated with the I/O channels. Data is loaded from source vector-register Rsa into special-register R0 using BusA and simultaneously from Rsb into R1 using BusB, in the same clock cycle. During succeeding clock cycles, the data loaded into R0/R1 is shifted out through ChA and ChB respectively, controlled by the explicit I/O shift-control bits IOSa/IOSb in those instructions. The source Rsa can be any register from the file (R8-R63), or any special-register accessible on BusA — R0, R2, R4, R6, R7d (Fig. 2). If the special-register R0 is specified as the source, then output along ChA is disabled. This makes it possible to selectively disable overwriting one of R0/R1 while the other is loaded for output. Restrictions on Rsb are similar to those on Rsa, as shown in Fig. 3c. The RVP does not have an explicit IN instruction — the WF instruction serves to transfer data from R0/R1 (previously shifted in from an input channel) to any of the registers of the file, and any of the other functional units can be directly loaded from R0/R1.

FPA (Floating-Point Add): During the first clock cycle, source operands from registers Rsa and Rsb are loaded and internally latched into the floating-point adder unit using BusA and BusB. The addition takes three further cycles and the result is latched into special-register R6, available using either BusA or BusB. The buses are thus used by an FPA operation only during the first cycle and are free during the last three cycles. The operand Rsa may be any register accessible along BusA — R0, R2, R4, R6, R7d or R8-R63, and Rsb may be any of R1, R3, R5, R6, R7, R8-R63. Even though faster adder designs were possible, a three cycle design was chosen due to VLSI area constraints.

FPS (Floating-Point Subtract): The operation is very similar to FPA, except that the data from Rsb is subtracted from that of Rsa.

FPM (Floating-Point Multiply): As with the FPA and FPS operations, the input operands are transferred along buses BusA and BusB during the first clock cycle and internally latched within the multiplier. The multiplier is a two-stage pipelined unit. The first stage takes four cycles after the initial data transfer cycle. The second stage takes one cycle and at the end of this cycle the result is latched in vector-register R7. The previous result, that resided in R7, is simultaneously transferred to register R7d. The availability of the last two multiplier results in this manner is very convenient for implementing composite matrix/vector operations using the base vector operations, as shown in Section 4.

LSR (Load Shift-Register): This instruction is used to load the shift-registers of the dual SBN, the mechanism for inter-FPP data communication. Register R2 is loaded from source vector-register Rsa via BusA while R3 is simultaneously loaded from Rsb via BusB. As explained for the OUT instruction, selective disabling of one of the dual loads is accomplished by using R2/R3 as the source for the register to be disabled. The shadow broadcast registers R4 and R5 in all three FPP's are automatically changed to be consistent with the contents of FPP1's R2 and R3 registers, respectively.

SH (SHift): The shift operation causes both the SBN shift-registers (R2,R3) to shift independently. The SBNa(SBNb) field in the instruction (Fig. 3b) specifies the shift mode — Right Circular (RC), Left Circular (LC) or No Shift (NS). Again, the broadcast operation is implicit after any shift — the broadcast registers R4 and R5 in all three FPP's are automatically changed to be consistent with the contents of FPP1's R2 and R3 registers, respectively, after the shift.

Fig. 4 illustrates the timing characteristics of the various vector instructions and also serves to emphasize the great degree of overlapped operation possible within the RVP. The single common RVP resource that is shared by various instructions is the dual-bus, but as shown in Fig. 4, considerable overlap in the execution of several instructions is nevertheless possible. The first instruction shows the timing sequence for an FPM. The instruction fetch (performed by a sequencer that drives the RVP, and hence shown lightly shaded) is followed by a read-register cycle where the input operands are fed via the buses and are internally latched within the multipliers. The first stage of each multiplier (FPM 1) takes four cycles, followed by a single-cycle second stage (FPM 2). The multiplier result is written into the output register R7 in the next cycle, when it can simultaneously be written into the file using a WF. The second instruction shown is a SH, that executes in one cycle, concurrent with the multiplier since they share no resources. The third instruction is an FPA (or FPS). The first cycle of the FPA is a read-register and can clearly overlap the multiplier execution since the buses are free. The execution of the addition takes three further cycles, completely overlapped with the multiplier execution, after which the result is latched into the special-register R6. During the cycle that R6 is being written, a WF can be used to simultaneously transfer the result to the register file. The OUT instruction that is executed next transfers data along the buses to load R0/R1; explicit shifting out of the data loaded into ChA/ChB is achieved through appropriate use of the IOSa/IOSb bits of succeeding instructions. Instruction 5 is another FPM, that is able to overlap the first FPM instruction due to the two-stage pipelined nature of the multiplier. Instruction 6 is an LSR and is able to overlap two FPM's, one FPA and one OUT instruction (during cycle 6) since none of the other instructions needs the use of the data

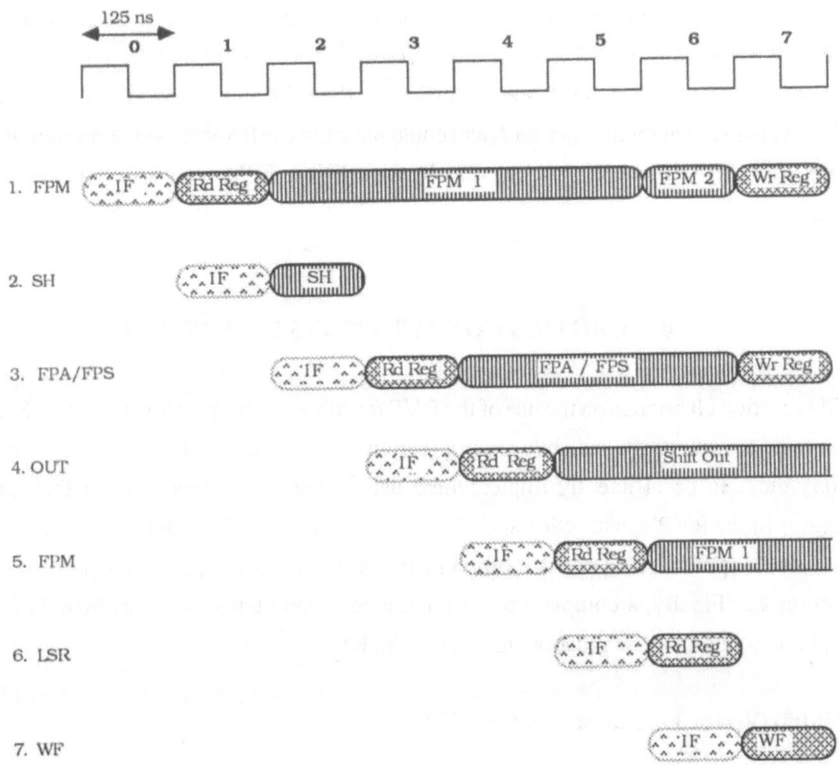

Figure 4. Illustration of Execution Timing on the RVP

buses in that cycle. Thus during cycle 6, 9 arithmetic operations may be simultaneously in progress (2 FPMs and one FPA on triadic vectors) as well as eight parallel word-level data transfers (6 along the three BusA's and BusB's for the LSR and 2 along ChA and ChB for the OUT). A WF is shown for instruction 7, and could possibly use R6 or R7 as the source even though they are just being written into, respectively, by the previous FPM or FPA/FPS instruction during cycle 7.

4. ROBOTICS COMPUTATIONS ON THE RVP

This section elaborates on the use of the RVP for robotics computations. First, in Section 4.1, it is shown how composite matrix/vector operations such as vector cross product, matrix-vector multiply etc. can be efficiently implemented using overlapped execution of the base vector operations of the RVP's instruction set. The execution of multiple such independent, composite matrix/vector operations on an RVP offers further scope for overlapped execution, as discussed in Section 4.2. Finally, a complete robot kinematics computation, the Jacobian, is used as an example in Section 4.3, to illustrate the use of the RVP.

4.1 Matrix/Vector Operations on the RVP

The implementation of more complex vector operations using the base vector instructions of the RVP will be described in this section. The matrix-vector (MV) product of a 3x3 matrix and 3x1 vector is used as an example of such a composite vector operation. Figure 5 shows a program for MV, timing relations for its execution and a reservation table that clearly brings out the overlap in the use of various components of FPP1.

The matrix M is assumed stored as three 3x1 column vectors M1, M2 and M3, distributed across the register-files of the three FPP's of the RVP. Thus M11 is stored as FPP1's component of vector-register "M1" (which is some register R8-R63 in the register-file), M21 in FPP2 and M31 in FPP3. The final result is stored in some vector-register Z in the register-file.

The vector V is first loaded into R2/R4 of the SBN (along BusA) through the first LSR instruction. Only shift-register A is effectively used; the second operand register is a "don't care", so that R3/R5 of the dual SBN may retain its previous value without loading any register along BusB. The LSR instruction causes V1 to be visible in all components of R4. The LSR instruction is executed in cycle 1; the instruction fetch for it is assumed to have occurred before the first cycle, and is not shown explicitly since it is implemented outside the RVP. This convention is followed

MV = [M1 M2 M3] [V] = Z

$$= \begin{bmatrix} M11 & M12 & M13 \\ M21 & M22 & M23 \\ M31 & M32 & M33 \end{bmatrix} \begin{bmatrix} V1 \\ V2 \\ V3 \end{bmatrix} = \begin{bmatrix} M11 * V1 + M12 * V2 + M13 * V3 \\ M21 * V1 + M22 * V2 + M23 * V3 \\ M31 * V1 + M32 * V2 + M33 * V3 \end{bmatrix}$$

a) Program*

1.	lsr	V, -		11.	nop	
2.	fpm	R4, M1		12.	fpa	R7d, R7
3.	nop			13.	nop	
4.	nop			14.	nop	
5.	sh	LC,NS		15.	nop	
6.	fpm	R4, M2		16.	fpa	R6, R7
7.	sh	LC,NS		17.	nop	
8.	nop			18.	nop	
9.	nop			19.	nop	
10.	fpm	R4, M3		20.	wf	R6, Z

* I/O control operands not shown

b) Timing chart

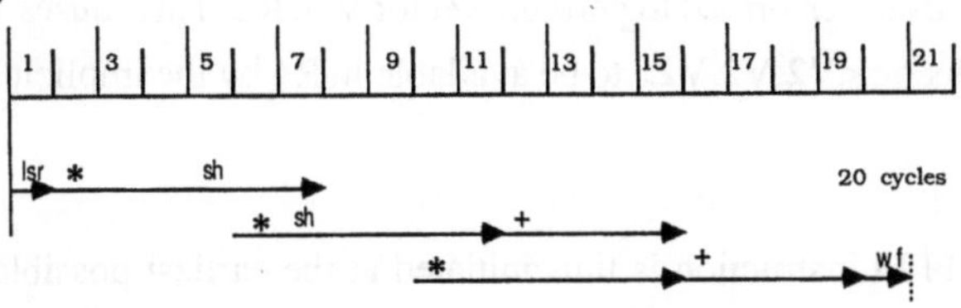

c) Reservation Table for FPP1

a = M11 * V1; b = M12 * V2; c = M13 *V3; d = a + b; Z = d + c

	1	2	3	4	5	6	7	8	9	10	11	12	13	14	15	16	17	18	19	20	21
fpm_{in} A / B		V1 / M11				V2 / M12				V3 / M13											
fpm_1			a'	a'	a'	a'	b'	b'	b'	b'	c'	c'	c'	c'							
fpm_2							a'				b'				c'						
R7								a	a	a	a	b	b	b	b	c	c	c	c		
R7d												a	a	a	a	b	b	b	b		
fpa_{in} A / B												a / b				c / d					
fpa													d'	d'	d'		Z'	Z'	Z'		
R6																d				Z	
Reg File																				Z	

Figure 5. Program Execution, Timing and Reservation Table for Matrix-Vector Multiply

for all the instructions. The next instruction (FPM) specifies R4 as one operand and M1 as the other. This results in the initiation of the computation for the intermediate vector **a** = V1*M1 = <V1*M11,V1*M21,V1*M31> by the FPM. This operation takes 7 cycles. During its first cycle (cycle 2), the input operands use the buses and are latched into internal FPM registers. The first stage of the multiplier FPM1 is used for the next four cycles. This is marked in the reservation table as **a'**, denoting the incomplete computation for **a** as occupying the stage. The next cycle uses the second multiplier stage FPM2, and the result is available in R7 after the last cycle (8).

The next two instructions are NOP's while the FPM executes. The next arithmetic operation is initiated in cycle 6 for the computation of intermediate vector **b** = V2*M2 = <V2*M12,V2*M22,V2*M32>. Just prior to this, in cycle 5, a left circular shift of the first register (R2/R4) of the SBN is performed to permute vector V in R2. This causes V2 to move from FPP2 to FPP1, thus causing <V2,V2,V2> to be available in R4 by the implicit broadcast operation of the SBN.

The second FPM instruction is thus initiated at the earliest possible time, overlapping its 'Register Read' cycle with the fourth cycle of the first stage execution of the first FPM instruction, as shown in the reservation table in Fig. 5c. Following the second FPM, another left circular shift is completed to broadcast V3, prior to initiation of computation for intermediate vector **c** = V3*M3 = <V3*M13,V3*M23,V3*M33>. At the end of cycle 12, intermediate vector **b** is output by the FPM and latched into R7, while simultaneously the intermediate result **a** that has been previously stored in R7 is moved into R7d. In cycle 12, a vector add for **d** = **a**+**b** is initiated by the adder. The intermediate vectors **c** and **d** are latched respectively at the outputs of the multiplier and adder (R7 and R6) at the end of cycle 16. The flow-through nature of the final write cycle of the adder/multiplier permit their results to be used in cycle 16, simultaneous with their latching in the special output registers. Thus the final vector addition **Z** = **d**+**c** is initiated in cycle 16 and the final result is written back into the register-file with the WF in cycle 20, again simultaneous with its latching in R6.

With a targeted clock cycle of 125ns, a matrix-vector multiply in 20 cycles corresponds to a realized performance of 6 Mflops on the single-chip RVP. In overlapped operation (see next section), only 12 cycles per matrix-vector multiply are required, i.e. 10 Mflops is achieved. Similarly 8Mflops (10 Mflops) performance is attained with matrix-matrix multiply in a nonoverlapped (overlapped) mode. Such performance for this computation is difficult to achieve today even on powerful mainframe computers, and further improvement in performance is possible here with the use of more sophisticated and area-intensive adder/multiplier designs in the RVP. It should also be noted that the short vector lengths involved make the achievable performance for such operations on general-purpose vector machines (such as the Cray-1) significantly lower than their peak rates.

4.2 Overlapped Execution of Matrix/Vector Operations

It is similarly possible to develop programs for other composite vector operations such as vector dot product, vector cross product etc. Table 1a summarizes the number of RVP cycles required for the various operations of relevance for robot kinematics and dynamics computations. When independent sets of vector operations are involved, further overlap in utilization of the RVP's functional units is often possible. For example, from Fig. 5 it may be observed that the first stage of the multiplier is free between cycles 15-20. Hence, if two independent matrix-vector multiplies were required, the second one could be initiated before the first one is completed.

Fig. 6 graphically displays the possible overlaps between a matrix-vector multiply and other succeeding operations. Consider first the overlap of a composite vector operation (VxV, V·V or MV) with MV. Any composite vector operation completes with the use of the adder, immediately preceded by the use of the multiplier and after an initial LSR instruction. Since there are more multiplication operations than additions in each of the composite vector operations and each of the multiplications takes longer to execute, the multiplier's use is the primary constraining factor limiting the amount of overlap possible. In particular, noting Fig. 6, the second vector operation (VxV, V·V or MV) can only begin such that there is no conflict in the use of the first stage of the multiplier. As shown, another multiplication may be initiated during cycle 14 so that the second vector operation may begin just prior to that. This overlappability is subject to the absence of any conflicts on the use of the buses for data movement. The regions of use of the bus by the previous MV operation are explicitly marked as the gray areas in the figure. When bus conflicts do occur, it is often possible to displace one of the uses of the bus slightly so that no loss of effective overlap occurs. This can be seen in the case of the VxV operation, where using the maximal overlap possible for the multiplier would have led to a conflict in the use of the bus by the read cycle for the adder by the MV operation and its use by the LSR instruction of the VxV. This is easily avoided by performing the LSR for the VxV one cycle earlier when the bus is free.

Considering now V+V operations, since the multiplier is the bottleneck resource in overlapping composite operations, there are regions where the adder is free, even after maximally overlapping composite operations. The hatched region in Fig. 6 corresponds to cycles of possible use of the adder at the end of a second previous composite operation, in maximal overlap with the previous one. Excluding all possible use of the adder by overlapped composite operations, the adder is seen to be free between cycles 8-11 and 20-23. Thus, V+V operations can be inserted within overlapped composite operations with no need for any additional cycles. Again, there may

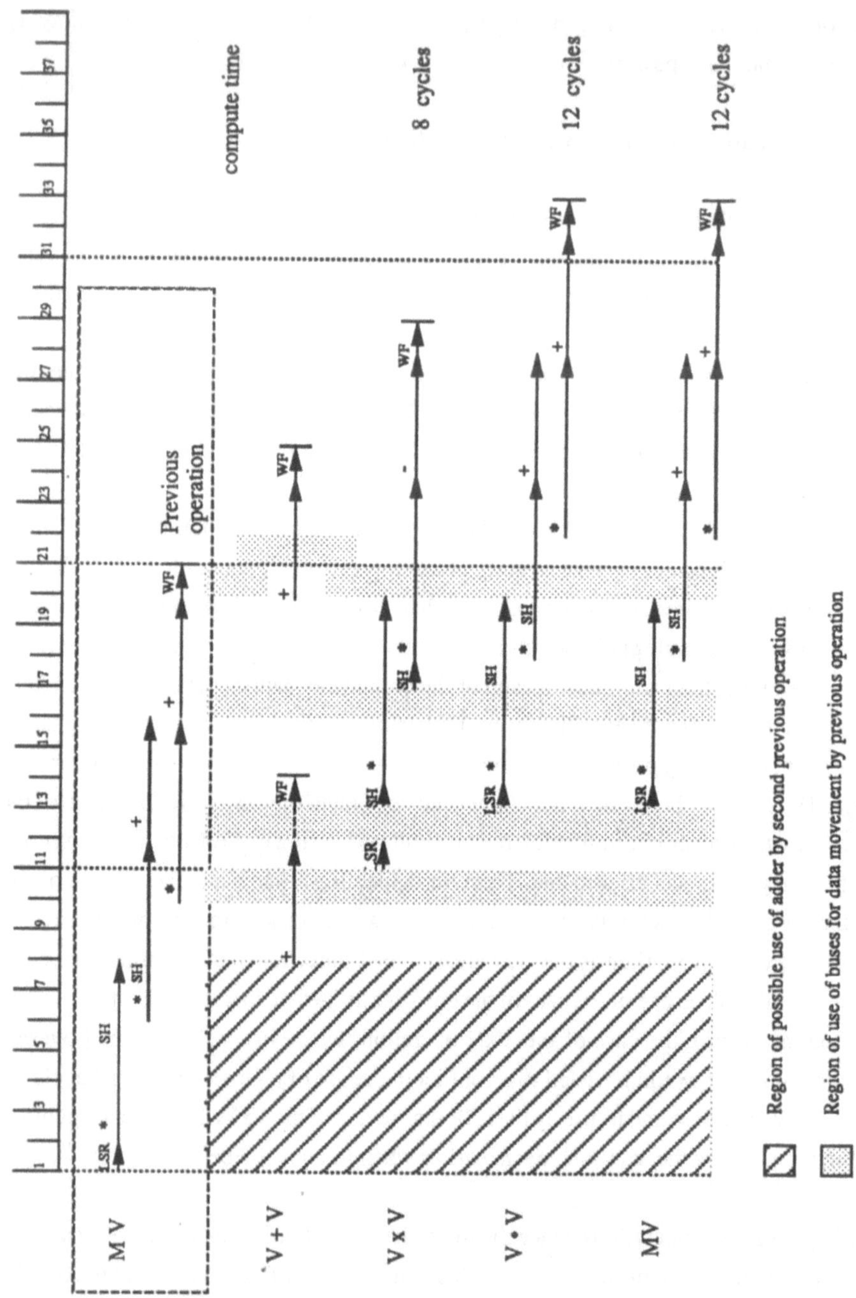

Figure 6. Overlapped Vector Operations with Matrix-Vector Multiply (MV) as Previous Operation

Table 1. Summary of Overlapped Execution Timing for RVP

(a) Sequential Execution

Operation	Cycles
V ∗ V	7
V + V	5
V x V	17
V • V	20
M V	20
M M	44

(b) Overlapped Execution of Composite Operations

Operation	Cycles*
V x V	8
V • V	12
M V	12
M M	36

*Cycles beyond end of previous composite operation

need to be slight adjustments to the sequences to avoid possible bus use conflicts, as can be seen to be the case for both possible positions where a V+V can be inserted in the schedule.

An analysis of the possible overlap of composite vector operations is summarized in Table 1b in terms of the finishing time of the operation relative to the end of the previous operation. The difference between an entry in Table 1b and the time for the same operation in Table 1a gives the amount of overlap possible. For example, by judiciously overlapping two independent vector cross products, it is possible to achieve a savings of 9 cycles — instead of 17 cycles each, the second overlapped operation can complete 8 cycles after the completion of the first. The information in Table 1 could be used by a compiler for the RVP in optimally scheduling the operations required to execute robotics computations. Its use in programming the Jacobian is illustrated in the next section.

4.3 Example: Jacobian Computation Using the RVP

The use of the RVP for robotics computations is now illustrated using the Jacobian as an example. Table 2 shows the Orin-Schrader formulation for the computation of the Jacobian, one of the six methods described in [18]. Only the case of revolute joints is shown; the treatment of the sliding joint case is somewhat simpler. In the equations, N is the number of degrees of freedom, ${}^{i}U_{i-1}$ is the 3x3 orientation matrix relating the ith link to the $i-1^{st}$ link, and ${}^{i}\mathbf{p}_i^*$ represents the position of the ith link with respect to the $i-1^{st}$ link. The 3x1 vectors γ_i and β_i form the ith column of the 6xN Jacobian matrix **J**.

A robot with six degrees of freedom is considered here. The computation required by the equations in Table 2 for this case can be expressed using a task graph, as seen in Fig. 7. The task graph shows the computation and its detailed scheduling on an RVP, including all input/output required. Each circle represents a component computational task at the matrix/vector operation level, with the operation explicitly indicated at the bottom of the circle. The number at the top of a task is its sequence number, representing the order of its scheduling on the RVP. The number range seen in the middle of a task gives the exact span of cycles that it actively executes on the

Table 2. Equations for Jacobian Computation [18]

$$\mathbf{J} = \begin{bmatrix} \gamma_1 & \gamma_2 & & \gamma_N \\ & & \cdots & \\ \beta_1 & \beta_2 & & \beta_N \end{bmatrix}$$

where

$$^{N+1}\mathbf{U}_{i-1} = {}^{N+1}\mathbf{U}_i \, {}^{i}\mathbf{U}_{i-1} \qquad i = N, \ldots 2, 1; \qquad (1)$$

$$^{N+1}\gamma_i = {}^{N+1}\mathbf{U}_i \begin{bmatrix} 0 \\ 0 \\ 1 \end{bmatrix} \qquad i = 1, 2, \ldots N; \qquad (2)$$

$$^{N+1}\mathbf{r}_{i-1} = {}^{N+1}\mathbf{r}_i - {}^{N+1}\mathbf{U}_i \, {}^{i}\mathbf{p}_i^* \qquad i = N, \ldots 2, 1; \qquad (3)$$

$$^{N+1}\beta_i = {}^{N+1}\gamma_i \times (-{}^{N+1}\mathbf{r}_{i-1}) \qquad i = 1, 2, \ldots N. \qquad (4)$$

RVP. For example, task 1 is a matrix-vector multiply (MV) and is scheduled for execution between cycles 7 and 26.

Directed edges in the task graph specify data flow and precedence constraints. Relevant intermediate quantities communicated between tasks are marked on the directed edges between tasks. The rectangular boxes explicitly show all input/output required by the RVP. The inputs for the Jacobian computation are the ${}^{i}\mathrm{U}_{i-1}$ matrices and the ${}^{i}\mathbf{p}_i^*$ vectors, $1 \leq i \leq 7$, while the outputs are the vectors γ_i and β_i, $1 \leq i \leq 6$, for each column of the Jacobian. At the bottom of each I/O box, the RVP channel used for I/O is displayed, and the actual cycles during which the operation takes place are specified by the numbers in the middle. Note that only ChA is used here.

The tasks in the task graph can be seen to correspond to the matrix/vector operations specified by the equations in Table 2. Eq. (1) requires a product of two 3x3 matrices for each of the six links, marked as MM in the task graph. Eq. (2) essentially specifies the extraction of the third column of a 3x3 matrix and thus does not require any computation. Eq. (3) requires a matrix-vector

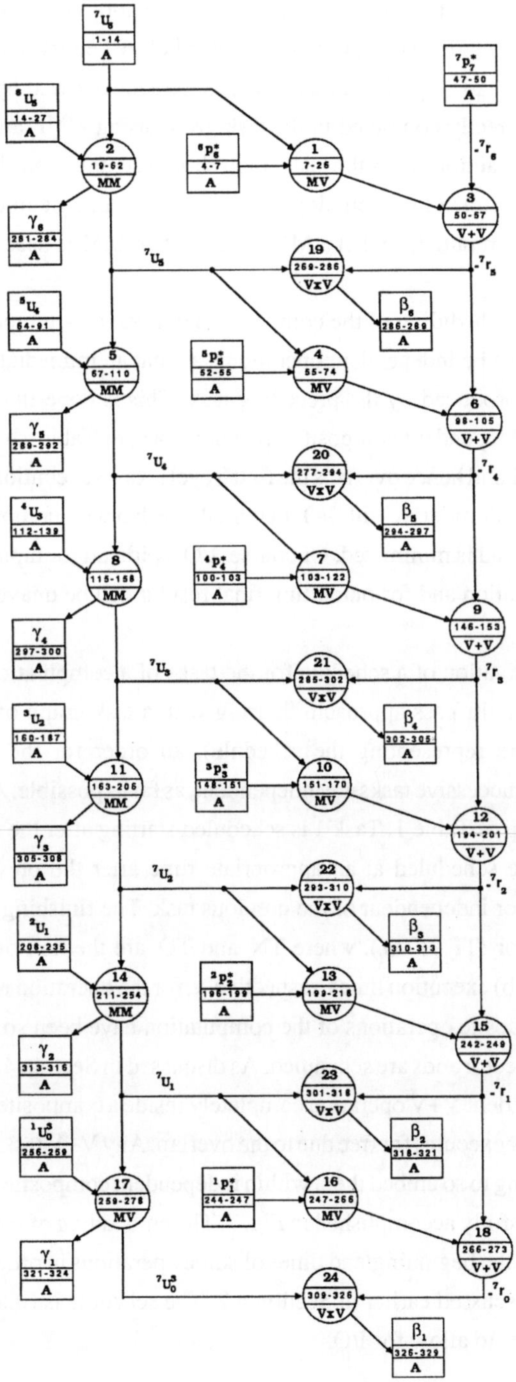

Figure 7. Task Graph and Scheduling of Jacobian Computation on an RVP

multiply (MV) as well as the subtraction of two 3x1 vectors to compute the intermediate 3x1 vectors ${}^{N+1}\mathbf{r}_{i-1}$. These vectors then have to be negated before being used in a vector cross product (VxV) in Eq. (4). Instead of performing a vector subtract, followed by a negation, the negated vectors - ${}^{N+1}\mathbf{r}_{i-1}$ are directly computed in the task graph using - ${}^{N+1}\mathbf{r}_i$ and vector addition (V+V) instead of vector subtraction. Thus the task graph should contain 6 MM's, 6 MV's, 6 V+V's and 6 VxV's. A minor optimization is employed — since only the last column of ${}^7U_0$ is needed, the entire matrix is not computed, so that a MV instead of an MM is used in this case.

Considering the scheduling of the computational tasks, an attempt is made to have successively scheduled tasks be independent operations, so that an immediately succeeding task does not use any results produced by the preceding task. This is done in order to capitalize on the overlappability of independent composite operations, as per Table 1b. Thus, in Fig. 7, task 2 is independent of task 1 and hence overlaps the first 8 cycles of its execution with the last eight cycles of task 1's execution. Similarly, with I/O, maximal overlap with computations is sought, so that the associated overhead is minimized. Of course, I/O overhead for inputting the initial quantities to begin the computation and for outputting final results will be unavoidable.

The manual derivation of a schedule for the tasks of a computation such as the Jacobian is completed using a multi-pass approach. Starting with a task graph of the computation (Fig. 7 without the numbers representing the schedule), an order for the composite operations is determined, so that successive tasks are independent, as far as possible. A first-cut schedule is then created with the help of Table 1. Task 1 is scheduled starting after the initial input required, and successive tasks are scheduled at an appropriate time after the previous task, depending on whether dependent or independent of the previous task. The finishing time of task i, TF_i is thus either (TF_{i-1}+TN_i) or (TF_{i-1}+TO_i), where TN_i and TO_i are the non-overlapped (Table 1a) and overlapped (Table 1b) execution times, respectively, for the operation represented by task i. After all constituent composite operations of the computation have been so scheduled, simple vector operations such as vector adds are scheduled. As discussed in Section 4.2, it is very often possible to embed an independent V+V operation completely inside a composite operation, so that the add operation essentially executes for free due to the overlap. Any V+V tasks in the task graph are then scheduled, attempting to so embed them within independent composite operations, as can be seen to have been successfully accomplished in Fig. 7. The embedding of the vector-adds may require slight adjustment of the beginning/end times of some operations in order to avoid conflicts on the use of the bus, as discussed earlier in Section 4.2. The schedule is finally refined to incorporate any necessary delays to allow for I/O.

For the Jacobian example, using the overlapped execution times of 8, 12 and 36 cycles respectively for VxV, MV and MM, and non-overlapped execution time of 20 cycles for the first

MV task, and assuming maximal overlap with all V+V's coming for free, a minimum possible finishing time may be calculated giving 20 + 5*36 + 6*12 + 6*8 = 320 cycles. Excluding the first 6 cycles for inputting two 3x1 vectors to get started, and a final 3 cycles for outputting the last output vector, this lower bound is actually achieved by the schedule shown. Thus, besides maximal overlap of computation, maximal possible overlap of all I/O has also been accomplished. The schedule shown in Fig. 7 requires 329 cycles for the Jacobian computation. The computation is comprised of 234 floating-point multiplications and 168 floating-point additions. Thus, with a cycle time of 125 ns, the computation takes 41 µs, providing an average performance of approximately 10 Mflops. The utilization of each FPM is approximately 95%, while the adders are 50% utilized. A conventional microprocessor with comparable FPA/FPM units would have required at least 234*7 + 168*5 = 2478 cycles; i.e. the RVP design provides a factor of improvement of at least 7.

The manual generation of such schedules as that for the Jacobian can be very tedious with complex robotics computations. Work is currently in progress on the development of an automated scheduler. For more complex computations, such a tool will be essential for the effective use of the RVP, especially so in the context of multiple-RVP systems that are considered next.

5. RESTRUCTURABLE MULTIPLE-RVP CONFIGURATIONS

The RVP design has been tailored to the effective exploitation of fine-grained parallelism in matrix/vector operations. Unlike the structured parallelism at this level, the parallelism available at higher levels of robot kinematics and dynamics computations has less regularity. However, when such a computation is expressed in terms of tasks at the level of matrix/vector operations, the total amount of data movement required between parallel tasks is much less than the intra-task data movement. Thus the impact of the data movement problem on the scheduling of tasks at the matrix/vector operation level is much less than if the scheduling was done at the primitive operation level. It is thus very attractive to use multiple RVP's to simultaneously execute different matrix/vector operations comprising a higher level computation such as the Jacobian. This section elaborates on the use of multiple-RVP configurations for robotics computations, using the Jacobian as an example to illustrate how a configuration can be structured to match the requirements of the computation.

When heterogenous task graphs are scheduled on a general-purpose multiprocessor, some dynamic synchronization mechanism has to be used to signal the end of tasks when they complete execution and to enable other tasks as soon as their predecessor tasks complete. Triggering mechanisms could of course be incorporated into a multiple-RVP system to enable scheduling of matrix/vector level task graphs. However, the conditional-free nature of the task graphs of robotics computations makes this unnecessary. Since the number of cycles required by an RVP to execute any of the component matrix/vector operations of the task graph is exactly known before execution, if additionally the time required for performing input/output between RVP's is explicitly factored into the scheduling, it is possible to determine a priori, the exact times at which these operations can be collectively scheduled on multiple RVP's to execute the entire task graph. Thus, given any particular cycle in this execution, the specific primitive instruction that needs to be executed by any RVP can be determined. The set of primitive instructions to be executed at a specific cycle by the various RVP's can in fact be considered to be a single very long horizontal micro-instruction. Hence, a multiple-RVP system may be viewed as a Very-Long-Instruction-Word (VLIW) processor [21], except that the conditional-free nature of the computations of interest means that none of the complications due to branches handled by trace-scheduling compilers for general-purpose VLIW architectures are encountered here.

The "length" of the VLIW instruction is proportional to the number of RVP's used. The actual number of RVP's used can be adapted to the specific task graph to be computed on the basis of a cost/performance trade-off. Using more RVP's implies greater exploitation of potential parallelism, but also means more inter-RVP communication overhead, and in general more idle time due to computational load-imbalance. Different configurations of RVP's will be appropriate for different computations. For example, the number of degrees of freedom of the robotics system being modeled will significantly impact the maximum number of RVP's effectively usable for the computation. The flexibility and variability of multiple RVP configurations leads us to label such multiple-RVP systems as Variable-Very-Long-Instruction-Word-Processors (V^2LIWP's). Section 5.1 details the structure of one such V^2LIWP configuration. In Section 5.2, a parallel formulation of the Jacobian computation is presented and mapped onto a 3-RVP configuration to illustrate the methodology for use of multiple RVP's. In Section 5.3, the application of multiple RVP's to other robotics computations, including the inertia matrix, is discussed.

5.1 Tightly Coupled RVP Configurations: V^2LIWP's

Fig. 8a illustrates the approach to controlling the tightly coupled RVP's of a V^2LIWP. The V^2LIWP is treated as an attached processor to a conventional host processor. The individual RVP program memories contain the component instructions for the RVP's. A single address needs to be provided to these memories to collectively read a "long instruction". A very simple organization for the RVP Controller is shown. The host computer is interconnected through a host interface with the RVP controller. In order to execute a computation on the V^2LIWP, the host processor first sets up input data (to be explained shortly) and then provides the RVP Controller with a starting address and instruction count for the relevant V^2LIWP program. These are loaded into an Address Register and a Size Register, respectively, in the RVP Controller through use of the 'Reg', 'Load' and 'Data' lines from the Host Interface. The host can then initiate the computation by causing the 'Start' line from the interface to be pulsed. This causes the contents of the Address Register and the Size Register to be loaded into a Program Counter (PC) and Count Register, respectively, in the RVP Controller. Following this, the PC is incremented and the Count Register decremented every clock cycle, until the Count Register goes to zero. When the Count Register reaches zero, the 'Done' flag is raised by the RVP Controller to notify the host processor of the completion of the computation. Also, the PC resets to zero, a predetermined address where all RVP program memories store a NOP.

Fig. 8b shows a 3-RVP V^2LIWP configuration structured for the parallel computation of the Jacobian for a 6 degree-of-freedom manipulator. The RVP's are interconnected in a ring fashion and interface to the host processor through FIFO (First-In-First-Out) buffers. An In-FIFO and an Out-FIFO are used with each RVP. The host processor places all input data for the computation into the In-FIFO's and expects all output results to be placed in the Out-FIFO's by the RVP's at the end of the computation. In the configuration shown, each FIFO pair is primarily associated with one of the RVP's (ChA), but is readable by another RVP. The FIFO's Read/Write controls are connected to the Channel control signals of the primary RVP, which is the only one that can write into it. The read strobes are also provided by the primary RVP, but due to the lock-step synchronized operation of the V^2LIWP, the other RVP, which connects to the data lines can simultaneously read the value. The three-way connection between a primary RVP, secondary RVP and a FIFO-pair thus permits R/W communication between the RVP's, R/W communication between the primary RVP and FIFO, and Read-Only communication between the secondary RVP and the In-FIFO. No control signals are used in inter-RVP communication, again due to the synchronized mode of operation. The flexible inter-RVP communication mechanism allows a multiple-RVP system to be easily structured into different configurations depending on the requirements of the algorithm being implemented. The following section shows the use of this V^2LIWP configuration in scheduling a parallel formulation of the Jacobian computation for a 6 degree-of-freedom manipulator.

5.2 Example: Parallel Jacobian Computation on a V^2LIWP

This section illustrates the approach to use of multiple-RVP systems, again considering the Jacobian example. While it is possible to map the computations for the equations in Table 1 onto

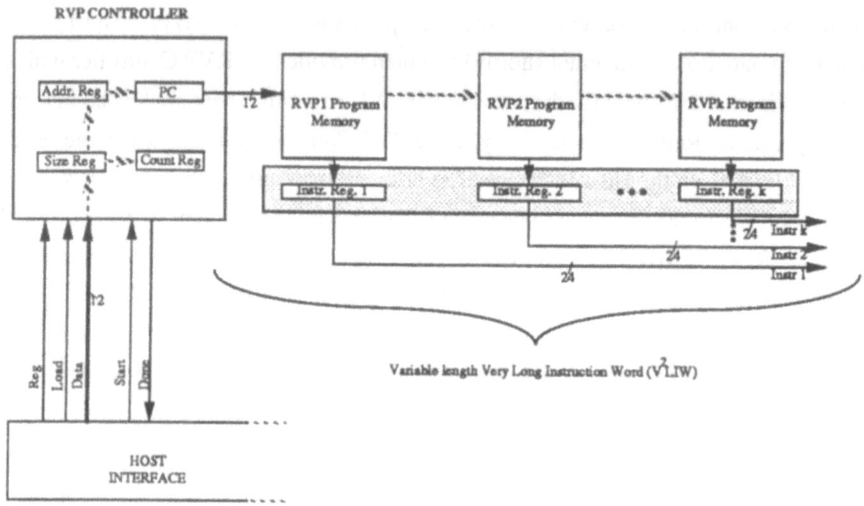

(a) V2LIW Control for Multiple-RVP System

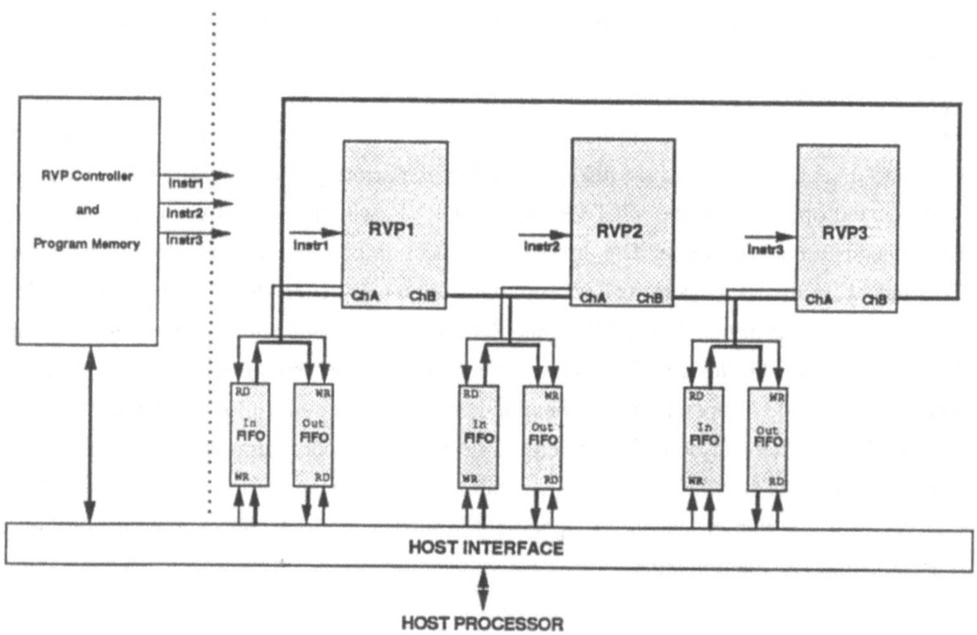

(b) 3-RVP System Structured for Parallel Jacobian Computation

Figure 8. Restructurable Variable-Very-Long-Instruction-Word (V^2LIW) Architecture a multiple-RVP system, it is preferable to use a formulation of the Jacobian computation that has greater inherent parallelism by virtue of its formulation. Hence a parallel formulation of the Jacobian computation for a six-degree-of-freedom manipulator is used [9]. The equations for the parallel Jacobian formulation are shown in Fig. 9a. Intermediate vectors **t** are used with this formulation, where $\mathbf{t}_{i-1,i} = \mathbf{p}_i^*$ and $\mathbf{t}_{i-1,N+1} = \mathbf{r}_{i-1}$. This parallel formulation uses a $\log_2 N$ stage computation for a manipulator with N degrees of freedom, so that there are three levels for the case considered here. The flow of computation for this parallel formulation is illustrated in Fig. 9b, where the partitioning of the computation among the processors of a three-RVP system (such as the one shown in Fig. 8b) is indicated. The boxes show quantities computed and the arrows show precedence and data flow. Thus the diagonal arrows represent the need for explicit interprocessor communication between the RVP's, while vertical arrows connect boxes to be allocated to the same processor and thus require no actual communication. The ring interconnection between the RVP's in Fig. 8b facilitates communication between any pair of processors. Certain input quantities are used by more than one processor, for example ${}^3U_2$, ${}^3\mathbf{p}_3^*$. In such cases, it may be possible (due to the synchronized model) for both the processors to simultaneously input it, if they are both connected to the FIFO along which the quantity is input.

After an initial partition of the overall computation, as shown in Fig. 9b at the intra-block level, a task graph at the lower matrix/vector operation level is generated to facilitate its efficient scheduling. This is shown in Fig. 10 along with the detailed schedule. Such a schedule can be generated using a procedure similar to the one explained for the earlier task graph in Fig. 7, except that additional explicit scheduling of inter-RVP communications is required. The computation takes 164 cycles to complete. When compared to the 329 cycles taken by the original serial algorithm on one RVP, an overall speedup of 329/164, or just over 2.0 is obtained. However, the parallel formulation inherently requires more computation than the serial form; if implemented on a single RVP, the best possible schedule for it takes 425 cycles. Thus, when comparing the 1-RVP implementation and the 3-RVP implementation of the same (parallel) formulation of the algorithm, the speedup obtained is actually 425/164 = 2.59, a figure that more closely reflects the effectiveness of the parallel schedule generated for the computation. The primary reason for the loss of speedup is processor idling due to inability to perfectly balance the total computational load among the RVP's. While interprocessor communication also adds to the overhead, its effect is only secondary — in the example discussed, only 4 cycles were added due to interprocessor synchronization.

5.3 Application to Other Robotics Computations

A number of other kinematics and dynamics computations have been considered for implementation on single or multiple-RVP configurations, with results obtained which are

$$
\left.\begin{aligned}
{}^k\mathbf{U}_i &= {}^k\mathbf{U}_j\,{}^j\mathbf{U}_i \\
{}^k\mathbf{t}_{i,k} &= {}^k\mathbf{t}_{j,k} + {}^k\mathbf{U}_j\,{}^j\mathbf{t}_{i,j}
\end{aligned}\right\} \qquad 0 \le (i,\, j,\, k) \le N+1
$$

$$
\left.\begin{aligned}
{}^{N+1}\boldsymbol{\gamma}_i &= {}^{N+1}\mathbf{U}_{i-1}\begin{bmatrix}0\\0\\1\end{bmatrix} \\
{}^{N+1}\boldsymbol{\beta}_i &= {}^{N+1}\,\boldsymbol{\gamma}_i \times (-{}^{N+1}\mathbf{t}_{i-1,N+1})
\end{aligned}\right\} \qquad i = 1,\, \ldots,\, N
$$

(a) Equations for Parallel Formulation

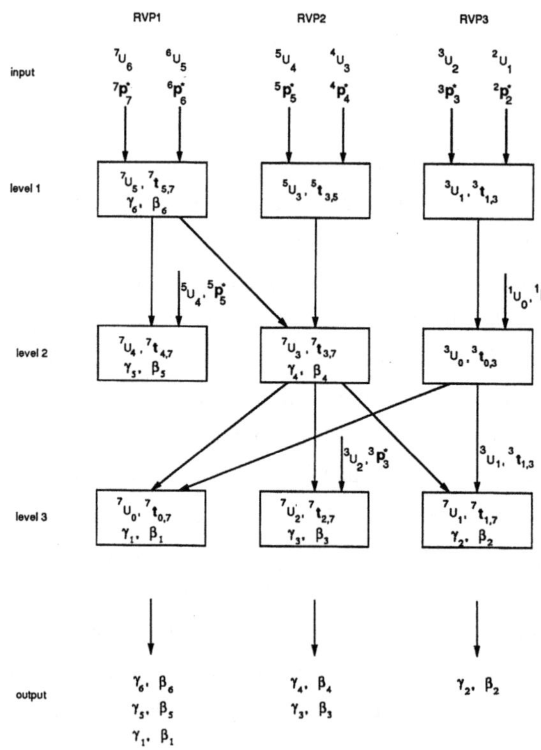

(b) Illustration of Flow of Computation and Data

Figure 9. Parallel Formulation of Jacobian Computation

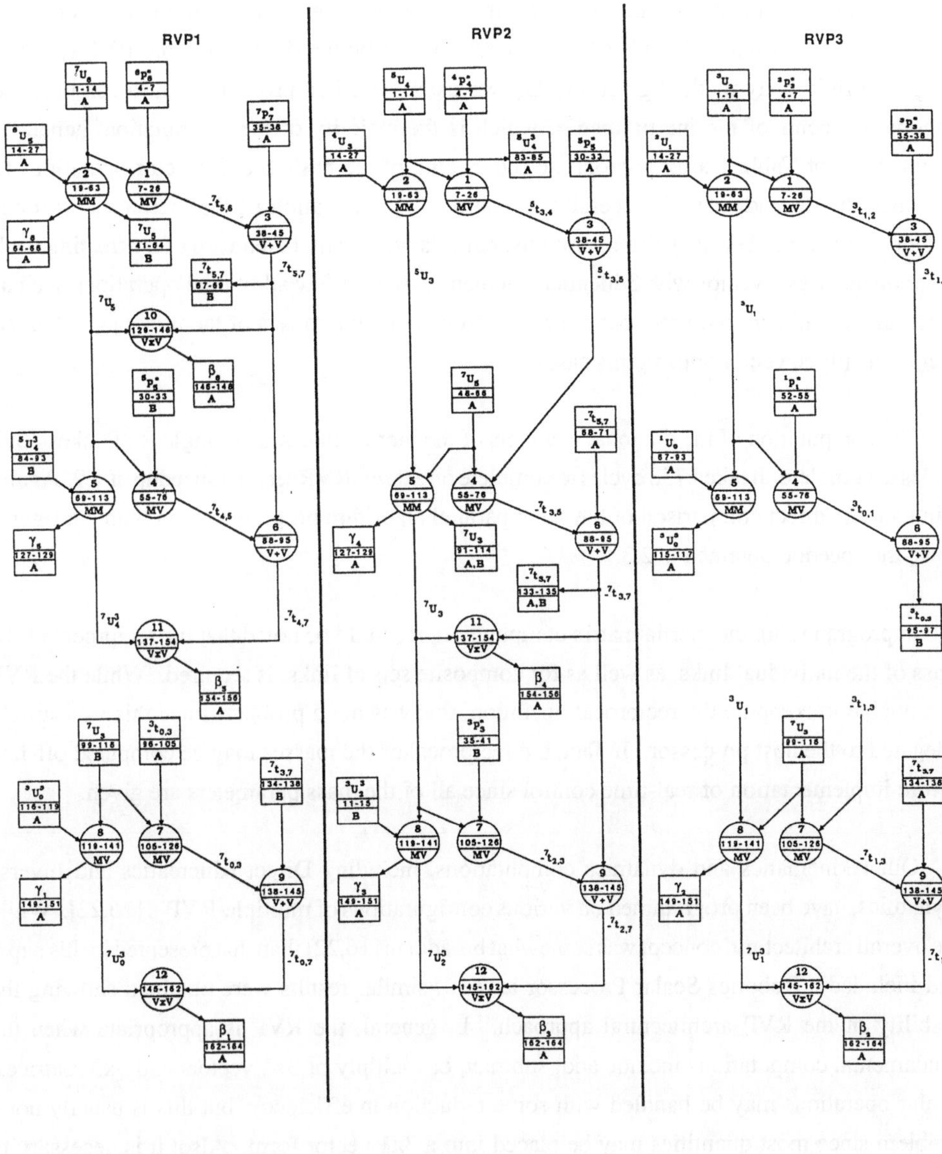

Figure 10. Task Graph and Scheduling of Parallel Jacobian Computation on a 3-RVP Architecture

similar to those for the Jacobian. In particular, the computation of the inertia matrix for a six-degree-of-freedom manipulator has been programmed on both a single and three-RVP set. The composite rigid body method given in [23] was used. Detailed programming to compute the diagonal elements of the inertia matrix, which is the majority of the computation, generally included use of Table 1 after appropriate sequencing of the basic matrix/vector operations to minimize the dependence of successive operations. Computation of the inertia matrix does include a small number of scalar operations, but this was easily taken care of by treating each scalar quantity as a vector with 3 identical elements. Also, a few additional operations such as matrix-matrix add are required, but are easily broken down into sets of the previously defined operations (three vector adds in this case).

The computation of the diagonal elements of the inertia matrix on a single RVP takes 1392 cycles to complete. It takes 733 cycles to complete on a three-RVP set for a speedup of 1.9. Again, using a more direct comparison of the same parallel algorithm programmed on both configurations, the speedup obtained is 2.3.

In programming the inertia matrix computation, it should be noted that the reciprocal of the mass of the individual links, as well as for composite sets of links, is required. While the RVP was unable to compute the reciprocal operation, this was not a problem since this was simply relegated to the host processor. In fact, the reciprocal of the masses may be computed off-line before implementation of real-time control since all of the mass parameters are given.

Other kinematics and dynamics computations, including Direct Kinematics and Inverse Dynamics, have been programmed on various configurations of multiple-RVP's [16,22]. While the overall architectural concept was somewhat broader in [16,22] than that presented in this paper and included a Robotics Scalar Processor as well, similar results were obtained showing the viability of the RVP architectural approach. In general, the RVP is appropriate when the fundamental computations include add, subtract, or multiply of 3x1 vectors and 3x3 matrices. Scalar operations may be handled with some reduction in efficiency, but this is usually not a problem since most quantities may be placed into a 3x1 vector form. Also, it is necessary to relegate trigonometric functions to the host processor, but this is in many cases not a problem since when a joint angle is sampled by the host processor, the sine and cosine may also be computed. In Direct Kinematics, Inverse Dynamics, the Jacobian, and Inertia Matrix, this is the only time that the need for trigonometric evaluation arises and is easily handled by the host processor. In other cases such as in evaluation using closed-form solutions to Inverse Kinematics, where a number of trigonometric evaluations are required, then use of a cordic processor may be effectively applied [15].

6. SUMMARY AND CONCLUSIONS

In this paper, the architecture for a Robotics Vector Processor (RVP) has been described. It consists of an SIMD array of three Floating-Point Processor (FPP) units, interconnected so as to facilitate implementation of robot kinematics and dynamics computations for real-time control. With considerable internal parallelism both within and between the FPP's, good speedup over conventional microprocessors may be obtained with a single RVP. Further exploitation of parallelism is possible by scheduling robotics computations expressed as task graphs at the matrix/vector operation level on restructurable arrays of RVP's. These tightly synchronized multiple-RVP configurations may be viewed as Variable-Very-Long-Instruction-Word Processors (V^2LIWP's) and achieve low overhead and high efficiencies due to the tight coupling. The Jacobian and Inverse Dynamics computations along with other computations necessary to implement Inverse Plant Plus Jacobian Control [17] have been evaluated on single and multiple-RVP systems, and real-time control is feasible [22].

A floor plan for the RVP has been completed (Fig. 11) and the entire processor, requiring approximately 90,000 transistors, can be fabricated on a single VLSI chip (7.9mm x 9.2mm) using 1.2 μm CMOS technology [22]. A 32-bit floating-point adder was fabricated using 3 micron scalable CMOS technology using the MOSIS facility and was successfully tested at a 12.5 Mhz. clock rate, faster than expected by approximately 50%. With improvements in fabrication technology in the future, increased performance can be expected both from higher usable clock rates, and more importantly, from the feasibility of incorporating area-intensive, faster arithmetic units in place of our current conservative but area-efficient designs. Design and simulation of the major elements of the RVP have been completed. Integration and control of all of these elements should not present any major difficulty since overall on-chip control functions have been greatly simplified (instruction fetch is off-chip).

Initial work has been completed on mapping multi-rate robotics control schemes [17] on multiprocessor systems comprised of a number of loosely-coupled V^2LIWP's [16,22]. The multiple levels of hardware parallelism exploitable, along with the great degree of flexibility and restructurability possible, make these loosely-coupled multiprocessor systems very powerful. Hopefully, with additional engineering work, such architectures for robotics will be realized and greatly expand the class of control schemes for which real-time implementation is possible.

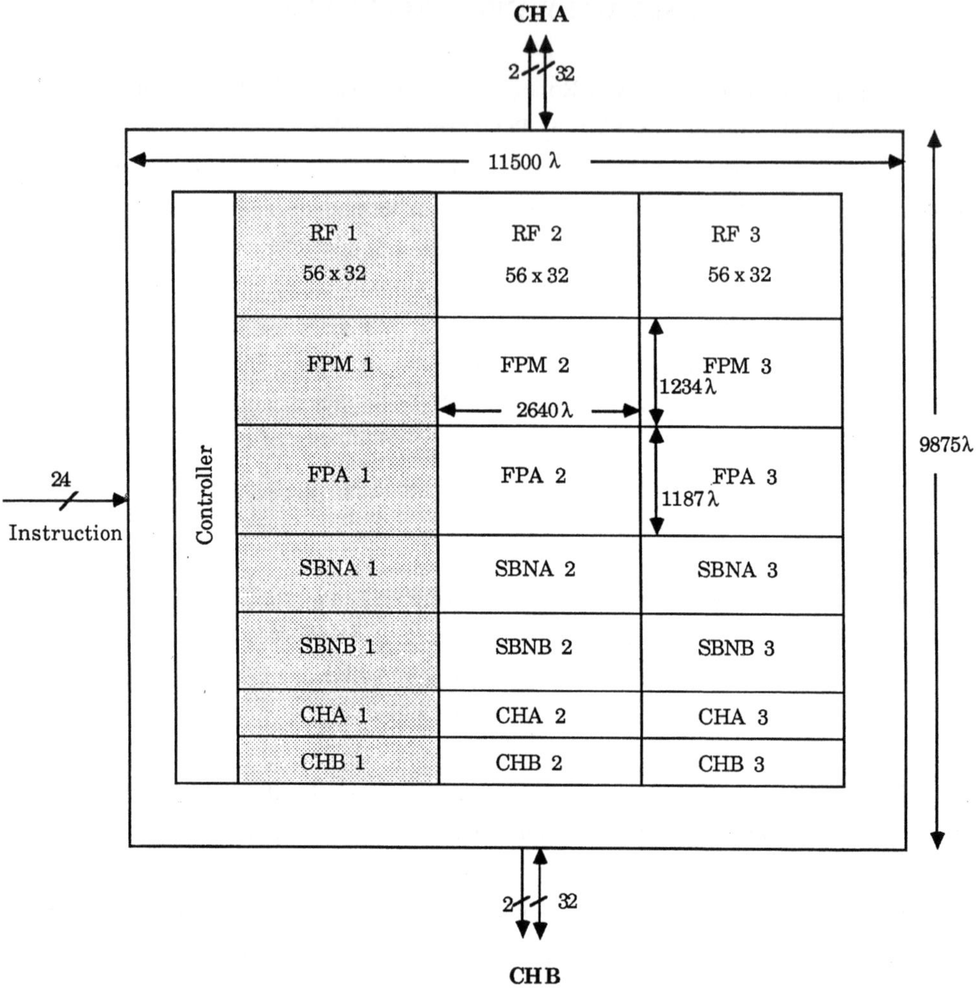

Figure 11. Robotics Vector Processor Floor Plan

ACKNOWLEDGEMENTS

The assistance of N. Jaksic in the preparation of this manuscript is gratefully acknowledged. This work was supported in part by the National Science Foundation, through Computer Engineering Grant No. DMC-8312677. The fabrication of the VLSI components was through the MOSIS facility with support from the Defense Advanced Research Projects Agency under Contract No. DAAE07-84-K-R001.

REFERENCES

[1] D.J. Kriegman, D.M. Siegel, S. Narasimhan, J.M. Hollerbach and G.E. Gerpheide, "Computational Architecture for the Utah/MIT Hand," Proc. of the IEEE International Conference on Robotics and Automation, pp. 918-924, St. Louis, MO, March 1985.

[2] J.B. Chen, R.S. Fearing, B.S. Armstrong and J.W. Burdick, "NYMPH: A Multiprocessor for Manipulation Applications," Proc. of the IEEE International Conference on Robotics and Automation, Vol. 3, pp. 1731-1736, San Francisco, CA, April 1986.

[3] R.B. McGhee, D.E. Orin, D.R. Pugh and M.R. Patterson, "A Hierarchically-Structured System for Computer Control of a Hexapod Walking Machine," Proc. of Symposium on Theory and Practice of Robots and Manipulators, Udine, Italy, June 1984.

[4] J.Y.S. Luh and C.S. Lin, "Scheduling of Parallel Computation for a Computer-Controlled Mechanical Manipulator," IEEE Transactions on Systems, Man and Cybernetics, Vol. SMC-12, pp. 214-234, March 1982.

[5] J. Barhen, "Robot Inverse Dynamics on a Concurrent Computation Ensemble," Proc. of 1985 ASME International Conference on Computers in Engineering, Vol. 3, pp. 415-429, Boston, MA, August 1985.

[6] R. Nigam and C.S.G. Lee, "A Multiprocessor-Based Controller for the Control of Mechanical Manipulators," IEEE Journal of Robotics and Automation, Vol. RA-1, No. 4, pp. 173-182, Dec. 1985.

[7] C.S.G. Lee and P.R. Chang, "Efficient Parallel Algorithm for Robot Inverse Dynamics Computation," IEEE Transactions on Systems, Man and Cybernetics, Vol. SMC-16, No. 4, pp. 532-542, July/August 1986.

[8] L.H. Lathrop, "Parallelism in Manipulator Dynamics," International Journal of Robotics Research, Vol. 4, No. 2, pp. 80-102, Summer 1985.

[9] D.E. Orin, K.W. Olson and H.H. Chao, "Systolic Architectures for Computation of the Jacobian for Robot Manipulators," in Computer Architectures for Robotics and Automation, pp. 39-67, Edited by J.H. Graham, Gordon and Breach Science Publishers, New York, 1987.

[10] M. Amin-Javaheri and D.E. Orin, "A Systolic Architecture for Computation of the Manipulator Inertia Matrix," Proc. of the IEEE International Conference on Robotics and Automation, Vol. 2, pp. 647-653, Raleigh, NC, April 1987.

[11] Y. Wang and S.E. Butner, "A New Architecture for Robot Control," Proc. of the IEEE International Conference on Robotics and Automation, Vol. 2, pp. 664-670, Raleigh, NC, April 1987.

[12] S.S. Leung and M.A. Shanblatt, "Real-Time DKS on a Single Chip," IEEE Journal of Robotics and Automation, Vol. RA-4, No. 3, pp. 281-290, August 1987.

[13] H.H. Chao, Parallel/Pipeline VLSI Computing Structures for Robotics Applications, Ph. D. dissertation, The Ohio State University, Columbus, OH, June 1985.

[14] Y.T. Tsai and D.E. Orin, "A Strictly Convergent Real-Time Solution for Inverse Kinematics of Robot Manipulators," Journal of Robotic Systems, Vol. 4, No. 4, pp. 477-501, 1987.

[15] C.S.G. Lee and P.R. Chang, "A Maximum Pipelined CORDIC Architecture for Inverse Kinematic Position Computation," IEEE Journal of Robotics and Automation, Vol. RA-3, No. 5, pp. 445-458, Oct. 1987.

[16] Y.L.C. Ling, K. Olson, D.E. Orin and P. Sadayappan, "A Layered Restructurable VLSI Architecture for Robotics Control," Proc. of 1987 IEEE International Conference on Computer Design, pp. 267-272, Port Chester, NY, October 1987.

[17] K.W. Lilly and D.E. Orin, "Multiprocessor Implementation of Dynamic Control Schemes for Robot Manipulators," Proceedings of 1986 ASME International Computers in Engineering Conference, Vol. 1, pp. 53-59, Chicago, IL, July 1986.

[18] D.E. Orin and W.W. Schrader, "Efficient Computation of the Jacobian for Robot Manipulators," International Journal of Robotics Research, Vol. 3, No. 4, pp. 66-75, Winter 1984.

[19] K. Hwang and F. Briggs, Computer Architecture and Parallel Processing, McGraw-Hill, New York, 1984.

[20] H.T. Kung, "Why Systolic Architectures?," IEEE Computer, Vol. 15, No. 1, pp. 37-46, Jan. 1982.

[21] J.A. Fisher, "Very Long Instruction Word Architectures and the ELI-512," Proceedings of 10th Annual Symposium on Computer Architecture, pp. 140-150, Stockholm, Sweden, June 1983.

[22] Y.L.C. Ling, Layered Multiprocessor Architecture Design in VLSI for Real-Time Robotic Control, Ph. D. dissertation, The Ohio State University, Columbus, OH, Dec. 1986.

[23] M.W. Walker and D.E. Orin, "Efficient Dynamic Computer Simulation of Robotic Mechanisms," ASME Journal of Dynamic Systems, Measurement, and Control, Vol. 104, pp. 205-211, September 1982.

ON THE PARALLEL ALGORITHMS FOR ROBOTIC COMPUTATIONS

C. S. George Lee
School of Electrical Engineering
Purdue University
West Lafayette, Indiana 47907, USA

ABSTRACT

The kinematics, dynamics, Jacobian, and their corresponding inverses are six major computational tasks in the real-time control of robot manipulators. The parallel algorithms for these computations are examined and analyzed. They are characterized based on six well-defined features that have greatest effects on the execution of parallel algorithms. These features include type of parallelism, degree of parallelism (granularity), uniformity of operations, fundamental operations, data dependency, and communication requirement. It is found that the inverse dynamics, the forward dynamics, the forward kinematics and the forward Jacobian computations possess highly regular properties and they are all in homogeneous linear recursive form. The inverse Jacobian is essentially the problem of solving a system of linear equations. The closed-form solution of the inverse kinematics problem is obviously non-uniform and robot dependent. The iterative solution for the inverse kinematics problem seems uniform and the parallel portions of the algorithm involve the forward kinematics, the forward Jacobian, and the inverse Jacobian computations. Suitable algorithms for the six basic robotics computations are selected and parallelized to make use of their common features. The characterization of the six basic robotics algorithms is tabulated for discussion and the results can be used to design better parallel architectures or a common architecture for the computation of these robotics algorithms.

This work was supported in part by a grant from the Ford Fund and in part by the National Science Foundation under Grant CDR 8803017 to the Engineering Research Center for Intelligent Manufacturing Systems.
Any opinions, findings, and conclusions or recommendations expressed in this article are those of the author and do not necessarily reflect the views of the funding agencies.

NATO ASI Series, Vol. F 66
Sensor-Based Robots: Algorithms and Architectures
Edited by C. S. George Lee

1. INTRODUCTION

Robot manipulators are highly nonlinear systems and their dynamic performance is directly dependent on the efficiency of the kinematic and dynamic models, the control schemes/algorithms, and the computer architecture for computing the control schemes. In general, robot manipulators are usually servoed in the joint-variable space while the objects to be manipulated are usually expressed in the world (or Cartesian) coordinate system. In order to control the position and orientation of the manipulator end-effector, the robot controller is required to compute, at a sufficient rate, such tasks as coordinate transformation between the joint-variable space and the Cartesian space, generalized forces/torques to drive the joint motors, the manipulator inertia matrix for model-based control schemes, and the Jacobian matrix which relates the joint velocity in the joint-variable space to the Cartesian space. These are the basic robotic computations for the control of robot manipulators. They are equivalent to the computations of kinematics, dynamics, Jacobian, and their corresponding inverses. These kinematics and dynamics computations reveal a basic characteristic and common problem in robot manipulator control — *intensive computations with a high level of data dependency*. They have become major computational bottlenecks in the control of robot manipulators. Despite their impressive speed, conventional general-purpose uniprocessor computers can not efficiently handle the kinematics and dynamics computations at the required computation rate because their architectures limit them to a mostly serial approach to computation, and therefore limit their usefulness for robotic computational problems. In addition, the processing of external sensory information for planning, coordination, and decision-making has further put a growing demand on the computational needs. Consequently, the quest for real-time robot arm control for better dynamic performance has resulted in a concerted effort and push for the development of *parallel algorithms of lower computational complexity as well as faster computational architectures* for their computations.

This paper addresses these computational problems in robot arm control and focuses on the investigation of various parallel algorithms in computing the robot arm kinematics, dynamics, Jacobian, and their corresponding inverses. The objective is to identify, analyze, and exploit the inherent parallelism of these robotics algorithms. The result of this study is to classify the common features among these parallel algorithms, which will inevitably aid the design of efficient parallel computational architectures for the control of robot manipulators.

2. BASIC ROBOTICS COMPUTATIONS

The six basic robotics computations in robot arm control are forward and inverse kinematics, forward and inverse dynamics, and forward and inverse Jacobian equations. These six basic robotics computations are required at various stages of robot arm control and computer simulation of robot motion. Unfortunately, they are all computational intensive tasks and constitute the major computational bottleneck in the real-time control of robot manipulators. In this section, we shall focus on various parallel algorithms for their computations.

2.1. Forward Kinematics Problem

The problem of computing the position and orientation (i.e., *pose*) of the manipulator end-effector from the measured data of angular/linear displacement of all the joints is known as the forward (or direct) kinematics problem [1-10] and can be stated as: Given the n measured joint variables $(q_1, q_2, \cdots, q_n)$, where $q_i \equiv \theta_i$ (joint angle) for a revolute joint or $q_i \equiv d_i$ (joint displacement) for a prismatic joint, find the position and orientation of the manipulator end-effector in the Cartesian space. Besides providing the pose information of the manipulator end-effector, this computation can also provide, to a certain extent, the location information of the manipulator's intermediate rigid links to better help the manipulator negotiating around obstacles when moving in an unpredictable environment.

To describe the translational and rotational relationship between adjacent robot links, an orthonormal coordinate frame ($\mathbf{x}_i, \mathbf{y}_i, \mathbf{z}_i$), based on the Denavit-Hartenberg matrix representation [1-7], is assigned to link i. Once a link coordinate system has been established for each link, the 4×4 homogeneous link transformation matrix, ${}^{i-1}\mathbf{A}_i$, can easily be developed relating the ith coordinate frame to the $(i-1)$th coordinate frame. Using the ${}^{i-1}\mathbf{A}_i$ matrix, one can relate a point $\mathbf{p}_i$ at rest in link i and expressed in homogeneous coordinates with respect to the ith coordinate system to the $(i-1)$th coordinate system established at link $(i-1)$ by

$$\mathbf{p}_{i-1} = {}^{i-1}\mathbf{A}_i \ \mathbf{p}_i \tag{1}$$

where $\mathbf{p}_{i-1} \triangleq (x_{i-1}, y_{i-1}, z_{i-1}, 1)^T$, $\mathbf{p}_i \triangleq (x_i, y_i, z_i, 1)^T$, and

$${}^{i-1}\mathbf{A}_i = \begin{bmatrix} \cos\theta_i & -\cos\alpha_i \sin\theta_i & \sin\alpha_i \sin\theta_i & a_i \cos\theta_i \\ \sin\theta_i & \cos\alpha_i \cos\theta_i & -\sin\alpha_i \cos\theta_i & a_i \sin\theta_i \\ 0 & \sin\alpha_i & \cos\alpha_i & d_i \\ 0 & 0 & 0 & 1 \end{bmatrix} . \tag{2}$$

The homogeneous transformation matrix ${}^0\mathbf{T}_i$, which specifies the position and orientation of the ith coordinate frame with respect to the base coordinate system, is the chain product of successive homogeneous link transformation matrices of ${}^{i-1}\mathbf{A}_i$ expressed as:

$${}^0\mathbf{T}_i = {}^0\mathbf{A}_1 \, {}^1\mathbf{A}_2 \cdots {}^{i-1}\mathbf{A}_i = \prod_{j=1}^{i} {}^{j-1}\mathbf{A}_j \tag{3}$$

$$= \begin{bmatrix} \mathbf{x}_i & \mathbf{y}_i & \mathbf{z}_i & \mathbf{p}_i \\ 0 & 0 & 0 & 1 \end{bmatrix} \quad \text{for } i = 1, 2, \cdots, n .$$

Specifically, for $i = n$, we obtain the manipulator hand matrix $\mathbf{T}$, $\mathbf{T} \equiv {}^0\mathbf{T}_n$, which specifies the position and orientation of the end-effector of a manipulator with respect to the base coordinate system. This Denavit-Hartenberg matrix representation reduces the direct kinematics problem to finding an equivalent 4×4 homogeneous transformation matrix which relates the spatial displacement of the "hand coordinate frame" to the reference coordinate frame. The advantage of using the Denavit-Hartenberg matrix representation is its algorithmic universality in deriving the kinematic equation of a robot arm.

Consider the **T** matrix to be of the form:

$$\mathbf{T} \triangleq \begin{bmatrix} \mathbf{R}_{3\times3} & | & \mathbf{p}_{3\times1} \\ - & | & - \\ \mathbf{0}_{1\times3} & | & 1\times1 \end{bmatrix} \triangleq \begin{bmatrix} n_x & s_x & a_x & p_x \\ n_y & s_y & a_y & p_y \\ n_z & s_z & a_z & p_z \\ 0 & 0 & 0 & 1 \end{bmatrix} \triangleq \begin{bmatrix} \mathbf{n} & \mathbf{s} & \mathbf{a} & \mathbf{p} \\ 0 & 0 & 0 & 1 \end{bmatrix} \tag{4}$$

and using the six ${}^{i-1}\mathbf{A}_i$ matrices of the PUMA robot arm in [2], the chain multiplication of these six homogeneous link transformation matrices yields a set of twelve equations, nine for orientation matrix and three for position information:

$$n_x = C_1[C_{23}(C_4C_5C_6 - S_4S_6) - S_{23}S_5C_6] - S_1[S_4C_5C_6 + C_4S_6]$$
$$n_y = S_1[C_{23}(C_4C_5C_6 - S_4S_6) - S_{23}S_5C_6] + C_1[S_4C_5C_6 + C_4S_6]$$
$$n_z = -S_{23}[C_4C_5C_6 - S_4S_6] - C_{23}S_5C_6 \tag{5.a}$$
$$s_x = C_1[-C_{23}(C_4C_5S_6 + S_4C_6) + S_{23}S_5S_6] - S_1[-S_4C_5S_6 + C_4C_6]$$
$$s_y = S_1[-C_{23}(C_4C_5S_6 + S_4C_6) + S_{23}S_5S_6] + C_1[-S_4C_5S_6 + C_4C_6]$$
$$s_z = S_{23}(C_4C_5S_6 + S_4C_6) + C_{23}S_5S_6 \tag{5.b}$$
$$a_x = C_1(C_{23}C_4S_5 + S_{23}C_5) - S_1S_4S_5$$
$$a_y = S_1(C_{23}C_4S_5 + S_{23}C_5) + C_1S_4S_5 \tag{5.c}$$
$$a_z = -S_{23}C_4S_5 + C_{23}C_5$$
$$p_x = C_1[d_6(C_{23}C_4S_5 + S_{23}C_5) + S_{23}d_4 + a_3C_{23} + a_2C_2] - S_1(d_6S_4S_5 + d_2)$$
$$p_y = S_1[d_6(C_{23}C_4S_5 + S_{23}C_5) + S_{23}d_4 + a_3C_{23} + a_2C_2] + C_1(d_6S_4S_5 + d_2)$$
$$p_z = d_6(C_{23}C_5 - S_{23}C_4S_5) + C_{23}d_4 - a_3S_{23} - a_2S_2 \tag{5.d}$$

where d_i and a_i are known PUMA's link parameters, and $C_i \equiv \cos\theta_i$, $S_i \equiv \sin\theta_i$, $C_{ij} \equiv \cos(\theta_i + \theta_j)$, and $S_{ij} \equiv \sin(\theta_i + \theta_j)$.

An examination of the above equations shows a large set of elementary operations: scalar multiplications, scalar additions, and transcendental functions. One can use microprocessors with co-processors and an appropriate table look-up technique for the transcendental functions to compute the end-effector location. Although this multiprocessor-based computing system is widely used in present-day robot controllers, it suffers from the solution accuracy, and lacks flexibility and modularity. The solution inaccuracy is due to the table look-up, while the flexibility is due to the need for changing the link coordinate frames of the manipulator, if desired. Furthermore, if one wants to obtain the pose of the manipulator end-effector with respect to a world coordinate frame instead of the robot's base coordinate frame, then an additional homogeneous transformation matrix relating the base coordinate frame to the external world coordinate frame must be included in Eq. (3), resulting in a different set of equations in Eqs. (5.a)-(5.d). Thus, computing a fixed set of equations as in Eqs. (5.a)-(5.d) does not present an attractive solution to the real-time computation of the manipulator end-effector location.

To speed up the computation in Eq. (3), most of the past research on this problem considered fine grain parallelism by designing a single chip processor with high vector computational speed and efficient trigonometric function generator. This includes the DKS chip with table generation of trigonometric function [8], the Taylor series expansion [9] with VLSI implementation, and the CORDIC pipelined architecture for computing the end-effector location [10].

Actually, a more cost-effective and efficient method for computing the successive matrix multiplication equation in Eq. (3) is to reformulate it into the homogeneous linear recursive form as

$$\begin{aligned} {}^0\mathbf{T}_1 &= {}^0\mathbf{A}_1 \\ {}^0\mathbf{T}_i &= {}^0\mathbf{T}_{i-1}\,{}^{i-1}\mathbf{A}_i \quad , \text{for } i = 2, \cdots, n , \end{aligned} \tag{6}$$

from which the configuration of all the coordinate frames can be obtained at the time complexity of $O(\lceil \log_2 n \rceil)$ by using the recursive doubling technique [38].

2.2. Inverse Kinematics Problem

The inverse kinematics problem [1-6],[8],[11-24] can be stated as: Given a desired position and orientation of the manipulator end-effector and the geometric link parameters with respect to a reference coordinate system, find the joint angles so that the manipulator can reach the desired prescribed manipulator hand position and orientation. (And if it can, how many different manipulator configurations will satisfy the same condition?) The inverse kinematics computation is often used to perform a kinematic path control in most industrial robots. The robot controller usually computes the joint angles of the manipulator from the desired pose of the manipulator end-effector at a sampling period of 20-30 milliseconds.

The joint solution computation often raises questions about existence, uniqueness, solvability, and computational efficiency of the solution [1-6],[8]. Due to the required computation time and solution accuracy, a closed-form joint solution, which yields a fixed computation time, is usually sought. A closed-form joint solution is a set of algebraic equations, each of which relates the given manipulator hand position and orientation to one of the unknown joint variables. Most industrial robots have simple geometry and satisfy one of the following sufficient conditions which make the closed-form joint solution possible: (1) Three adjacent joint axes intersecting at a point or (2) Three adjacent joint axes parallel to one another. PUMA† robot arm satisfies the first condition while MiniMover‡ robot arm satisfies the second condition for finding the closed-form joint solution. However, if the geometry of a

†PUMA is a trademark of Unimation, Inc.
‡MiniMover is a trademark of Microbot, Inc.

robot does not satisfy one of these conditions, then some efficient iterative solutions [20-24] must be used to find the joint solution in real-time.

In general, the inverse kinematic position solution can be obtained by various techniques such as inverse transform [3,11], screw algebra [12], dual matrices [13], dual quaternion [14], iterative [20-24], geometric approach [2,15], and the VLSI architecture approach [18-19]. Pieper [8] presented the kinematics solution for any six degree-of-freedom manipulator which has revolute or prismatic pairs for the first three joints and the joint axes of the last three joints intersect at a point. The inverse transform technique yields a set of explicit, non-iterative joint angle equations which involve multiplications, additions, square root, and transcendental function operations. The iterative methods can obtain robot independent joint solution, but they usually have some disadvantages: more computations than the closed-form solution, variable computation time and, more important, convergence problem, especially in the singular and degenerate cases. Furthermore, as with the inverse transform technique, there is no indication on how to choose the correct solution for a particular arm configuration.

Traditionally, the joint solution is computed by the robot controller which is a multiprocessor-based system. In order to obtain a fixed computation time for the joint angle solution, time-consuming transcendental functions (sine, cosine, and arc tangent) are implemented as table look-up at the expense of the solution accuracy. Based on an actual implementation on a multiprocessor system having a circuit to synchronize the CPUs and software scheduling for computing the joint solution [16-17], the best reported computation time was 3.6 milliseconds for a six-jointed manipulator versus 20 milliseconds running on a uniprocessor system.

A closed-form solution for PUMA-type robots is listed in Table 1 (Tables are placed at the end of the paper). These equations appear highly non-uniform since each joint angle is obtained through very distinct computations which contain a large set of elementary operations including scalar addition, multiplication, reciprocal, square root, trigonometric and inverse trigonometric functions (sine, cosine, and arctangent). One possible way to parallelly process these equations is to perform a functional decomposition such that each computational module can be processed by a CORDIC processor [18]. Unfortunately, this decomposition shows only a limited amount of parallelism with a large amount of sequentialism in the flow of computation, and the CORDIC pipelined architecture presents a very expensive solution in computing the joint solution.

To achieve higher parallelism for the inverse kinematics problem, the iterative method provides a better approach, since nearly every presented iterative method contains the computations of forward kinematics, Jacobian, and inverse Jacobian [20-24], which have been shown to be highly parallelized. Tsai and Orin [24] presented a parallel iterative algorithm for the inverse kinematics problem (see Figure 1). The basic algorithm of their method is based on the integration of the joint velocity and can be expressed by the following equations:

$$\dot{\mathbf{q}}(t) = \mathbf{J}^{-1}(\mathbf{q})\dot{\mathbf{x}}(t) \tag{7}$$

$$\mathbf{q}(t) = \int \dot{\mathbf{q}}(\tau) d\tau \tag{8}$$

where $\dot{\mathbf{x}}(t)$ is the linear and angular velocity vector of the manipulator end-effector, $\dot{\mathbf{q}}(t)$ is the n-dimensional joint velocity vector of the manipulator, and $\mathbf{J}^{-1}(\mathbf{q})$ is the $n \times 6$ inverse Jacobian. The inputs to this algorithm are the values of the desired manipulator end-effector position and velocity $\mathbf{x}_d$ and $\dot{\mathbf{x}}_d$, respectively, and the outputs are the joint variables q_i's, $i = 1, 2, \cdots, n$. Details about this iterative method can be found in [24].

2.3. Inverse Dynamics Problem

Since robot manipulators are highly nonlinear systems, in order to achieve better performance in path tracking control, the dynamic model of the robot must be utilized in computing the generalized forces/torques to servo the joint motors. To perform a dynamic path-tracking control, the robot controller must repeatedly compute the required generalized forces/torques to drive all the joint motors. The problem of computing manipulator joint torques based on a manipulator dynamic model is known as the inverse dynamics problem [25-42] and can be stated as: Given the joint positions and velocities as $\{ q_i(t), \dot{q}_i(t) \}_{i=1}^{n}$ which describe the state of the manipulator at time t, expressed in the base (or reference) coordinate system, together with the desired joint accelerations $\{ \ddot{q}_j(t) \}_{j=1}^{n}$ at that time, solve the dynamic equations of motion for the joint torques $\{ \tau_j(t) \}_{j=1}^{n}$ as follows:

$$\boldsymbol{\tau}(t) = \mathbf{f}(\mathbf{q}(t), \dot{\mathbf{q}}(t), \ddot{\mathbf{q}}(t)) \tag{9}$$

where $\boldsymbol{\tau}(t) = (\tau_1, \tau_2, \cdots, \tau_n)^T$, $\mathbf{q}(t) = (q_1, q_2, \cdots, q_n)^T$, $\dot{\mathbf{q}}(t) = (\dot{q}_1, \dot{q}_2, \cdots, \dot{q}_n)^T$, $\ddot{\mathbf{q}}(t) = (\ddot{q}_1, \ddot{q}_2, \cdots, \ddot{q}_n)^T$, and the superscript "$T$" denotes transpose operation on vectors and matrices. Obviously, the execution time for computing the generalized forces/torques partially determines the feasibility of implementing the control scheme in real time.

There are a number of ways to compute the generalized forces/torques $\boldsymbol{\tau}(t)$ applied to the joint motors [25-42], among which the computation of joint torques from the Newton-Euler (NE) equations of motion (see Table 2) is the most efficient and has been shown to possess the time lower bound of $O(n)$ running in uniprocessor computers [27,38], where n is the number of degrees of freedom (DOF) of the manipulator. Based on the study of [27], it requires $(150n - 48)$ multiplications and $(131n - 48)$ additions per trajectory set point for an n-jointed manipulator with rotary joints. It seems unlikely that further substantial improvements in computational efficiency can be achieved, since the recursive NE equations are efficiently computing the minimum information needed to compute the generalized forces/torques: angular velocity and acceleration, linear accelerations at the center of mass of the link, and joint forces and torques. Nevertheless, some improvements could be achieved by taking advantage of particular computation structures [29-30], customized algorithms/architectures for specific manipulators [31], parallel computations [37-42], and scheduling algorithms for multiprocessor systems [32-36].

The recursive structure of the NE equations of motion is obviously well suited to standard single-instruction-stream single-data-stream (SISD) computers. It is, however, not an efficient parallel processing for single-instruction-stream multiple-data-stream (SIMD) computers that are capable of performing many simultaneous operations. In order to overcome this inherent recurrence problem of the NE formulation, the NE equations of motion can be reformulated in a homogeneous linear recurrence form, and the "recursive doubling" algorithm [71-72] for parallel solution of the linear recurrence problem can be utilized to compute the joint torques in SIMD computers.

This reformulation of the NE equations of motion in a homogeneous linear recurrence form leads to an immediate question: What is the limitation of speeding up the computation of the inverse dynamics problem of a manipulator (based on the NE equations of motion) running on p processors, where $1 \le p \le n$? This can be answered by the following time lower bound theorem for computing the inverse dynamics of an n-jointed robot manipulator parallelly using p processors, where $1 \le p \le n$.

> **Theorem 1 (Time Lower Bound [38].)** The shortest parallel time to evaluate the joint torques $\{\tau_j(t)\}_{j=1}^{n}$ in equation (9) using p processors, $1 \le p \le n$, is bounded below by $O(k_1 \lceil n/p \rceil + k_2 \lceil \log_2 p \rceil)$, where k_1 and k_2 are specified constants.

The proof of Theorem 1 can be found in [38]. Two extreme cases follow from Theorem 1:

(a) If $p = 1$ (i.e., using a uniprocessor computer), then the shortest computing time to calculate the joint torques of an n-jointed manipulator is not lower than $O(n)$.

(b) If $p = n$, then the shortest parallel computing time is not lower than $O(\lceil \log_2 n \rceil)$, which is the information theoretic lower bound [73-74].

Theorem 1 indicates that the shortest parallel computing time for the inverse dynamics problem is bounded below by and not necessary equal to $O(k_1 \lceil n/p \rceil + k_2 \lceil \log_2 p \rceil)$. In other words, an efficient algorithm with p-fold parallelism may not have a computing time of the same order as $O(k_1 \lceil n/p \rceil + k_2 \lceil \log_2 p \rceil)$. However, if a parallel algorithm possesses the time lower bound, then it must be the most efficient algorithm of evaluating the inverse dynamics. Since the NE equations of motion in Table 2 possess a time complexity order of $O(n)$, it is the most efficient algorithm of evaluating the inverse dynamics running on uniprocessor computers. However, Theorem 1 also indicates that a better solution is to find an efficient parallel algorithm, running on p processors, that possesses a time complexity order of $O(k_1 \lceil n/p \rceil + k_2 \lceil \log_2 p \rceil)$. To achieve this time order, the NE equations of motion must be reformulated in a homogeneous linear recurrence form, and the recursive doubling algorithm for parallel solution of linear recurrence problems can be utilized to compute the joint torques in SIMD computers.

The NE equations of motion are very efficient in evaluating the inverse dynamics whether they are formulated in the base coordinate frame or in the link coordinate frames. The clear advantage of referencing both the dynamics and kinematics to the link coordinates is to obviate a great deal of coordinate transformations and to allow the inertia tensor to be fixed in each link coordinate frame, which results in a much faster computation in a uniprocessor computer. However, the recursive structure of this formulation is in an inhomogeneous linear recursive form (see Table 2), e.g., $\boldsymbol{\omega}_i = a_i \boldsymbol{\omega}_{i-1} + b_i$, where $a_i \equiv {}^i\mathbf{R}_{i-1}$ and $b_i \equiv {}^i\mathbf{R}_{i-1}\,\mathbf{z}_0 \dot{q}_i$†. This inhomogeneous linear recursive form requires more calculations and arrangements for parallel processing than the homogeneous linear recursive form. The NE formulation in the base coordinates can be re-arranged and transformed into a homogeneous linear recurrence form (see Table 3), e.g., $\boldsymbol{\omega}_i = \boldsymbol{\omega}_{i-1} + \dot{q}_i\,\mathbf{z}_{i-1}$, which is more suitable for parallel processing on an SIMD computer, yielding a much shorter computing time.

Kogge and Stone [71-72] developed an efficient technique, called recursive doubling, to solve a large class of recurrence problems on parallel (SIMD) computers. For the manipulator dynamics problem (forward and inverse dynamics problems), we are only interested in the first-order linear recurrence problem. In general, the first-order linear recurrence problem can be stated as: Given $a_i \neq 0$ and b_i, $0 \le i \le n$, and the recursive equation

$$x_i = a_i * x_{i-1} + b_i \tag{10}$$

where * and + may be scalar (or matrix) multiplication and scalar (or vector or matrix) addition, respectively, find $x_1, x_2, \cdots, x_n$ by a parallel algorithm running on an n-processor computer. If either a_i's are identities or b_i's are null for all i, then this is the first-order *homogeneous* linear recurrence (HLR) problem. Otherwise, it becomes the first-order *inhomogeneous* linear recurrence (IHLR) problem. On a serial (SISD) computer, we first use the initial conditions to compute one new x_i, then using the new x_i to compute x_{i+1}, and so on until all n x_i's elements are computed. The computation time is proportional to n (or of order $O(n)$). The recursive doubling technique is especially suited for solving the linear recurrence problems on SIMD computers. It involves the splitting of the computation of such problem into two equally complex subproblems. The evaluation of these subproblems can be performed simultaneously in two separate processors. By repeating the same procedure, each subproblem can be further split and spread over more processors. For the computation of a sequence of n elements, there will be $(n+1)/2^k$ parallel operations at the kth splitting until $\lceil \log_2(n+1) \rceil$ splittings. The resulting algorithm computes the entire series $x_0, x_1, \cdots, x_n$ in time complexity order of $O(\lceil \log_2(n+1) \rceil)$ on an n-processor computer. Thus, recursive doubling achieves the time lower bound and hence is the best solution for the linear recurrence problem.

For the inverse dynamics problem, we are only interested in the homogeneous linear recurrence problem of size $(n+1)$; that is, Eq. (10) with all the b_i's being null and $a_0 \neq$ identity. This can be solved by the recursive doubling algorithm as depicted in Figure 2 using the First Order Homogeneous Recurrence Algorithm (FOHRA). Similarly, the inhomogeneous linear recurrence problem can be solved by the First-Order Inhomogeneous Recurrence Algorithm (FOIHRA). Both FOHRA and FOIHRA algorithms are summarized in Figure 3, and

† a_i and b_i are used here as either matrices or vectors which must be computed in advance.

both have a computational complexity of $O(\lceil \log_2(n+1) \rceil)$ which is the time lower bound of the linear recurrence problem.

In order to apply the recursive doubling algorithm to the NE equations of an n-jointed manipulator, the equations of motion must be reformulated in the base coordinates and re-arranged to a homogeneous linear recursive form. The procedure in computing the inverse dynamics from the NE equations of motion formulated in the base coordinates as a linear recurrence problem is given in Table 4. Applying this parallel NE (PNE) procedure to the PUMA robot arm in [2], the total computational complexity of the parallel algorithm indicates $27\lceil \log_2 n \rceil + 116$ scalar multiplications and $24\lceil \log_2 n \rceil + 9\lceil \log_2(n+1) \rceil + 84$ scalar additions. If $n = 6$, then the complexity of the parallel NE algorithm is 197 multiplications and 183 additions as compared with the complexity of the NE algorithm running on a uniprocessor [38]: 852 multiplications and 738 additions. Moreover, even if n becomes large, say $n = 12$ (for redundant robots), then the number of multiplications and additions increases only by 27 and 33, respectively.

2.4. Forward Dynamics Problem

The manipulator forward dynamics problem [43-47] concerns the determination of the motion of the manipulator from a set of applied joint forces/torques and can be stated as [45]: Given an input force/torque vector $\boldsymbol{\tau}(t)$ and a vector of external forces/torques exerted on the last link of the manipulator $\mathbf{k}(t)$, compute the joint acceleration vector $\ddot{\mathbf{q}}(t)$, based on an appropriate manipulator dynamic model, from values of $\boldsymbol{\tau}(t)$, $\mathbf{k}(t)$, the joint position $\mathbf{q}(t)$, and the joint velocity $\dot{\mathbf{q}}(t)$. The resultant $\ddot{\mathbf{q}}(t)$ is then integrated to give new values of $\dot{\mathbf{q}}(t)$ and $\mathbf{q}(t)$, and the process is repeated for the next input force/torque vector. Thus, the forward dynamics problem is essential for the real-time dynamic simulation of robot arm motion.

The forward dynamics problem is more computational intensive than the inverse dynamics problem. At the present time, there are several methods available to solve this problem [43-44]. Among these, the composite rigid-body method [43] (see Table 5), based on the computation of the NE equations of motion, is widely used to develop efficient parallel algorithms [45-47]. All these methods are analyzed and compared in [45]. Several reasons can be accounted for this popular use of the composite rigid-body method. First, the composite rigid-body method is the most efficient for a reasonable number of degrees of freedom of the manipulator (i.e., for most industrial robots, $n \leq 12$) [45] and is suitable for parallel processing. Second, the composite rigid-body method also computes the inertia matrix of the manipulator. Hence, the inertia matrix and the joint acceleration vector can be obtained at the same time using the same algorithm. This is of great advantage because various model-based control schemes utilize the inertia matrix [48-49]. Another advantage for the composite rigid-body method is that efficient parallel algorithms for the inverse dynamics computation have been well developed and can be used to speed up the computation time. In fact, most of the forward dynamics parallel algorithms are all based on the composite rigid-body method and are much alike [45-47]. Essentially, they all utilize the recursive doubling technique

[71-72],[38] to reformulate the serial algorithm of the composite rigid-body method into the linear recursive equation form to reduce the order of computation. What make these parallel algorithms to have different computation complexities are the designed computer architectures and/or the number of processors used in their architectures.

Let us consider this composite rigid-body method here and try to characterize the algorithm. The dynamic equations of motion of a manipulator can be written as

$$\mathbf{H}(\mathbf{q})\ddot{\mathbf{q}}(t) + \mathbf{C}(\mathbf{q}, \dot{\mathbf{q}})\dot{\mathbf{q}}(t) + \mathbf{G}(\mathbf{q}) = \boldsymbol{\tau}(t) \tag{11}$$

and can be re-written as

$$\mathbf{H}(\mathbf{q})\,\ddot{\mathbf{q}}(t) = \boldsymbol{\tau}(t) - \mathbf{b} \tag{12}$$

$$\mathbf{b} = \mathbf{C}(\mathbf{q}, \dot{\mathbf{q}})\dot{\mathbf{q}}(t) + \mathbf{G}(\mathbf{q}) \tag{13}$$

where $\mathbf{H}(\mathbf{q})$ is an $n \times n$ symmetric, positive-definite inertia matrix, $\mathbf{G}(\mathbf{q})$ is the gravity vector, $\mathbf{C}(\mathbf{q}, \dot{\mathbf{q}})$ is the Coriolis and centrifugal force/torque vector, $\boldsymbol{\tau}(t)$ is the applied force/torque vector, and $\mathbf{b}$ is the bias torque vector due to the gravity $\mathbf{G}(\mathbf{q})$ and the velocity term $\mathbf{C}(\mathbf{q}, \dot{\mathbf{q}})$. The biased vector $\mathbf{b}$ may be solved by simply setting $\ddot{\mathbf{q}}$ equal to $\mathbf{0}$ when computing the joint torques from the inverse dynamics, using the parallel NE procedure in Table 4. That is, from Eq. (11), $\mathbf{b} = \boldsymbol{\tau}|_{\ddot{\mathbf{q}}=\mathbf{0}}$. The algorithm to solve for the joint acceleration $\ddot{\mathbf{q}}(t)$ from the above equations consists of three parts. First, the inertia matrix $\mathbf{H}(\mathbf{q})$ is computed based on the composite rigid-body method. Second, the biased vector $\mathbf{b}$ is obtained from the parallel computation of the inverse dynamics using the parallel NE procedure in Table 4, and finally, the system of linear equations for $\ddot{\mathbf{q}}(t)$ is solved. The algorithm for computing the joint accelerations is shown in Table 5 which is very similar to that of the inverse dynamics.

2.5. Forward Jacobian Problem

The Jacobian matrix [50-59], which is often used in the velocity analysis of a mechanism, specifies the mapping from the joint velocities of the manipulator in the joint-variable space to the linear and angular velocities of the manipulator end-effector in the Cartesian space as in

$$\dot{\mathbf{x}}(t) \triangleq \begin{bmatrix} \mathbf{v}(t) \\ \boldsymbol{\omega}(t) \end{bmatrix} = \mathbf{J}(\mathbf{q})\dot{\mathbf{q}}(t) \tag{14}$$

where $\mathbf{v}(t)$ and $\boldsymbol{\omega}(t)$ are, respectively, the linear and angular velocities of the manipulator end-effector, $\dot{\mathbf{q}}(t)$ is the n-dimensional joint velocity vector of the manipulator, and $\mathbf{J}(\mathbf{q})$ is the $6 \times n$ manipulator Jacobian matrix and it is a function of the joint variables. Furthermore, the transpose of the Jacobian matrix relates static contact forces and moments to the set of joint torques as

$$\boldsymbol{\tau}(t) = \mathbf{J}^T(\mathbf{q})\mathbf{F}(t) \tag{15}$$

where $\mathbf{F}(t) \triangleq (F_x, F_y, F_z, M_x, M_y, M_z)^T$ is a 6-dimensional static force/moment vector, and

$\tau(t)$ is an n-dimensional joint torque vector. From Eq. (14), the forward Jacobian problem can be stated as: Given the 3×3 rotation transformation matrices ${}^{i-1}\mathbf{R}_i$,

$$
{}^{i-1}\mathbf{R}_i = \begin{bmatrix} \cos\theta_i & -\cos\alpha_i \sin\theta_i & \sin\alpha_i \sin\theta_i \\ \sin\theta_i & \cos\alpha_i \cos\theta_i & -\sin\alpha_i \cos\theta_i \\ 0 & \sin\alpha_i & \cos\alpha_i \end{bmatrix} \quad \text{for } i = 1, \cdots, n,
$$

and the position vector from the origin of link i coordinate frame to the origin of link $(i-1)$ coordinate frame with respect to the link i coordinate frame ${}^{i}\mathbf{p}_i^*$, $i = 1, \cdots, n$, determine the Jacobian matrix.

Existing methods in computing the Jacobian are mostly confined to uniprocessor computers. In particular, Renaud [50], Waldron [51], Orin/Schrader [56], and Yeung/Lee [57] exploited the linear recurrence characteristics of the Jacobian equations. These methods differed from each other only by a different selection of the reference coordinate frame for computation. The reference coordinate frame is selected such that all the vectors and matrices and the Jacobian computed are referred to that reference coordinate system. Specifically, for an n-jointed robot manipulator, Waldron's method corresponds to the reference coordinate frame k being selected at the base coordinates $k = 0$, Renaud's method corresponds to $k = n/2$, Orin/Schrader's method corresponds to selecting the reference coordinate frame at the end-effector coordinates $k = n$, and Yeung/Lee's method is the Generalized-k, where $0 \le k \le n$. They all have the computational order of $O(n)$ for an n-jointed manipulator. The Generalized-k algorithm developed by Yeung and Lee [57] is shown in Table 6.

Other methods for computing the Jacobian include Uicker [20], Whitney [53], Paul et al. [55], Vukobratovic and Potkonjak [52], and Fu et al. [2]. Uicker [20] obtained the Jacobian in terms of the differential change of transform elements. Whitney [17] computed the Jacobian based on vector methods. Paul et al. [55] developed the differential matrix approach to obtain the Jacobian directly from the homogeneous transformation matrices. Vukobratovic and Potkonjak [52] calculated the n individual columns of the Jacobian matrix from the base coordinates to the end-effector coordinates expressed with respect to the end-effector coordinates. In Fu et al. [2], the Jacobian is computed as a by-product from the computation of the Newton-Euler equations of motion in solving the inverse dynamics problem.

Recently, Orin et al. [58] developed pipeline and parallel algorithms for configuring a systolic array of processors to implement the Jacobian. $(n + 1)/2$ time units are taken to obtain the Jacobian for the pipeline algorithm, and $2n$ processors are used with an initiation rate equals to 2. The parallel algorithm, which is based on a "divide-and-conquer" strategy, reduces the computational order to $O(\log_2(n+1))$. Hypercube interconnection network is selected for connecting the processors. Yeung and Lee [57] also designed two VLSI systolic pipelines for the computation of the Jacobian matrix. A linear VLSI systolic pipeline is designed to implement the Generalized-k algorithm, and a parallel VLSI systolic pipeline is designed to implement the Parallel Forward And Backward Recursive Doubling algorithm.

2.6. Inverse Jacobian Problem

The inverse Jacobian problem [60-69] can be stated as: Given the linear and angular velocity vector of the manipulator end-effector in the Cartesian space, find the corresponding joint rates. The computation of inverse Jacobian is useful in the resolved motion rate control (RMRC). In the RMRC, the manipulator end-effector (or hand) is commanded to move along a desired Cartesian direction in a coordinated rate control, the motions of the various joint motors are combined and resolved into separately controllable hand motions along the world coordinate axes. This implies that several joint motors must run simultaneously at different time-varying rates in order to achieve the desired, coordinated hand motion along any one world coordinate axis. This RMRC enables the user to specify the direction and speed along any arbitrarily oriented path for the manipulator to follow and greatly simplifies the specification of the sequence of motion for completing a task. Unfortunately, the computational bottleneck of this kinematic control scheme is the real-time computation of the inverse Jacobian.

The most direct method of obtaining the inverse Jacobian is to calculate the Jacobian matrix and then invert the matrix. Unfortunately, the Jacobian matrix becomes singular at the robot's deadpoints and the inverse of the Jacobian is not defined because the Jacobian matrix is not of full rank. Moreover, since the Jacobian is a $6 \times n$ matrix, its inverse does not exist when n is not equal to six. For these unpleasant cases, the concept of generalized inverse has been applied [60]. The inverse Jacobian algorithms for a general manipulator can be divided into two categories. One is to calculate the inverse or the generalized inverse Jacobian explicitly [22],[61]. The other is to consider the inverse Jacobian problem as a system of linear equations and solve the joint rate from the Cartesian velocity implicitly [21],[24],[62]. For practical purposes, the latter approach is easier to be parallelized due to the use of some standard techniques to solve a system of linear equations such as the Gaussian elimination method. Like the inverse kinematics problem, few parallel algorithms for the inverse Jacobian computation have been developed.

The algorithm to compute the joint rates is different for manipulators with different number of degrees of freedom. If $n = 6$ and the manipulator is at a non-singular location, then the Jacobian matrix is square and Eq. (14) can be solved directly as a system of linear equations by simply applying the Gaussian Elimination method [69]. If n is less than six, then Eq. (14) is an overdetermined system and the Moore-Penrose pseudo-left inverse may be employed to give the least-squares solution

$$\mathbf{J}^{-L} = (\mathbf{J}^T \mathbf{J})^{-1} \mathbf{J}^T \tag{16}$$

where $\mathbf{J}^{-L}$ is the Moore-Penrose pseudo-left inverse of the Jacobian matrix $\mathbf{J}$. If n is greater than six, then Eq. (14) is an underdetermined system and the Moore-Penrose pseudo-right inverse can provide the minimum energy solution

$$\mathbf{J}^{-R} = \mathbf{J}^T (\mathbf{J}\mathbf{J}^T)^{-1} \tag{17}$$

where $\mathbf{J}^{-R}$ is the Moore-Penrose pseudo-right inverse of the Jacobian matrix $\mathbf{J}$. For practical purposes, $\mathbf{J}^{-1}$, $\mathbf{J}^{-L}$ or $\mathbf{J}^{-R}$ are not computed explicitly through the above equations. Rather, they are implicitly used to obtain the joint rates, for example, via Householder reflections [63] or singular value decomposition (SVD) technique [64-66] or residue arithmetics [61,67-68]. Hence, the inverse Jacobian computation reduces to just a general type of system of linear equations.

The solution of a system of linear equations is the most common computational problem in linear algebra [69,75-76] and there has been a long tradition of research on parallel algorithms for solving various types of systems of linear equations whether the matrix is square or rectangular. The techniques to solve the system of linear algebraic equations are all rather regular and homogeneous and can map quite naturally onto many types of multiprocessor architecture, especially one- or two-dimensional arrays of processors. Generally speaking, there are two approaches for solving linear systems [75-76]. The first approach is the direct method which usually includes two processes: factorization of the matrix into a triangular matrix and solving the triangular systems of equations. The second approach is the iterative method which is amenable for parallel processing.

3. CHARACTERISTICS OF PARALLEL ALGORITHMS

The above six basic robotics computations are required at various stages of robot arm control and computer simulation of robot motion. Unfortunately, they are all computational intensive tasks and constitute the major computational bottleneck in the real-time control of robot manipulators. In the past decade, there has been a growing interest in speeding up these computations. Most of the past research focused on the use and/or design of parallel architectures such as multiprocessor-based systems and special-purpose VLSI chips with parallel processing and pipelined processing techniques [9-10], [16-19],[29-42],[45-47],[57-59],[61]. Since efficient utilization of parallel architectures rests on the development of proper parallel algorithms, common features and characteristics of these robotics parallel algorithms must be identified, which can be used to understand their inherent parallelism and architectural requirements. This in turn leads to the design of algorithmically-specialized parallel architectures for the efficient computation of these robotics algorithms.

In order to examine the characteristics of the six basic robotics parallel algorithms, a set of features which have the greatest effects on the execution of parallel algorithms is defined [70].

- **Type of parallelism.** Two levels of parallelism can be identified.

  (a) *Job-level parallelism.* The original algorithm is reformulated to a parallel processable form. In this level, the variables carrying the same kind of information but with different indices (e.g., for different links or joints of a manipulator) are processed parallelly. Due to the nature of the robot's serial link structure, variables representing the same physical meaning are defined for each link such as joint velocities, joint accelerations, and joint torques. Usually, the same class of variables are produced through an identical computational procedure but with different set of data. This property is called uniformity of operations as defined below. So the job-level parallelism will often be amenable to the single-instruction-stream multiple-data-stream (SIMD) implementation and usually the required number of processors depends on the number of degrees of freedom of the manipulator (i.e., one processor for each joint).

(b) *Task-level parallelism.* The original algorithm is decomposed into multiple subtasks. While the computation within a subtask is serial, the number of subtasks that can be processed concurrently is maximized by using some scheduling techniques. Obviously, this implies multiple-instruction-stream multiple-data-stream (MIMD) operations. Furthermore, for this level of parallelism, a subtask usually performs the same computation for different set of data, and hence the operation can be pipelined. An advantage of this task-level parallelism is that the required number of processors is independent of the number of degrees of freedom of the manipulator.

- **Degree of parallelism (Granularity).** Three levels of granularity are distinguished. In the *large grain granularity*, the parallelism is performed at the algorithmic level. That is, only the parallelism between different segments or subtasks is considered. For the *medium grain granularity*, the concurrency is considered at the operation level and the parallelism is performed based on some basic mathematical operations such as vector cross product and matrix-vector multiplication. If we consider the implementation of parallelism within the basic arithmetic operations, then the *fine grain granularity* is achieved. For example, a 4×4 systolic array can be designed for basic matrix operations in robotics application. Different degrees of parallelism often imply different synchronization requirements. The finer the granularity is, the more frequent synchronization is required. Most researchers consider the degree of parallelism an important feature for characterizing the basic robotics parallel algorithms.

- **Uniformity of operations.** A robotics algorithm is said to possess uniformity of operations if the required computations for some set of variables, especially the joint variables, are uniform. An algorithm with operation uniformity can be implemented on an SIMD machine with higher efficiency.

- **Fundamental operations.** Two classes of operations are used in robotics algorithms. One is basic matrix-vector operations including vector addition, inner product, and cross product. The other is trigonometric function. The identification of basic operations performed in the algorithm will dictate the processor capabilities needed.

- **Data dependency.** Three kinds of data dependency are classified for robotics algorithms: *local dependency* or *neighborhood dependency*, *special type dependency*, and *global dependency*. The local dependency means that the required operands in an operation come from its neighborhood; for example, from the results of last operation or using the same operands of last operation. The special type dependency is defined for some special equation or problem. There are some special types of data dependency that are peculiar and inherent to the robotics algorithms. Among them, the homogeneous (or inhomogeneous) linear recursive type of dependency which describes the data dependency in a homogeneous (or inhomogeneous) linear recursive equation appears most frequently. This linear recurrence structure plays a major role in the robotics algorithms because the variables of a joint are usually related to the corresponding variables of its adjacent joint due to the robot's serial link structure. Other special types of data dependency are defined for some well-known problems; for example, system of linear equations and Column-Sweeping algorithm for a triangular linear system. Besides the above two data dependencies, the global dependency means that the results of some operations may be required by other operations or equations that may appear in other places of the algorithm. Since few algorithms possess absolutely one kind of data dependency, we can just identify whether an algorithm is local data dependency oriented or not. The data dependency in an algorithm usually dictates memory organization, data allocation and communication requirements.

- **Communication requirement.** The communication requirement decides the required interconnection type between processor and processor or between processor and memory. Three types of interconnection are considered: *one-to-one* connection, *permutation* and *broadcast* connections. It will be shown that the permutation capability is the most important for the efficient computation of robotics parallel algorithms and most of the permutation requirements are of regular types; for example, uniform module shifts connection. Of course, the exact required interconnection type for each computation in an algorithm depends on many factors such as task assignment of each processor, data allocation in the memories, and data dependency of each computation. Hence, the exact required interconnection type can only be decided at the time of the algorithm-architecture mapping process. In examining robotics parallel algorithms, only rough connection requirements can be observed.

4. CHARACTERIZATION OF BASIC ROBOTICS ALGORITHMS

If we consider these six basic robotics computations as a set of tasks that we need to compute for the control of robot manipulators, then based on the set of features defined in the last section, the parallel algorithm of each basic robotics algorithm will be examined and analyzed to find the common features and characteristics among them. The results will be useful for better understanding of the inherent parallelism within robotics computations and for designing suitable parallel architectures for their computations.

For the forward kinematics computation, considering the type of parallelism for the matrix multiplication equation in Eq. (3), it is natural to perform the job-level parallelism. That is, apply the recursive doubling technique to n linear array processors to solve the homogeneous linear recursive equations at the time complexity of $O(\lceil \log_2 n \rceil)$. The fundamental operations of this problem are 4×4 matrix multiplication, and more importantly, the sine and cosine functions (trigonometric function calculations). The important data dependency is only the HLR type and the communication requirement is the permutation capability to provide the recursive doubling interconnection.

For the inverse kinematics computation, the closed-form solution in Table 1 shows that the data dependencies among these equations are irregular and global, and the required communication capabilities are one-to-one and broadcast. These irregular and inhomogeneous equations are obviously not suitable for the job-level parallelism. The parallel iterative algorithm presented by Tsai and Orin [24] provides a higher parallelism and is amenable for parallel processing at the job-level parallelism due to its step by step nature, although some steps can only be processed serially; e.g., convergence testing. As the general property of iterative methods, the data dependency tends to be local; however, convergence testing necessarily involves gathering global information and this requires global communication capability.

For the inverse dynamics computation, using the NE equations of motion in Tables 2 and 3, one possible approach to parallelly process these equations is to decompose them into some computational modules, each of which calculates the kinematic and dynamic variables for all the joints such as angular velocities, angular and linear accelerations, joint forces and moments. For example, equations (3-a) to (3-i) in Table 3 are treated as consisting of nine computational modules. Then each one of these computational modules is assigned to one processor and different modules are processed concurrently. This type of parallelism is the task-level parallelism as defined in section 3. However, if we observe the structure of these equations more carefully, one can find that equations (3-a), (3-b), (3-c), (3-g), and (3-h) are all in HLR form, and solving a linear recursive equation on a uniprocessor computer has a time complexity of $O(n)$ as indicated by Theorem 1. This time complexity can be reduced to $O(\lceil \log_2 n \rceil)$ if the recursive doubling technique is applied to n processors as discussed above. Moreover, due to the data precedency, some processor(s) will be idle for waiting the required data calculated by other processors. For example, not until the processor which is assigned to compute the angular velocities $\boldsymbol{\omega}_i$, $i = 1, 2, \cdots, n$, completes its work, no other processors can start initiating their computations. This means that the distribution of tasks among processors will lead to low processor utilization. So a better parallelization approach is to perform the job-level parallelism, that is, n processors collaborate to compute each equation in Table 3. Hence, final results of each equation are produced at the same time and no data precedency problem needs to be worried about as long as these equations are processed in order. The above parallelization is considered at the algorithmic level; that is, at the large grain granularity. Moreover, since the procedures to calculate the variables with the same kinematic or dynamic meaning but for different links, e.g., $\boldsymbol{\omega}_i$, $i = 1, 2, \cdots, n$, are identical, the concurrency can be achieved at each basic operation (i.e., each matrix-vector operation). This means that the NE equations of motion possess the uniformity of operations. So, as far as the degree of parallelism is concerned, these NE equations can be processed with parallelism at the medium grain granularity. Furthermore, parallelism at the fine grain granularity is also possible because the fundamental operations of the NE equations are some basic matrix-vector operations with fixed data size (e.g., 3×3 matrix or 3×1 vector), so a special-purpose processor can be designed to perform these operations parallelly.

Considering their data dependencies, a number of the NE equations are in homogeneous linear recursive form and so possess the so called dependency of HLR type. Further observations indicate that although the outputs or operands of one equation are usually used as operands of the next equation, which means local data dependency, these equations have global data dependency because some variables appear in many equations irregularly. For instance, $\boldsymbol{\omega}_i$ and $\dot{\boldsymbol{\omega}}_i$ are required in equations (3-c), (3-d), and (3-f), and $\mathbf{z}_{i-1}$ appears in equations (3-a), (3-b), (3-c), and (3-i). A rough observation indicates that the most important communication requirement is permutation and no broadcast capability is necessary because there is no constant or variable which is common to all the manipulator links. Moreover, the required permutation capabilities are the one which can provide the recursive doubling technique to solve HLR equations and the one which connects a processor to its corresponding memory module (straight connection) or to its neighboring processor and corresponding memory module (nearest neighborhood connection) because some joint variables relate to the variables of its own joints or its adjacent joint. For example, equation (3-d) in Table 3, $\mathbf{a}_i$ is a function of $\boldsymbol{\omega}_i$, $\dot{\boldsymbol{\omega}}_i$, $\mathbf{s}_i$, and $\dot{\mathbf{v}}_i$, all from its own link i; equation (3-b) in Table 3, the angular acceleration of link i, $\dot{\boldsymbol{\omega}}_i$, is a function of $\dot{\boldsymbol{\omega}}_{i-1}$, $\mathbf{z}_{i-1}$, $\ddot{q}_i$, $\boldsymbol{\omega}_{i-1}$, and $\dot{q}_i$, among which $\mathbf{z}_{i-1}$ and $\boldsymbol{\omega}_{i-1}$ are variables of link $(i-1)$, and $\dot{\boldsymbol{\omega}}_{i-1}$ forms the linear recursive relation. So the required permutation types are regular.

For the forward dynamics computation, the equations in Table 5 possess uniformity of operations. Equations (5-a) to (5-f) are applied to different joint j. That means the variables with different index j which corresponds to the jth joint, e.g., $\mathbf{c}_1, \mathbf{c}_2, \cdots, \mathbf{c}_{n-1}$, are obtained through identical computation with distinct operands. Equation (5-i) is solved via the inverse dynamics algorithm which has been shown to be highly operational uniform. Equations (5-g) and (5-h) are equations with two indices i and j, so they can be viewed as "two-dimensional" problems. They also posses uniform computation because for each j, $1 \le j \le n$, all n_{ij}'s (or h_{ij}'s), $1 \le i \le j$, are obtained through identical computation. We identify this uniformity because parallel architectures can be designed to make use of this operational uniformity. Finally, equation (5-j) in Table 5 is a system of linear equations which is a very common computational problem discussed by many researchers [63],[69],[75-76]. This system of linear equations has also shown its regularity.

As the type of parallelism is concerned, it is natural to perform the job-level parallelism based on the above discussion. Equations (5-a) and (5-c) in Table 5 are in HLR form. Equation (5-b) is in IHLR form. Equation (5-g) can be considered as a set of homogeneous linear recurrence equations (SHLR) [45]. Equation (5-i) is actually an inverse dynamics problem which can be solved by the parallel NE procedure in Table 4. Each of these equations is suitable to be implemented on $O(n)$ or $O(n^2)$ array processors [45-47].

The fundamental operations of the forward dynamics problem are basic matrix-vector computations and a number of scalar additions, subtractions, multiplications, and reciprocal especially for the linear system solver. The degree of parallelism is considered at the large grain granularity in most of the past research on this problem. However, the forward dynamics equations can also be parallelized at the medium or fine grain granularities as the inverse dynamics problem can.

For the data dependency, they are more complex than the inverse dynamics problem due to some special types of data dependency. Besides the HLR type, there are the IHLR type, the set of homogeneous linear recursive equations type, and the linear system solver (LSS) type dependencies. Correspondingly, the communication requirements are also stricter. The major requirement is still the permutation capability. However, in order to implement the linear system solver, the broadcast and one-to-one connections are inevitable. The parallel composite rigid-body algorithm for the forward dynamics problem is shown in Table 7.

For the forward Jacobian computation, the equations of the the Generalized-k algorithm shown in Table 6 show high uniformity of operations. In particular, equations (6-a)-(6-d) are all in homogeneous linear recursive form, among which equations (6-a) and (6-b) are the forward recursive equations, and equations (6-c) and (6-d) are the backward recursive equations. In fact, the forward and backward recursive equations can be processed together. For example, equations (6-a) and (6-c) can be computed simultaneously to produce all the matrices ${}^k\mathbf{R}_i$, $i = 1, \cdots, n$, and so do equations (6-b) and (6-d). Although the forward and backward recursive equations can be processed together, this will require some irregular data dependencies [57]. We call this special type as *forward and backward homogeneous linear recursive* (FBHLR) type data dependency. This requires the communication requirement to include irregular permutation and broadcast. The parallel forward Jacobian algorithm is shown in Table 8.

For the inverse Jacobian computation, it involves solving a system of linear equations. The approach of factorization of the matrix into a triangular matrix and solving the triangular systems of equations possesses global data dependency because each successive column, row, or submatrix of the matrix factors depends on all the preceding columns, rows, or submatrices and the broadcast communication capability is often required. The second approach is the iterative method which has the advantage of local data dependency and is also amenable for parallel processing. However, in general, the iterative method can not be faster than the direct method. The common fundamental operations include scalar adds, subtracts, multiplication, vector operations (e.g., inner products), and reciprocal. Parallelism can be performed at the fine grain granularity with subtasks of complexity $O(1)$ as on systolic and wavefront processor array [77], or at the medium grain granularity with subtasks of complex $O(n)$ as on a linear array of processors [77].

The characteristics of the six basic robotics algorithms are tabulated in Table 9. This table shows that these six basic robotics algorithms do possess some important common features and characteristics. This is especially true for the inverse dynamics, the forward dynamics, the forward kinematics and the forward Jacobian computations for the following three reasons. First, they are all suitable to be parallelized at the job-level and the parallelization can be performed at the large, medium and fine grain granularities simultaneously, although different granularities are emphasized in each individual algorithm. Second, their operations are all uniform for the variables corresponding to each joint, and the most important fundamental operation is matrix-vector operation. Finally, the strongest common feature is that they are all in homogeneous linear recursive form, for which the recursive doubling technique can be applied to achieve the time lower bound of $O(\lceil \log_2 n \rceil)$. The communication requirement indicates that one-to-one and some regular or irregular permutation capabilities are required for these four computational problems and the broadcast capability is necessary for the forward dynamics and the forward Jacobian algorithms. This indicates that some efficient, versatile network is required in the parallel architecture for their computations.

The inverse Jacobian and the inverse kinematics computations may seem less common to the former four algorithms. However, if proper or less efficient methods to solve these two problems are chosen individually, then these two algorithms may possess some common features to the other four algorithms, and a common parallel architecture can be designed to match all these common characteristics for the computation of these six robotics algorithms. From Table 9 and discussions in the previous sections, we found that either the direct method or the iterative method for the inverse Jacobian is a proper candidate for parallel processing, while the direct method is more efficient with somewhat complex data dependencies. For the inverse kinematics problem, only the iterative method possesses regular properties similar to the other four computations.

A common feature of today's research on robotics computational problems is that a specific problem, mostly the inverse dynamics or the inverse kinematics, is studied at a time and usually an algorithmically-specialized architecture or machine is developed for this particular algorithm. Of course, the special architecture designed by this approach can make the most use of the parallel properties of a specific algorithm. However, most advanced robot control schemes always require to solve the combination of *some* or *all* of the six basic robotic computation problems. Hence, the above characterization of the six basic robotics algorithms can be used to design a parallel architecture that best matches the common features of these robotics algorithms [70].

5. CONCLUSION

The six basic computations in the control of robot manipulators are kinematics, dynamics, Jacobian, and their corresponding inverse. The parallel algorithms for these computations were examined and analyzed. Their common features were characterized based on six well-defined features that have greatest effects on the execution of parallel algorithms. These features include type of parallelism, degree of parallelism (granularity), uniformity of operations, fundamental operations, data dependency, and communication requirement. Their characteristics are summarized in Table 9. It is found that the inverse dynamics, the forward dynamics, the forward kinematics and the forward Jacobian computations possess highly regular properties and they are all in homogeneous linear recursive form. The inverse Jacobian is essentially the problem of a system of linear equations. Its two general approaches, the direct and iterative methods, are both uniform. The closed-form solution of the inverse kinematics problem is obviously non-uniform and robot dependent. The iterative algorithm for the inverse kinematics problem seems uniform and the parallel portions of the algorithm involve the forward kinematics, the forward Jacobian, and the inverse Jacobian computations. The characterization of the six basic robotics algorithms can be used to design a common architecture for the computation of these robotics algorithms.

6. REFERENCES

1. C. S. G. Lee, "Robot Arm Kinematics, Dynamics, and Control," *IEEE Computer,* Vol. 15, No. 12, pp. 62-80, Dec. 1982.
2. K. S. Fu, R. C. Gonzalez, and C. S. G. Lee, *Robotics: Control, Sensing, Vision, and Intelligence,* New York: McGraw-Hill, 1987.
3. R. P. Paul, *Robot Manipulator: Mathematics, Programming and Control,* the MIT Press, Cambridge, MA, 1981.
4. Y. Koren, *Robotics for Engineers,* New York: McGraw-Hill, 1985.
5. J. J. Craig, *Introduction to Robotics,* Addison-Wesley, 1986.
6. J. Denavit and R. B. Hartenberg, "A Kinematic notation for Lower-pair Mechanisms based on Matrices," *Trans. of ASME, J. Appl. Mech.,* Vol. 23, 1955.
7. D. L. Pieper, "The Kinematics of Manipulators under Computer Control," *Artificial Intelligence Project Memo No. 72.,* Computer Science Department, Stanford University, Palo Alto, Calif.
8. S. S. Leung and M. A. Shanblatt, "Real-Time DKS on a Single Chip," *IEEE J. of Robotics and Automation,* Vol. RA-3, No. 4, pp. 281-290, August 1987.
9. V. Seshadri, "A Real-time VLSI Architecture for Direct Kinematics," *Proc. of 1987 IEEE Int'l Conf. on Robotics and Automation,* Raleigh, North Carolina, pp. 1116-1120, March 30-April 3, 1987.
10. C. S. G. Lee, "CORDIC-Based Architectures for Robot Direct Kinematics and Jacobian Computations," to appear in *IEEE Transactions on Systems, Man, and Cybernetics.*
11. R. P. Paul, B. E. Shimano, and G. Mayer, "Kinematic Control Equations for Simple Manipulators," *IEEE Trans. Syst. Man, Cybern.,* Vol. SMC-11, No. 6, pp. 456-460, 1981.
12. D. Kohli and A. H. Soni, "Kinematic Analysis of Spatial Mechanisms via Successive Screw Displacements," *Trans. of ASME, Journal of Engineering for Industry,* Vol. 2, Series B, pp. 739-747, May 1975.

13. G. R. Pennock and A. T. Yang, "Application of Dual-Number Matrices to the Inverse Kinematics Problem of Robot Manipulators," *Trans. of ASME, J. of Mechanism, Transmissions, and Automation in Design,* Vol. 107, pp. 201-208, June 1985.

14. A. T. Yang, "Displacement Analysis of Spatial Five-link Mechanisms Using (3x3) Matrices with Dual-number Elements," *Trans. of ASME, J. of Engineering for Industry,* Vol. 91, No. 1, Series B, pp. 152-157, Feb. 1969.

15. C. S. G. Lee and M. Ziegler, "A Geometric Approach in Solving the Inverse Kinematics of PUMA Robots," *IEEE Trans. on Aerospace and Electronic Systems,* Vol. AES-20, No. 6, pp. 695-706, 1984.

16. T. Watanabe et al., "Improvement of the Computing Time of Robot Manipulators Using a Multiprocessor," *Proc. of ASME Winter Annual Meeting,* Miami, Florida, pp. 13-22, 1985.

17. M. Kametani and T. Watanabe, "Hardware and Software of a Multiprocessor System Applied for Robot Control," *1984 Proc. of Industrial Electronic Conf.,* pp. 749-758, 1984.

18. C. S. G. Lee and P. R. Chang, "A Maximum Pipelined CORDIC Architecture for Robot Inverse Kinematic Position Computation," *IEEE J. Robotics and Automation,* Vol. RA-3, No. 5, pp. 445-458, Oct. 1987.

19. R. G. Harber, J. Li, X. Hu, and S. C. Bass, "The Application of Bit-Serial CORDIC Computational Units to the Design of Inverse Kinematics Processors," *Proc. of 1988 IEEE Int'l Conf. on Robotics and Automation,* pp. 1152-1157, 1988.

20. J. J. Uicker, J. Denavit, and R. S. Hartenberg, "An Iterative Method for the Displacement Analysis of Spatial Mechanisms," *Trans. of ASME, Journal of Applied Mechanics,* pp. 309-314, June 1964.

21. J. Angeles, "On the Numerical Solution of the Inverse Kinematic Problem," *Int. J. Robotics Res.,* Vol. 4, No. 2, pp. 21-37, Summer 1985.

22. A. A. Goldenberg, B. Benhabib, and R. G. Fenton, "Complete Generalized Solution to the Inverse Kinematics of Robots," *IEEE J. Robotics Automation,* Vol. RA-1, No. 1, pp. 14-20, March 1985.

23. L. W. Tsai and A. P. Morgan, "Solving the Kinematics of the Most General Six- and Five-DOF Manipulators by Continuation Methods," *Trans. of ASME, J. of Mechanism, Transmissions, and Automation in Design,* Vol. 107, pp. 189-200, June 1985.

24. Y. T. Tsai and D. E. Orin, "A Strictly Convergent Real-time Solution for Inverse Kinematics of Robot Manipulators," *J. of Robotic Systems,* Vol. 4, No. 4, pp. 477-501, 1987.

25. A. K. Bejczy, "Robot Arm Dynamics and Control," Jet Propulsion Lab Technical Memo 33-669, Pasadena, CA., Feb. 1974.

26. D. E. Orin, R. B. McGhee, M. Vukobratovic, and G. Hartoch, "Kinematic and Kinetic Analysis of Open-Chain Linkages Utilizing Newton-Euler Methods," *Math. Biosci.,* Vol. 43, pp. 107-130, 1979.

27. J. Y. S. Luh, M. W. Walker, and R. P. C. Paul, "On-line Computational Scheme for Mechanical Manipulator," *Trans. of ASME, J. Dynam. Syst., Meas. Contr.,* Vol. 102, pp. 69-76, June 1980.

28. J. Hollerbach, "A Recursive Lagrange Formulation of Manipulator Dynamics and a Comparative Study of Dynamics Formulation Complexity," *IEEE Trans. on Syst. Man. Cybern.,* Vol. SMC-10, No. 11, pp. 730-736, Nov. 1980.

29. C. S. G. Lee et al., "Hierarchical Control Structure Using Special Purpose Processors for the Control of Robot Arms," *Proc. of 1982 Pattern Recognition and Image Processing Conf.*, Las Vegas, Nevada, pp. 634-640, June 14-17, 1982.

30. R. Nigam and C. S. G. Lee, "A Multiprocessor-Based Controller for the Control of Mechanical Manipulators," *IEEE J. of Robotics and Automation*, Vol. RA-1, No. 4, pp. 173-182, Dec. 1985.

31. T. Kanade, P. K. Khosla, and N. Tanaka, "Real-Time Control of the CMU Direct Arm II Using Customized Inverse Dynamics," *Proc. of 1984 IEEE Conf. on Decision and Contr.*, pp. 1345-1352, Dec. 1984.

32. J. Y. S. Luh and C. S. Lin, "Scheduling of Parallel Computation for a Computer-controlled Mechanical Manipulator," *IEEE Trans. on Syst. Man. Cybern.*, Vol. SMC-12, No. 2, pp. 214-234, March 1982.

33. H. Kasahara and S. Narita, "Parallel Processing of Robot-arm Control Computation on a Multimicroprocessor System," *IEEE J. of Robotics and Automation*, Vol. RA-1, No. 2, pp. 104-113, June 1985.

34. J. Barhen, "Robot Inverse Dynamics on a Concurrent Computation Ensemble," *Proc. of 1985 ASME Int'l Conf. on Computers in Engineering*, Vol. 3, pp. 415-429, 1985.

35. C. L. Chen, C. S. G. Lee, and E. S. H. Hou, "Efficient Scheduling Algorithms for Robot Inverse Dynamics Computation on a Multiprocessor System," *IEEE Trans. on Syst. Man. Cybern.*, Vol. SMC-18, No. 5, pp. 729-743, Sep./Oct. 1988.

36. C. S. G. Lee and C. L. Chen, "Efficient Mapping Algorithms for Scheduling Robot Inverse Dynamics Computation on a Multiprocessor System," Engineering Research Center Technical Report, TR-ERC 88-20, Schools of Engineering, Purdue University, West Lafayette, Indiana, September 1988.
Also to appear in *IEEE Trans. on Syst. Man. Cybern.*

37. L. H. Lathrop, "Parallelism in Manipulator Dynamics," *Int'l J. of Robotics Res.*, Vol. 4, No. 2, pp. 80-102, Summer 1985.

38. C. S. G. Lee and P. R. Chang, "Efficient Parallel Algorithm for Robot Inverse Dynamics Computation," *IEEE Trans. on Syst. Man. Cybern.*, Vol. SMC-16, No. 4, pp. 532-542, July/Aug. 1986.

39. D. E. Orin, H. H. Chao, and W. W. Schrader, "Pipeline/Parallel Algorithms for the Jacobian and Inverse Dynamics Computations," *Proc. 1985 IEEE Int'l Conf. on Robotics*, pp. 785-789, St. Louis, MO, 1985.

40. E. Binder and J. Herzog, "Distributed Computer Architecture and Fast Parallel Algorithms in Real-time Robot Control," *IEEE Trans. on Syst. Man. Cybern.*, Vol. SMC-16, No. 4, pp. 543-549, July/Aug. 1986.

41. M. Rahman and D. G. Meyer, "A Cost-efficient High-Performance Bit-serial Architecture for Robot Inverse Dynamics Computation," *IEEE Trans. on Syst. Man. Cybern.*, Vol. SMC-17, No. 6, pp. 1050-1058, Nov./Dec. 1987.

42. P. K. Khosla and S. Ramos, "A Comparative Analysis of the Hardware Requirements for the Lagrange-Euler and Newton-Euler Dynamics Formulations," *Proc. of 1988 IEEE Int'l Conf. on Robotics and Automation*, Philadelphia, PA, pp. 291-296, April 24-29, 1988.

43. M. W. Walker and D. E. Orin, "Efficient Dynamic Computer Simulation of Robot Mechanisms," *Trans. of ASME, J. Dynam. Syst. Meas. and Contr.*, Vol. 104, pp. 205-211, Sept. 1982.

44. R. Featherstone, "The Calculation of Robot Dynamics Using Articulated-body Inertia," *Int. J. Robotics Res.*, Vol. 2, No. 1, pp. 13-30, Spring 1983.

45. C. S. G. Lee and P. R. Chang, "Efficient Parallel Algorithms for Robot Forward Dynamics Computation," *IEEE Trans. on Syst. Man. Cybern.*, Vol. SMC-18, No. 2, pp. 238-251, Mar./Apr. 1988.

46. M. Amin-Javaheri and D. E. Orin, "A Systolic Architecture for Computation of the Manipulator Inertia Matrix," *Proc. of 1987 IEEE Int'l Conf. on Robotics and Automation,* Raleigh, North Carolina, pp. 647-653, March 30-April 3, 1987.

47. A. Fijany and A. K. Bejczy, "An Efficient Method for Computing the Manipulator Inertia Matrix," *2nd Int. Symp. on Robotics and Manufacturing Research,* Albuqurque, Nov., 1988.

48. A. K. Bejczy, T. J. Tarn, and X. Yun, "Robust Robot Arm Control with Nonlinear Feedback," *Proc. of IFAC Symp. on Robot Control,* Barcelona, 1985.

49. O. Khatib, "Dynamic Control of Manipulator in Operational Space," *6th CISM-IFTOmn,* 1983.

50. M. Renaud, "Geometric and Kinematic Models of a Robot Manipulator: Calculation of the Jacobian Matrix and its Inverse," *Proc. of 11th Int'l Symp. Industrial Robots,* pp. 757-763, Oct. 1981.

51. K. J. Waldron, "Geometrically Based Manipulator Rate Control Algorithms," *Mechanism and Theory,* Vol. 17, No. 6, pp. 379-385, 1982.

52. M. Vukobratovic and V. Potkonjak, "Contribution of the Forming of Computer Methods for Automatic Modelling of Spatial Mechanisms Motions," *Mechanism and Machine Theory,* Vol. 14, pp. 179-200, 1979.

53. D. E. Whitney, "The Mathematics of Coordinated Control of Prosthetic Arms and Manipulators," *Trans. of ASME, J. Dyn. Sys. Measurement, Contr.*, Vol. 94, pp. 303-309, 1979.

54. C. H. Wu and R. P. Paul, "Resolved Motion Force Control of Robot Manipulators," *IEEE Trans. Sys. Man Cyber.*, Vol. SMC-12, No. 3, pp. 266-275, May/June 1982.

55. R. P. Paul, B. Shimano, and G. E. Mayer, "Differential Kinematic Control Equations for Simple Manipulators," *IEEE Trans. on Syst. Man. Cybern.*, Vol. SMC-11, No. 6, pp. 456-460, June 1981.

56. D. E. Orin and W. W. Schrader, "Efficient Computation of the Jacobian for Robot Manipulators," *the Int. J. of Robotics Research,* Vol. 3, No. 4, pp. 66-75, Winter 1984.

57. T. B. Yeung and C. S. G. Lee, "Efficient Parallel Algorithms and VLSI Architectures for Manipulator Jacobian Computation," *IEEE Trans. on Systems, Man, and Cybernetics,* Vol. SMC-19, No. 5, pp. 1154-1166, September/October 1989.

58. D. E. Orin, et al., "Systolic Architectures for Computation of the Jacobian for Robot Manipulators," *Computer Architectures for Robotics and Automation,* (J. H. Graham, editor) Gordon and Breach Science Publishers, pp. 39-68, 1987.

59. A. Fijany and J. G. Pontnau, "Parallel Computation of the Jacobian for Robot Manipulators," *Proc. IASTED,* Santa Barbara, May 1987.

60. C. R. Rao and S. K. Mitra, *Generalized Inverse: Theory and Applications,* Wiley, NY, 1974.

61. P. R. Chang and C. S. G. Lee, "Residue Arithmetic VLSI Array Architecture for Manipulator Pseudo-Inverse Jacobian Computation," to appear in *IEEE Transactions on*

Robotics and Automation, Vol. RA-5, No. 5, pp. 569-582, October 1989.

62. R. Featherstone, "Position and Velocity Transformations Between Robot End-effector Coordinates and Joint Angles," *the Int. J. of Robotics Research,* Vol. 2, No. 2, Summer 1983, pp. 35-45.

63. P. Businger and G. H. Golub, "Linear Least Square Solution by Householder Transformations," in *Handbook for Automatic Computation,* ed. J. H. Wilkinson and C. Reinsch, Berlin: Springer-Verlag, Vol. 2, pp. 111-118, 1971.

64. V. Klema and A. Laub, "The Singular Value Decomposition: Its Computation and Some Applications," *IEEE Trans. on Automatic Control,* Vol. AC-25, No. 2, April 1980, pp. 164-176.

65. G. H. Golub and W. Kahan, "Calculating the Singular Values and Pseudo-Inverse of a Matrix," *J. SIAM Ser. B: Number. Anal.,* Vol. 2, pp. 205-224, 1965.

66. G. H. Golub and C. Reinsch, "Singular Value Decomposition and Least Squares Solutions," in *Handbook for Automatic Computation,* ed. J. H. Wilkinson and C. Reinsch, Berlin: Springer-Verlag, Vol. 2, 1971, pp. 134-151.

67. W. T. Stallings and T. L. Boullion, "Computation of Pseudo-Inverse Matrices Using Residue Arithmetic," *SIAM Review,* Vol. 14, 1972, pp. 152-163.

68. T. M. Rao et al., "Residue Arithmetic Algorithms for Exact Computation of Generalized-Inverse of Matrices," *SIAM J. Numer. Anal.,* Vol. 13, 1976, pp. 155-171.

69. J. R. Rice, *Matrix Computations and Mathematical Software,* New York: McGraw-Hill, 1981, pp. 46-48.

70. C. T. Lin, "Parallel Algorithms and Reconfigurable Architecture for Robotics Computations," MSEE Thesis, School of Electrical Engineering, Purdue University, West Lafayette, IN, August 1989.

71. P. M. Kogge, "Parallel Solution of Recurrence Problems," *IBM J. Res. Develop.,*

72. P. M. Kogge and H. S. Stone, "A Parallel Algorithm for the Efficient Solution of a General Class of Recurrence Equations," *IEEE Trans. on Computer,* Vol. C-22, pp. 789-793, Aug. 1973.

73. E. Horowitz and S. Sahni, *Fundamentals of Computer Algorithms,* Computer Science Press Inc., 1978, pp. 488-494.

74. K. Hwang and F. A. Briggs., *Computer Architecture and Parallel Processing,* New York: McGraw Hill, 1984.

75. A. H. Sameh, "Solving the Linear Least Squares Problem on a Linear Array of Processors," *Proc. of Purdue Workshop on Algorithmically-Specialized Computer Organizations,* 1982.

76. G. A. Geist, M. T. Heath, and E. Ng, "Parallel Algorithms for Matrix Computations," in *The Characteristics of Parallel Algorithms,* The MIT press, 1987.

77. S. Y. Kung, *VLSI Array Processor,* Prentice-Hall International Inc., 1988.

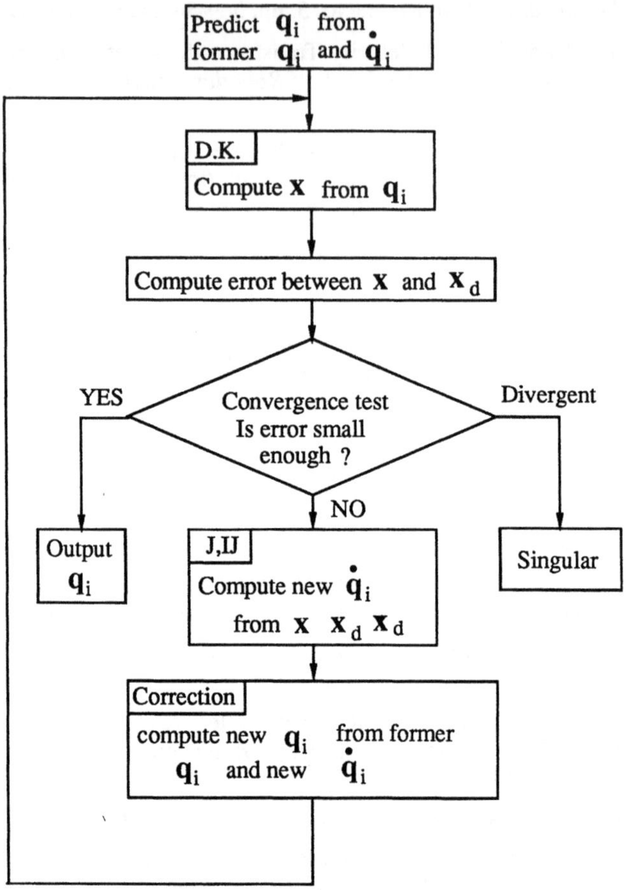

NOTE : $\mathbf{x}_d$ $\mathbf{x}_d$ are the desired end effector position and velocity respectively. $\mathbf{x}$ is the predicted end effector position. $\mathbf{q}_i$ is joint positions. $\dot{\mathbf{q}}_i$ is joint rate. DK mean direct kinematics algorithm. J means forward Jacobian algorithm. IJ means inverse Jacobian algorithm.

Figure 1. Iterative method for inverse kinematics solution.

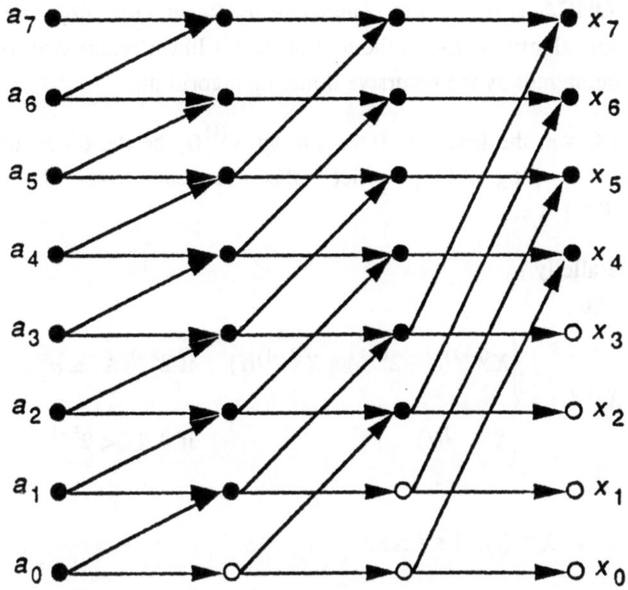

Figure 2. Parallel computation of $x_i = x_{i-1} * a_i$.
Dark circles represent operations; open circles represent null operations.

Algorithm FOHRA (*First-Order Homogeneous Recurrence Algorithm*). Given the terms a_i, $0 \le i \le n$, this algorithm solves the homogeneous linear recurrence form in equation (10) with all the b_i being null by the recursive doubling algorithm.

F1. [Initialization.] Given the terms a_i, $0 \le i \le n$, let $X^{(k)}(i)$ be the ith sequence at the kth splitting and $s = \lceil \log_2(n+1) \rceil$. Set the sequence at the initial step, $X^{(0)}(i) \leftarrow a_i$, $0 \le i \le n$.

F2. [Compute x_i parallelly.]
for $k \leftarrow 1$ **to** s, **do**

$$X^{(k)}(i) = \begin{cases} X^{(k-1)}(i-2^{k-1}) * X^{(k-1)}(i) \;, & \text{if } 2^{k-1} \le i \le n \\ X^{(k-1)}(i) & , \text{if } 0 \le i < 2^{k-1} \end{cases}$$

end {for}

Set $x_i \leftarrow X^{(s)}(i)$, $1 \le i \le n$.

END FOHRA.

Algorithm FOIHRA (*First-Order Inhomogeneous Recurrence Algorithm*). Given a_i, b_i, $0 \le i \le n$, this algorithm computes the first-order inhomogeneous linear recurrence equation (10) using the recursive doubling technique.

I1. [Initialization.] Given a_i, b_i, $0 \le i \le n$, let $X^{(k)}(i)$, $Y^{(k)}(i)$ be the ith sequences at the kth level, and set $X^{(0)}(i) = a_i$, $Y^{(0)}(i) = b_i$, for $0 \le i \le n$, and $s = \lceil \log_2(n+1) \rceil$.

I2. [Compute x_i parallelly.]
for $k \leftarrow 1$ **to** s, **do**

$$X^{(k)}(i) = \begin{cases} X^{(k-1)}(i-2^{k-1}) * X^{(k-1)}(i), & \text{if } 2^{k-1} \le i \le n \\ X^{(k-1)}(i), & \text{if } 0 \le i < 2^{k-1} \end{cases}$$

$$Y^{(k)}(i) = \begin{cases} X^{(k-1)}(i) * Y^{(k-1)}(i-2^{k-1}) + Y^{(k-1)}(i) \,, & \text{if } 2^{k-1} \le i \le n \\ Y^{(k-1)}(i) \,, & \text{if } 0 \le i < 2^{k-1} \,. \end{cases}$$

end {for}

Set $x_i \leftarrow Y^{(s)}(i)$, $1 \le i \le n$.

End FOIHRA.

where "*" in step I2 denotes an associative operator.

Figure 3. First-order homogeneous and inhomogeneous recurrence algorithms.

Table 1. Inverse Kinematic Position Solution for PUMA-Type Robots

Given the position/orientation of the manipulator hand as:

$$\mathbf{T} = \begin{bmatrix} n_x & s_x & a_x & p_x \\ n_y & s_y & a_y & p_y \\ n_z & s_z & a_z & p_z \\ 0 & 0 & 0 & 1 \end{bmatrix} = \begin{bmatrix} \mathbf{n} & \mathbf{s} & \mathbf{a} & \mathbf{p} \\ 0 & 0 & 0 & 1 \end{bmatrix}$$

The joint angle equations for the PUMA-type robot to position its end-effector as desired are [11]:

$$r = (p_x^2 + p_y^2)^{1/2}$$

$$\theta_1 = \tan^{-1}\left[\frac{p_y}{p_x}\right] - \tan^{-1}\left[\frac{d_3}{\pm\sqrt{r^2 - d_3^2}}\right]$$

$$f_{11p} = p_x C_1 + p_y S_1 \;\; ; \;\; f_{11s} = s_x C_1 + s_y S_1 \;\; ; \;\; f_{11a} = a_x C_1 + a_y S_1$$

$$f_{13p} = -p_x S_1 + p_y C_1 \;\; ; \;\; f_{13s} = -s_x S_1 + s_y C_1 \;\; ; \;\; f_{13a} = -a_x S_1 + a_y C_1$$

$$d = f_{11p}^2 + f_{12p}^2 - d_4^2 - a_3^2 - a_2^2 \;\; ; \;\; e = 4a_2^2 a_3^2 + 4a_2^2 d_4^2 \quad \text{(constant)}$$

$$\theta_3 = \tan^{-1}\left[\frac{a_3}{-d_4}\right] - \tan^{-1}\left[\frac{d}{\pm\sqrt{e - d^2}}\right]$$

$$\theta_{23} = \tan^{-1}\left[\frac{w_2 f_{11p} - w_1 p_z}{w_1 f_{11p} + w_2 p_z}\right]$$

$$\theta_2 = \theta_{23} - \theta_3$$

$$\theta_4 = \tan^{-1}\left[\frac{-S_1 a_x + C_1 a_y}{C_{23}(C_1 a_x + S_1 a_y) - S_{23} a_z}\right] = \tan^{-1}\left[\frac{f_{13a}}{C_{23} f_{11a} - S_{23} a_z}\right]$$

$$\theta_5 = \tan^{-1}\left[\frac{C_4[C_{23}(C_1 a_x + S_1 a_y) - S_{23} a_z] + S_4[-S_1 a_x + C_1 a_y]}{S_{23}(C_1 a_x + S_1 a_y) + C_{23} a_z}\right]$$

$$= \tan^{-1}\left[\frac{C_4(C_{23} f_{11a} - S_{23} a_z) + S_4 f_{13a}}{S_{23} f_{11a} + C_{23} a_z}\right]$$

$$\theta_6 = \tan^{-1}\left[\frac{-C_5[C_4(C_{23} f_{11s} - S_{23} s_z) + S_4 f_{13s}] + S_5(S_{23} f_{11s} + C_{23} s_z)}{-S_4(C_{23} f_{11s} - S_{23} s_z) + C_4 f_{13s}}\right]$$

where $w_1 = a_2 C_3 + a_3$, $w_2 = d_4 + a_2 S_3$, $C_i \equiv \cos\theta_i$, $S_i \equiv \sin\theta_i$, $C_{ij} \equiv \cos(\theta_i + \theta_j)$, and $S_{ij} \equiv \sin(\theta_i + \theta_j)$.

Table 2. Recursive Newton-Euler Equations of Motion Referenced to the Link Coordinate Systems

Forward (or Outward) Equations: $i = 1, 2, \cdots, n$

$$ {}^i\mathbf{R}_0\boldsymbol{\omega}_i = {}^i\mathbf{R}_{i-1}[\,{}^{i-1}\mathbf{R}_0\boldsymbol{\omega}_{i-1} + \dot{q}_i\mathbf{z}_0\,(1-\lambda_i\,)] $$

$$ {}^i\mathbf{R}_0\dot{\boldsymbol{\omega}}_i = {}^i\mathbf{R}_{i-1}[\,{}^{i-1}\mathbf{R}_0\dot{\boldsymbol{\omega}}_{i-1} + (\ddot{q}_i\mathbf{z}_0 + {}^{i-1}\mathbf{R}_0\boldsymbol{\omega}_{i-1}\times\dot{q}_i\mathbf{z}_0\,)(1-\lambda_i\,)] $$

$$ \begin{aligned} {}^i\mathbf{R}_0\dot{\mathbf{v}}_i = {} & (\,{}^i\mathbf{R}_0\dot{\boldsymbol{\omega}}_i\,)\times(\,{}^i\mathbf{R}_0\mathbf{p}_i^*\,) + (\,{}^i\mathbf{R}_0\boldsymbol{\omega}_i\,)\times[\,(\,{}^i\mathbf{R}_0\boldsymbol{\omega}_i\,)\times(\,{}^i\mathbf{R}_0\mathbf{p}_i^*\,)] \\ & + {}^i\mathbf{R}_{i-1}(\ddot{q}_i\mathbf{z}_0\lambda_i + {}^{i-1}\mathbf{R}_0\dot{\mathbf{v}}_{i-1}\,) + 2\,(\,{}^i\mathbf{R}_0\boldsymbol{\omega}_i\,)\times(\,{}^i\mathbf{R}_{i-1}\mathbf{z}_0\dot{q}_i\,)\lambda_i \end{aligned} $$

$$ {}^i\mathbf{R}_0\mathbf{a}_i = (\,{}^i\mathbf{R}_0\dot{\boldsymbol{\omega}}_i)\times(\,{}^i\mathbf{R}_0\mathbf{s}_i\,) + (\,{}^i\mathbf{R}_0\boldsymbol{\omega}_i\,)\times[(\,{}^i\mathbf{R}_0\boldsymbol{\omega}_i\,)\times(\,{}^i\mathbf{R}_0\mathbf{s}_i\,)] + {}^i\mathbf{R}_0\dot{\mathbf{v}}_i $$

Backward (or Inward) Equations: $i = n, n-1, \cdots, 1$

$$ {}^i\mathbf{R}_0\mathbf{f}_i = {}^i\mathbf{R}_{i+1}(\,{}^{i+1}\mathbf{R}_0\mathbf{f}_{i+1}\,) + m_i\,{}^i\mathbf{R}_0\mathbf{a}_i $$

$$ \begin{aligned} {}^i\mathbf{R}_0\mathbf{n}_i = {} & {}^i\mathbf{R}_{i+1}[\,{}^{i+1}\mathbf{R}_0\mathbf{n}_{i+1} + (\,{}^{i+1}\mathbf{R}_0\mathbf{p}_i^*\,)\times(\,{}^{i+1}\mathbf{R}_0\mathbf{f}_{i+1}\,)] \\ & + (\,{}^i\mathbf{R}_0\mathbf{p}_i^* + {}^i\mathbf{R}_0\mathbf{s}_i\,)\times(m_i\,{}^i\mathbf{R}_0\mathbf{a}_i\,) \\ & + (\,{}^i\mathbf{R}_0\mathbf{I}_i\,{}^0\mathbf{R}_i\,)(\,{}^i\mathbf{R}_0\dot{\boldsymbol{\omega}}_i\,) + (\,{}^i\mathbf{R}_0\boldsymbol{\omega}_i)\times[(\,{}^i\mathbf{R}_0\mathbf{I}_i\,{}^0\mathbf{R}_i\,)(\,{}^i\mathbf{R}_0\boldsymbol{\omega}_i\,)] \end{aligned} $$

$$ \tau_i = (\,{}^i\mathbf{R}_0\mathbf{n}_i\,)^T(\,{}^i\mathbf{R}_{i-1}\mathbf{z}_0\,)(1-\lambda_i\,) + (\,{}^i\mathbf{R}_0\mathbf{f}_i\,)^T(\,{}^i\mathbf{R}_{i-1}\mathbf{z}_0\,)\lambda_i + b_i\dot{q}_i $$

where $\mathbf{z}_0 = (0, 0, 1)^T$, and all the variables are referenced to link i coordinate frame:

λ_i : equal to 0 if link i is rotational; equal to 1 if link i is translational.

b_i : viscous damping coefficient for joint i.

τ_i : torque exerted by the actuator at joint i if rotational, force if translational.

q_i : joint variable of joint i (θ_i if rotational and d_i if translational)

m_i : total mass of link i.

${}^i\mathbf{R}_0\,\boldsymbol{\omega}_i$: angular velocity of link i.

${}^i\mathbf{R}_0\,\dot{\boldsymbol{\omega}}_i$: angular acceleration of link i.

${}^i\mathbf{R}_0\,\dot{\mathbf{v}}_i$: linear acceleration of link i.

${}^i\mathbf{R}_0\,\mathbf{a}_i$: linear acceleration of the center of mass of link i.

$^i\mathbf{R}_0\,\mathbf{f}_i$: force exerted on link i by link $i-1$.

$^i\mathbf{R}_0\,\mathbf{n}_i$: moment exerted on link i by link $i-1$.

$^i\mathbf{R}_0\,\mathbf{p}_i^*$: origin of the ith coordinate frame the origin of the $(i-1)$th coordinate system expressed with respect to the link i coordinate frame.

$^i\mathbf{R}_0\,\mathbf{s}_i$: position of the center of mass of link i from the origin of the ith coordinate system expressed with respect to the link i coordinate frame.

$^i\mathbf{R}_0\,\mathbf{I}_i\,{}^0\mathbf{R}_i$: inertia matrix of link i about its center of mass with reference to the link i coordinate system (it is a constant).

Initial conditions:

$^0\mathbf{R}_0\,\boldsymbol{\omega}_0 = {}^0\mathbf{R}_0\,\dot{\boldsymbol{\omega}}_0 = \mathbf{0},\ {}^0\mathbf{R}_0 = \mathbf{I}_{3\times3}$.

$^0\mathbf{R}_0\,\dot{\mathbf{v}}_0 = \mathbf{g} = (g_x\,, g_y\,, g_z\,)^T$ to include gravity, and $|\mathbf{g}| = 0.98062\ m/s^2$.

$^{n+1}\mathbf{R}_0\,\mathbf{f}_{n+1} = \mathbf{f}_e$ = external force $\mathbf{f}_e$ exerting on link n.

$^{n+1}\mathbf{R}_0\,\mathbf{n}_{n+1} = \mathbf{n}_e$ = external moment $\mathbf{n}_e$ exerting on link n.

Table 3. Recursive Newton-Euler Equations of Motion Referenced to the Base Coordinate System

Forward (or Outward) Equations: $i = 1, 2, \cdots, n$

$$\boldsymbol{\omega}_i = \boldsymbol{\omega}_{i-1} + \dot{q}_i \mathbf{z}_{i-1} (1 - \lambda_i) \tag{3-a}$$

$$\dot{\boldsymbol{\omega}}_i = \dot{\boldsymbol{\omega}}_{i-1} + (\ddot{q}_i \mathbf{z}_{i-1} + \boldsymbol{\omega}_{i-1} \times \dot{q}_i \mathbf{z}_{i-1})(1 - \lambda_i) \tag{3-b}$$

$$\begin{aligned} \dot{\mathbf{v}}_i = {} & \dot{\boldsymbol{\omega}}_i \times \mathbf{p}_i^* + \boldsymbol{\omega}_i \times (\boldsymbol{\omega}_i \times \mathbf{p}_i^*) + \dot{\mathbf{v}}_{i-1} \\ & + (2\boldsymbol{\omega}_i \times \dot{q}_i \mathbf{z}_{i-1} + \ddot{q}_i \mathbf{z}_{i-1}) \lambda_i \end{aligned} \tag{3-c}$$

$$\mathbf{a}_i = \dot{\boldsymbol{\omega}}_i \times \mathbf{s}_i + \boldsymbol{\omega}_i \times (\boldsymbol{\omega}_i \times \mathbf{s}_i) + \dot{\mathbf{v}}_i \tag{3-d}$$

Backward (or Inward) Equations: $i = n, n-1, \cdots, 1$

$$\mathbf{F}_i = m_i \mathbf{a}_i \tag{3-e}$$

$$\mathbf{N}_i = \mathbf{I}_i \dot{\boldsymbol{\omega}}_i + \boldsymbol{\omega}_i \times (\mathbf{I}_i \boldsymbol{\omega}_i) \tag{3-f}$$

$$\mathbf{f}_i = \mathbf{F}_i + \mathbf{f}_{i+1} \tag{3-g}$$

$$\mathbf{n}_i = \mathbf{n}_{i+1} + \mathbf{p}_i^* \times \mathbf{f}_{i+1} + (\mathbf{p}_i^* + \mathbf{s}_i) \times \mathbf{F}_i + \mathbf{N}_i \tag{3-h}$$

$$\tau_i = (\mathbf{n}_i^T \mathbf{z}_{i-1})(1 - \lambda_i) + (\mathbf{f}_i^T \mathbf{z}_{i-1}) \lambda_i + b_i \dot{q}_i \tag{3-i}$$

where all the variables are referenced to the base coordinate frame:

λ_i : equal to 0 if link i is rotational; equal to 1 if link i is translational.

b_i : viscous damping coefficient for joint i.

$\boldsymbol{\omega}_i$: angular velocity of link i.

$\dot{\boldsymbol{\omega}}_i$: angular acceleration of link i.

$\dot{\mathbf{v}}_i$: linear acceleration of link i.

$\mathbf{a}_i$: linear acceleration of the center of mass of link i.

$\mathbf{F}_i$: total external force exerted on link i at the center of mass.

$\mathbf{N}_i$: total external moment exerted on link i at the center of mass.

$\mathbf{f}_i$: force exerted on link i by link $i-1$.

$\mathbf{n}_i$: moment exerted on link i by link $i-1$.

τ_i : torque exerted by the actuator at joint i if rotational, force if translational.

$\mathbf{z}_i$: $\mathbf{z}$ component of the origin of the ith coordinate frame.

q_i : joint variable of joint i (θ_i if rotational and d_i if translational)

$\mathbf{p}_i^*$: origin of the ith coordinate frame the origin of the $(i-1)$th coordinate system.

$\mathbf{s}_i$: position of the center of mass of link i from the origin of the ith coordinate system.

m_i : total mass of link i.

$\mathbf{I}_i$: inertia matrix of link i about its center of mass with reference to the base coordinate system.

Initial conditions:

$\boldsymbol{\omega}_0 = \dot{\boldsymbol{\omega}}_0 = \mathbf{0}$.

$\dot{\mathbf{v}}_0 = \mathbf{g} = (g_x, g_y, g_z)^T$ to include gravity, and $|\mathbf{g}| = 0.98062\ m/s^2$.

$\mathbf{f}_{n+1} = \mathbf{f}_e$ = external force $\mathbf{f}_e$ exerting on link n.

$\mathbf{n}_{n+1} = \mathbf{n}_e$ = external moment $\mathbf{n}_e$ exerting on link n.

Table 4. Parallel Newton-Euler (PNE) Algorithm.

Given $\dot{q}_i$, m_i, ${}^i\mathbf{p}_i^*$, ${}^i\mathbf{s}_i$, ${}^i\mathbf{I}_i$, and ${}^{i-1}\mathbf{R}_i$, for $1 \le i \le n$, this algorithm computes the generalized joint torque/force τ_i parallelly.

N1. For $i = 1$ to n parallelly do /*HLR equations*/

$${}^0\mathbf{R}_i = {}^0\mathbf{R}_{i-1}\ {}^{i-1}\mathbf{R}_i$$

N2. For $i = 1$ to n parallelly do /*parallelly compute*/

$$\mathbf{z}_i = {}^0\mathbf{R}_i\ \mathbf{z}_0 \quad , \quad \mathbf{z}_0 = [0, 0, 1]^T$$

$$\mathbf{p}_i^* = {}^0\mathbf{R}_i\ {}^i\mathbf{p}_i^*$$

$$\mathbf{s}_i = {}^0\mathbf{R}_i\ {}^i\mathbf{s}_i$$

N3. For $i = 1$ to n parallelly do /*parallelly compute*/

$$b_i = \mathbf{z}_{i-1}\dot{q}_i\ (1 - \lambda_i)$$

For $i = 1$ to n parallelly do /*HLR equations*/

$$\boldsymbol{\omega}_i = \boldsymbol{\omega}_{i-1} + b_i$$

N4. For $i = 1$ to n parallelly do /*parallelly compute*/

$$b_i = (\mathbf{z}_{i-1}\ \ddot{q}_i + \boldsymbol{\omega}_{i-1} \times \mathbf{z}_{i-1}\ \dot{q}_i)\ (1 - \lambda_i)$$

For $i = 1$ to n parallelly do /*HLR equations*/

$$\dot{\boldsymbol{\omega}}_i = \dot{\boldsymbol{\omega}}_{i-1} + b_i$$

N5. For $i = 1$ to n parallelly do /*parallelly compute*/

$$b_i = \dot{\boldsymbol{\omega}}_i \times \mathbf{p}_i^* + \boldsymbol{\omega}_i \times (\boldsymbol{\omega}_i \times \mathbf{p}_i^*) + (\mathbf{z}_{i-1}\ \ddot{q}_i + 2\ \boldsymbol{\omega}_i \times (\mathbf{z}_{i-1}\ \dot{q}_i))\ \lambda_i$$

For $i = 1$ to n parallelly do /*HLR equations*/

$$\dot{\mathbf{v}}_i = \dot{\mathbf{v}}_{i-1} + b_i$$

N6. For $i = 1$ to n parallelly do /*parallelly compute*/

$$\mathbf{a}_i = \dot{\boldsymbol{\omega}}_i \times \mathbf{s}_i + \boldsymbol{\omega}_i \times (\boldsymbol{\omega}_i \times \mathbf{s}_i) + \dot{\mathbf{v}}_i$$

N7. For $i = 1$ to n parallelly do /*parallelly compute*/

$$\mathbf{F}_i = m_i \mathbf{a}_i$$

N8. For $i = 1$ to n parallelly do /*parallelly compute*/

$${}^i\boldsymbol{\omega}_i = {}^i\mathbf{R}_0\ \boldsymbol{\omega}_i = ({}^0\mathbf{R}_i)^T\ \boldsymbol{\omega}_i$$

$${}^i\dot{\boldsymbol{\omega}}_i = {}^i\mathbf{R}_0\ \dot{\boldsymbol{\omega}}_i = ({}^0\mathbf{R}_i)^T\ \dot{\boldsymbol{\omega}}_i$$

$$^{i}\mathbf{N}_i = {}^{i}\mathbf{I}_i\ {}^{i}\dot{\boldsymbol{\omega}}_i + {}^{i}\boldsymbol{\omega}_i \times ({}^{i}\mathbf{I}_i\ {}^{i}\boldsymbol{\omega}_i)$$

$$\mathbf{N}_i = {}^{0}\mathbf{R}_i\ {}^{i}\mathbf{N}_i$$

N9. For $i = 1$ to n parallelly do /*HLR equations*/

$$\mathbf{f}_i = \mathbf{f}_{i+1} + \mathbf{F}_i$$

N10. For $i = 1$ to n parallelly do /*parallelly compute*/

$$b_i = \mathbf{N}_i + (\mathbf{p}_i^* + \mathbf{s}_i) \times \mathbf{F}_i + \mathbf{p}_i^* \times \mathbf{f}_{i+1}$$

For $i = 1$ to n parallelly do /*HLR equations*/

$$\mathbf{n}_i = \mathbf{n}_{i+1} + b_i$$

N11. For $i = 1$ to n parallelly do /*parallelly compute*/

$$\tau_i = \begin{cases} (\mathbf{n}_i)^T \mathbf{z}_{i-1} & , \text{ if } \lambda_i = 0 \\ (\mathbf{f}_i)^T \mathbf{z}_{i-1} & , \text{ if } \lambda_i = 1 \end{cases}$$

END

where

$^{i-1}\mathbf{R}_i$: 3×3 rotation matrix indicate the orientation of link i coordinates referenced to link $(i-1)$ coordinates.

$^{i}\mathbf{p}_i^*$: the origin of the ith coordinate frame with respect to the $(i-1)$th coordinate system, expressed with respect to link i coordinates.

$^{i}\mathbf{s}_i$: position of the center of mass of link i from the origin of the ith coordinate system, expressed with respect to link i coordinates.

$^{i}\mathbf{I}_i$: inertia matrix of link i about its center of mass, expressed with respect to link i coordinates.

$^{i}\boldsymbol{\omega}_i$: angular velocity of link i with respect to the ith coordinate frame.

$^{i}\mathbf{N}_i$: total external moment exerted on link i at the center of mass, expressed with respect to link i coordinates.

λ_i equal to 0 if link i is rotational; equal to 1 if link i is translational.

Table 5. The Composite Rigid-Body Method for Forward Dynamics Computation.

/* Compute the inertia matrix $\mathbf{H}(q)$ */

$$M_j = M_{j+1} + m_j \tag{5-a}$$

$$\mathbf{c}_j = \frac{1}{M_j}\{m_j(\mathbf{s}_j + \mathbf{p}_j^*) + M_{j+1}(\mathbf{c}_{j+1} + \mathbf{p}_j^*)\} \tag{5-b}$$

$$\begin{aligned}\mathbf{E}_j = \mathbf{E}_{j+1} &+ M_j[(\mathbf{c}_{j+1} + \mathbf{p}_j^* - \mathbf{c}_j)^T \cdot (\mathbf{c}_{j+1} + \mathbf{p}_j^* - \mathbf{c}_j)\mathbf{I}_{3\times3} \\ &- (\mathbf{c}_{j+1} + \mathbf{p}_j^* - \mathbf{c}_j)(\mathbf{c}_{j+1} + \mathbf{p}_j^* - \mathbf{c}_j)^T] + \mathbf{I}_j \\ &+ m_j[(\mathbf{s}_j + \mathbf{p}_j^* - \mathbf{c}_j)^T \cdot (\mathbf{s}_j + \mathbf{p}_j^* - \mathbf{c}_j)\mathbf{I}_{3\times3} \\ &- (\mathbf{s}_j + \mathbf{p}_j^* - \mathbf{c}_j)(\mathbf{s}_j + \mathbf{p}_j^* - \mathbf{c}_j)^T]\end{aligned} \tag{5-c}$$

where $1 \le j \le n-1$.

$$\mathbf{F}_j = \begin{cases} \mathbf{z}_{j-1} \times (M_j \mathbf{c}_j), & \text{if joint } j \text{ is rotational} \\ M_j \mathbf{z}_{j-1}, & \text{if joint } j \text{ is translational} \end{cases} \tag{5-d}$$

$$\mathbf{N}_j = \begin{cases} \mathbf{E}_j \mathbf{z}_{j-1}, & \text{if joint } j \text{ is rotational} \\ \mathbf{0}, & \text{if joint } j \text{ is translational} \end{cases} \tag{5-e}$$

$$\mathbf{f}_{jj} = \mathbf{F}_j, \quad \mathbf{n}_{jj} = \mathbf{N}_j + \mathbf{c}_j \times \mathbf{F}_j, \quad 1 \le j \le n \tag{5-f}$$

$$\begin{aligned}\mathbf{f}_{i,j} &= \mathbf{f}_{(i+1),j} \\ \mathbf{n}_{i,j} &= \mathbf{n}_{i+1,j} + \mathbf{p}_i^* \times \mathbf{f}_{(i+1),j}\end{aligned} \qquad 1 \le i \le j-1, \ 2 \le j \le n \tag{5-g}$$

$$h_{ij} = \begin{cases} \mathbf{z}_{i-1}^T \, \mathbf{n}_{i,j}, & \text{if joint } j \text{ is rotational} \\ \mathbf{z}_{i-1}^T \, \mathbf{f}_{i,j}, & \text{if joint } j \text{ is translational} \end{cases} \tag{5-h}$$

where $1 \le i \le j$, $1 \le j \le n$.

/* Compute the bias torque vector $\mathbf{b}$ */

$$\mathbf{b} = \mathbf{C}(\mathbf{q}, \dot{\mathbf{q}})\dot{\mathbf{q}}(t) + \mathbf{G}(\mathbf{q}) \tag{5-i}$$

/* Solve the system of linear equations */

$$\mathbf{H}(\mathbf{q})\ddot{\mathbf{q}}(t) = \boldsymbol{\tau}(t) - \mathbf{b} \tag{5-j}$$

where

M_j: total mass of links j through n.

$\mathbf{c}_j$: the location of the composite center of mass of links j through n with reference to the $(j-1)$th coordinate frame.

$\mathbf{E}_j$: the moment of inertia matrix of the composite system of links j through n.

$\mathbf{F}_j$: total force exerted on link i.

$\mathbf{N}_j$: total moment exerted on link i.

$\mathbf{f}_{ij}$: the force exerted on joint i due to the motion of the composite system of links j through n.

$\mathbf{n}_{ij}$: the moment exerted on joint i due to the motion of the composite system of link j through n.

h_{ij}: the (i,j)th component of inertia matrix.

$\mathbf{C}(\mathbf{q}, \dot{\mathbf{q}})$: $n \times n$ matrix specifying Centrifugal and Coriolis effects.

$\mathbf{G}(\mathbf{q})$: $n \times 1$ vector specifying the effects due to gravity.

$\boldsymbol{\tau}(t)$: generalized applied force/torque vector.

Initial conditions:

$M_n = m_n$

$\mathbf{c}_n = \mathbf{s}_n + \mathbf{p}_n^*$

$\mathbf{E}_n = \mathbf{I}_n$

Table 6. Forward Jacobian Computation

Given a desired reference coordinate frame k, $0 \le k \le n$, this algorithm computes the Jacobian matrix with respect to the desired reference coordinate frame k for an n-jointed manipulator.

J1. [Forward Recurrence.]

for $i = k+1$ step 1 **to** n, **do**

$$ {}^{k}\mathbf{R}_i = {}^{k}\mathbf{R}_{i-1}\ {}^{i-1}\mathbf{R}_i \tag{6-a}$$

$$ {}^{k}\mathbf{p}_i = {}^{k}\mathbf{p}_{i-1} - {}^{k}\mathbf{R}_i\ {}^{i}\mathbf{p}_i^{*} \tag{6-b}$$

end {for}

J2. [Backward Recurrence.]

for $i = k-1$ step -1 **to** 1, **do**

$$ {}^{k}\mathbf{R}_i = {}^{k}\mathbf{R}_{i+1}\ {}^{i}\mathbf{R}_{i+1}^{T} \tag{6-c}$$

$$ {}^{k}\mathbf{p}_i = {}^{k}\mathbf{p}_{i+1} + {}^{k}\mathbf{R}_{i+1}\ {}^{i+1}\mathbf{p}_{i+1}^{*} \tag{6-d}$$

end {for}

J3. [Jacobian Computation.]

for $i = 1$ step 1 **to** n, **do**

$$ {}^{k}\boldsymbol{\beta}_i = (1-\lambda_i)\,{}^{k}\mathbf{R}_i\ \mathbf{z}_0 \tag{6-e}$$

$$ {}^{k}\boldsymbol{\mu}_i = (1-\lambda_i)({}^{k}\boldsymbol{\beta}_i \times (-{}^{k}\mathbf{p}_i)) + \lambda_i({}^{k}\mathbf{R}_i\ \mathbf{z}_0) \tag{6-f}$$

end {for}

where initially $\mathbf{z}_0 = (0,0,1)^T$, ${}^{k}\mathbf{R}_k = \mathbf{I}_{3\times3}$, ${}^{k}\mathbf{p}_k = \mathbf{0}_{3\times1}$.

Table 7. Parallel Composite-Rigid-Body Algorithm.

Given $\boldsymbol{\tau}(t)$, $\mathbf{q}$, $\dot{\mathbf{q}}$, m_i, ${}^i\mathbf{p}_i^*$, ${}^i\mathbf{s}_i$, ${}^i\mathbf{I}_i$, and ${}^{i-1}\mathbf{R}_i$, for $1 \le i \le n$, this algorithm computes the joint acceleration vector $\ddot{\mathbf{q}}(t)$ parallelly.

C1. For $i = 1$ to n parallelly do /*HLR equations*/

$$ {}^0\mathbf{R}_i = {}^0\mathbf{R}_{i-1}\, {}^{i-1}\mathbf{R}_i $$

C2. For $i = 1$ to n parallelly do /*parallelly compute*/

$$ \mathbf{z}_i = {}^0\mathbf{R}_i \mathbf{z}_0 \quad , \quad \mathbf{p}_i^* = {}^0\mathbf{R}_i\, {}^i\mathbf{p}_i^* \quad , \quad \mathbf{s}_i = {}^0\mathbf{R}_i\, {}^i\mathbf{s}_i $$

$$ \mathbf{I}_i = {}^0\mathbf{R}_i\, {}^i\mathbf{I}_i\, {}^i\mathbf{R}_0 = {}^0\mathbf{R}_i\, {}^i\mathbf{I}_i ({}^0\mathbf{R}_i)^T $$

where $\mathbf{z}_0 = (0\,,0\,,1)^T$.

C3. Initialize $M_n = m_n$.
For $j = 1$ to $n-1$ parallelly do /*HLR equations*/

$$ M_j = M_{j+1} + m_j $$

C4. (i) Initialize $\mathbf{c}_n = \mathbf{s}_n + \mathbf{p}_n^*$.
(ii) For $j = 1$ to $n-1$ parallelly do /*parallelly compute*/

$$ a_j = \frac{M_{j+1}}{M_j} $$

$$ \mathbf{b}_j = \frac{m_j}{M_j}(\mathbf{s}_j + \mathbf{p}_j^*) + a_j \mathbf{p}_j^* $$

(iii) For $j = 1$ to $n-1$ parallelly do /*HLR equations*/

$$ \mathbf{c}_j = a_j \mathbf{c}_{j+1} + \mathbf{b}_j $$

C5. (i) Initialize $\mathbf{E}_n = \mathbf{I}_n$.
(ii) For $j = 1$ to n parallelly do /*parallelly compute*/

$$ \begin{aligned} \mathbf{b}_j = M_{j+1}[&(\mathbf{c}_{j+1} + \mathbf{p}_j^* - \mathbf{c}_j)^T (\mathbf{c}_{j+1} + \mathbf{p}_j^* - \mathbf{c}_j)\mathbf{I}_{3\times3} \\ &- (\mathbf{c}_{j+1} + \mathbf{p}_j^* - \mathbf{c}_j)(\mathbf{c}_{j+1} + \mathbf{p}_j^* - \mathbf{c}_j)^T] + \mathbf{I}_j \\ &+ m_j[(\mathbf{s}_j + \mathbf{p}_j^* - \mathbf{c}_j)^T (\mathbf{s}_j + \mathbf{p}_j^* - \mathbf{c}_j)\, \mathbf{I}_{3\times3} \\ &- (\mathbf{s}_j + \mathbf{p}_j^* - \mathbf{c}_j)(\mathbf{s}_j + \mathbf{p}_j^* - \mathbf{c}_j)^T] \end{aligned} $$

(iii) For $j = 1$ to $n-1$ parallelly do /*HLR equations*/

$$ \mathbf{E}_j = \mathbf{E}_{j+1} + \mathbf{b}_j $$

C6. For $j = 1$ to n parallelly do /*parallelly compute*/

$$\mathbf{F}_j = \mathbf{z}_{j-1} \times (M_j \mathbf{c}_j)(1-\lambda_j) + \lambda_j M_j \mathbf{z}_{j-1}$$

$$\mathbf{N}_j = \mathbf{E}_j \mathbf{z}_{j-1}(1-\lambda_j)$$

C7. (i) Initialize $\mathbf{f}_{ij} = \mathbf{F}_j$, $1 \le i \le j$, $1 \le j \le n$.
(ii) For $j = 1$ to n parallelly do /*parallelly compute*/

$$\mathbf{n}_{jj} = \mathbf{N}_j + \mathbf{c}_j \times \mathbf{F}_j$$

(iii) For $j = 2$ to n do
For $i = 1$ to $j-1$ parallelly do /*parallelly compute*/

$$\mathbf{b}_{ij} = \mathbf{p}_i^* \times \mathbf{F}_j \;, \; 1 \le i \le j-1 \,, 2 \le j \le n \,.$$

(iv) $1 \le i \le j-1, 2 \le j \le n$ parallelly compute /*HLR equations*/

$$\mathbf{n}_{ij} = \mathbf{n}_{(i+1),j} + \mathbf{b}_{ij}$$

C8. $1 \le i \le j$, $1 \le j \le n$, parallelly compute /*parallelly compute*/

$$h_{ij} = \begin{cases} \mathbf{z}_{i-1}^T \,\mathbf{n}_{ij} & \text{, if joint } i \text{ is rotational} \\ \mathbf{z}_{i-1}^T \,\mathbf{f}_{ij} & \text{, if joint } i \text{ is translational} \end{cases}$$

C9. Parallelly compute the bias vector $\mathbf{b}$ /*HLR equations*/

$$\mathbf{b} = \mathbf{C}(\mathbf{q}, \dot{\mathbf{q}})\, \dot{\mathbf{q}}(t) + \mathbf{G}(\mathbf{q})$$

C10. Parallelly solve the system equation for $\ddot{\mathbf{q}}(t)$ /*parallel Cholesky factorization */

$$\mathbf{H}(\mathbf{q})\ddot{\mathbf{q}}(t) = \mathbf{y} = \boldsymbol{\tau}(t) - \mathbf{b}$$

Table 8. Parallel Forward Jacobian Computation

Given a desired reference coordinate frame k, $0 \le k \le n$, this algorithm parallelly computes the Jacobian matrix with respect to the desired reference coordinate frame k for an n-jointed manipulator.

J1. [Compute ${}^k\mathbf{R}_i$.]
for $i = k+1$ to n and $j = k-1$ **step** (-1) to 1 **parallelly do** /*HLR equations*/

$${}^k\mathbf{R}_i = {}^k\mathbf{R}_{i-1}\ {}^{i-1}\mathbf{R}_i \quad \text{(forward)}$$

$${}^k\mathbf{R}_j = {}^k\mathbf{R}_{j+1}\ {}^j\mathbf{R}_{j+1}^T \quad \text{(backward)}$$

end {for}

J2. [Compute ${}^k\mathbf{p}_i$.]
for $i = k+1$ to n and $j = k-1$ **step** (-1) to 1 **parallelly do** /*HLR equations*/

$${}^k\mathbf{p}_i = {}^k\mathbf{p}_{i-1} - {}^k\mathbf{R}_i\ {}^i\mathbf{p}_i^* \quad \text{(forward)}$$

$${}^k\mathbf{p}_j = {}^k\mathbf{p}_{j+1} + {}^k\mathbf{R}_{j+1}\ {}^{j+1}\mathbf{p}_{j+1}^* \quad \text{(backward)}$$

end {for}

J3. [Compute ${}^k\boldsymbol{\beta}_i$ in one time step.]
for $i = 1$ to n **parallelly do** /*parallelly compute*/

$${}^k\boldsymbol{\beta}_i = (1-\lambda_i)\,{}^k\mathbf{R}_i\,\mathbf{z}_0$$

J4. [Compute ${}^k\boldsymbol{\mu}_i$ in one time step.]
for $i = 1$ to n **parallelly do** /*parallelly compute*/

$${}^k\boldsymbol{\mu}_i = (1-\lambda_i)({}^k\boldsymbol{\beta}_i \times (-{}^k\mathbf{p}_i)) + \lambda_i({}^k\mathbf{R}_i\,\mathbf{z}_0)$$

Table 9. Characteristics of Six Basic Robotics Algorithms.

Algorithms	CHARACTERISTICS					
	Type of Parallelism	Degree of Parallelism	Uniformity of Operations	Fundamental Operations	Data Dependency	Communication Requirement
Inverse Dynamics	Job level	Large grain	Yes	Matrix-Vector	HLR	(Regular) Permutation
Forward Dynamics	Job level	Large grain	Yes	Scalar ops. Reciprocal Matrix-Vector	HLR,IHLR SHLR, PNE System of Linear Eqs.	one-to-one Permutation Broadcast
Forward Kinematics	Job level	Medium or Fine grain	Yes	Matrix Mult. Trigonometric	HLR	(Regular) Permutation
Forward Jacobian	Job level	Medium or Fine grain	Yes	Matrix-Vector	HLR (Forward & Backward)	(Irregular) Permutation Broadcast
Inverse Jacobian (Direct)	Job level	Medium or Fine grain	Yes	Scalar ops. Reciprocal Vector ops.	Global	Permutation Broadcast
Inverse Jacobian (iterative)	Job level	Medium or Fine grain	Yes	Scalar ops. Reciprocal Vector ops.	Local	Permutation Broadcast
Inverse Kinematics (Direct)	Task level	Fine grain	No	Scalar ops. Reciprocal Square root Trigonometric	Global	one-to-one Broadcast
Inverse Kinematics (iterative)	Job level	Medium or Fine grain	Yes	Scalar ops. Reciprocal Matrix-Vector Trigonometric	Local	one-to-one Permutation Broadcast

Report on the Group Discussion about NEURAL NETWORKS IN ROBOTICS

Carme Torras
Institut De Cibernetica
Diagonal 647, 2s planta
08028 - Barcelona
Spain

The discussion developed around the following four topics: advancements that have led to the resurgence of neural network (NN), operations that NN are best suited for, main applications, and key problems that need to be solved.

Concerning the advancements, the contributions of Hopfield, Hinton and Sejnowski were mentioned. Hopfield (*Proc. Natl. Acad. Sci.*, USA, vol. 79, pp. 2554-2558, April 1982) characterized a NN as a dynamical system and showed that stored patterns correspond to attractors of such a system. Hinton and Sejnowski (*Proc. IEEE Conf. Comp. Vision and Pat. Recog.*, pp. 448-453, 1983) pushed this analogy further in characterizing a NN as a thermodynamical system, in which patterns are retrieved through a simulated annealing process; these authors also endowed their model network (named "Boltzmann machine") with a learning algorithm based on the Boltzmann equation. Besides these contributions, what has made the real difference between the 60's — when NN started to develop — and the 80's — in which this field has flourished — is that the technology needed to build truly parallel NN has become available.

Neural networks seem to be well-suited to carry out constraint satisfaction, constrained optimization, pattern association, generalization and learning. It was pointed out that generalization in a fuzzy world is interesting, but the application of neural generalization to the construction of exact mappings (e.g., inverse kinematics) is not. Everybody agreed on the general statement, but not on the example chosen, since, under some hypotheses, there is no conventional algorithm to find inverse kinematics.

Several successful applications of NN were cited along the discussion. Besides the conventional ones, mainly in the signal processing field, three more were mentioned: space-ship orientation, strategic planning, and path planning. The former one has been developed at JPL and has been tested against another procedure to accomplish the same task based on artificial intelligence techniques; while the AI-based product requires 1MB of memory, the NN-based one requires 4KB to achieve a 100% retrieval performance with 90% accuracy. At the

NATO ASI Series, Vol. F 66
Sensor-Based Robots: Algorithms and Architectures
Edited by C. S. George Lee

Imperial College of Science and Technology, UK, there is work in progress on the application of NN to strategic planning; this is accomplished by formulating such kind of planning as a constraint optimization problem, the main difficulty being the establishment of a metric in a symbolic space. An application of NN to path planning has been developed at Princeton University, which is reminiscent of the potential-field approach to this problem. It is important to mention that DARPA has allocated 40M dollars for a period of two years to assess the specific areas on which NN are more likely to have impact.

Three key problems were identified and an open question was raised, all of which can be posed in an interrogative form as follows. What is a useful representation of a problem for a NN to be able to solve it efficiently? Which analog/digital neural models scale up? How can the branching underlying if-then rules be implemented in NN? Are NN adequate for closed-chain robot control? Answers to these questions will help to clarify the field of NN and its impact on robotics.

List of Lecturers and Participants

Professor Igor Aleksander (L)†
Department of Electrical Engineering
Imperial College of Science, Technology and Medicine
Exhibition Road
London SW7 2BZ
ENGLAND
phone: 1-589-5911 Ext. 4985

Dr. Jacob Barhen (L)
Jet Propulsion Lab
California Institute of Technology
4800 Oak Grove Drive
Pasadena, CA 91109
phone: 818-354-9218

Professor Luis Basanez (P)
Institut de Cibernètica (UPC/CSIC)
Diagonal 647, 2a planta
08028 Barcelona
SPAIN
phone: (3) 249 28 42
Telex: 52 821 UPC E

Dr. Antonio Bicchi (P)
University of Pisa
Scuola Superiore "S. Anna"
Via Carducci, 40
56100 Pisa
ITALY

Dr. Giorgio Buttazzo (L)
Center "E. Piaggio",
Faculty of Engineering
University of Pisa
Via Diofisalvi, 2
56100 Pisa
ITALY
phone: (50) 444780

Dr. James L. Crowley (P)
Institut Polytechnique de Grenoble
LIFIA-IMAG, 46 Ave. Felix Viallet
38031 Grenoble
FRANCE

Professor Hugh G. Durrant-Whyte (P)
Department of Engineering Science,
University of Oxford,
Parks Road,
Oxford, OX1 3PJ,
United Kingdom
phone: (0865) 273146

Professor Eddie Grant (L)
The Turing Institute
George House
36 N Hanover St.
Glasgow, G1 2AD
United Kingdom

Dr. Thomas Henderson (P)
INRIA (Sophia-Antipolis)
2004 Route des Lucioles
06565 Valbonne Cedex
FRANCE

Professor Yorgo Istefanopulos (P)
Bogazici University
Electrical-Electronic Engineering Dept.
P.K.2 Bebek, Istanbul
TURKEY
phone: (90)-1-1631500 Ext. 422

Dr. Juan Juan (P)
Institut de Technologia, S.A.
Parc Tecnologic del Valles
08290 Cerdanyola
Barcelona, SPAIN
phone: 343.580.10.00

† L = Lecturer; P = Participant; C = Program Committee Member.

Professor Robert B. Kelley (L)
Electrical, Computer, and
Systems Engineering Department
Rensselaer Polytechnic Institute
Troy, New York 12180-3590
phone: 518-276-2653

Dr. Francis G. King (L)
Supervisor
Manufacturing Development Center
Ford Motor Company
24500 Glendale Avenue
Detroit, Michigan 48239
phone: 313-592-2367

Professor S. Y. Kung (L)
Department of Electrical Engineering
Princeton University
Princeton, NJ 08544
phone: 609-452-3780

Professor C. S. George Lee (C, L)
School of Electrical Engineering
Purdue University
West Lafayette, Indiana 47907-0501
phone: 317-494-1384

Professor P. Levi (P)
Universität Karlsruhe, Fakultät für Informatik
Institut für Prozeβrechentechnik und Robotik
P.O. Box 69 80, D-7500 Karlsruhe 1, FRG
phone: 721/608-3957

Dr. J. Majumdar (L)
Universität Karlsruhe, Fakultät für Informatik
Institut für Prozeβrechentechnik und Robotik
P.O. Box 69 80, D-7500 Karlsruhe 1, FRG
phone: 721/608-3957

Professor D. E. Orin (L)
Department of Electrical Engineering
Ohio State University
2015 Neil Avenue
Columbus, Ohio 43210
phone: 614-292-3064

Professor Ulrich Rembold (C, L)
Universität Karlsruhe, Fakultät für Informatik
Institut für Prozeβrechentechnik und Robotik
P.O. Box 69 80, D-7500 Karlsruhe 1, FRG
phone: 721/608-3957

Professor George N. Saridis (P)
Electrical, Computer, and
Systems Engineering Department
Rensselaer Polytechnic Institute
Troy, New York 12180-3590
phone: 518-276-6076

Professor Joao Jose dos Santos Sentieiro (P)
Centro de Analise e Processamento de Sinais -
Complexo 1 do INIC
Instituto Superior Technico
Av. Rovisco Pais
1096 Lisboa Codex
PORTUGAL
phone: 011 3511 572399 - ext. 359

Dr. Thierry Simeon (P)
L.A.A.S.
7 Avenue du colonel Roche
31400 Toulouse
FRANCE
phone: 61.33-63-49

Dr. Sharon Stansfield (P)
Sandia National Labs
P. O. Box 5800
Albuquerque, New Mexico

Professor Harry E. Stephanou (C, P)
Center for Advanced Technology
in Automation and Robotics
CII Building, Room 8005
Rensselaer Polytechnic Institute
Troy, NY 12180-3590
phone: 518-276-6156

Professor Carme Torras (P)
Institut de Cibernètica
Diagonal 647, 2a planta
08028 Barcelona
SPAIN
phone: (3) 249.28.42

Professor Spyros Tzafestas (P)
Control and Robotics Group,
Computer Engineering Div.
National Technical University
Zografou 15773
Athens, GREECE
phone: 0030/1/7757504

Professor Andrew K. C. Wong (L)
PAMI Lab, Systems Design Engineering
University of Waterloo
Waterloo, Ontario,
Canada N2L 3G1
phone: 519-999-4649

NATO ASI Series F

Including Special Programmes on Sensory Systems for Robotic Control (ROB) and on Advanced Educational Technology (AET)

Vol. 1: Issues in Acoustic Signal – Image Processing and Recognition. Edited by C. H. Chen. VIII, 333 pages. 1983.

Vol. 2: Image Sequence Processing and Dynamic Scene Analysis. Edited by T. S. Huang. IX, 749 pages. 1983.

Vol. 3: Electronic Systems Effectiveness and Life Cycle Costing. Edited by J. K. Skwirzynski. XVII, 732 pages. 1983.

Vol. 4: Pictorial Data Analysis. Edited by R. M. Haralick. VIII, 468 pages. 1983.

Vol. 5: International Calibration Study of Traffic Conflict Techniques. Edited by E. Asmussen. VII, 229 pages. 1984.

Vol. 6: Information Technology and the Computer Network. Edited by K. G. Beauchamp. VIII, 271 pages. 1984.

Vol. 7: High-Speed Computation. Edited by J. S. Kowalik. IX, 441 pages. 1984.

Vol. 8: Program Transformation and Programming Environments. Report on a Workshop directed by F. L. Bauer and H. Remus. Edited by P. Pepper. XIV, 378 pages. 1984.

Vol. 9: Computer Aided Analysis and Optimization of Mechanical System Dynamics. Edited by E. J. Haug. XXII, 700 pages. 1984.

Vol. 10: Simulation and Model-Based Methodologies: An Integrative View. Edited by T. I. Ören, B. P. Zeigler, M. S. Elzas. XIII, 651 pages. 1984.

Vol. 11: Robotics and Artificial Intelligence. Edited by M. Brady, L. A. Gerhardt, H. F. Davidson. XVII, 693 pages. 1984.

Vol. 12: Combinatorial Algorithms on Words. Edited by A. Apostolico, Z. Galil. VIII, 361 pages. 1985.

Vol. 13: Logics and Models of Concurrent Systems. Edited by K. R. Apt. VIII, 498 pages. 1985.

Vol. 14: Control Flow and Data Flow: Concepts of Distributed Programming. Edited by M. Broy. VIII, 525 pages. 1985.

Vol. 15: Computational Mathematical Programming. Edited by K. Schittkowski. VIII, 451 pages. 1985.

Vol. 16: New Systems and Architectures for Automatic Speech Recognition and Synthesis. Edited by R. De Mori, C.Y. Suen. XIII, 630 pages. 1985.

Vol. 17: Fundamental Algorithms for Computer Graphics. Edited by R.A. Earnshaw. XVI, 1042 pages. 1985.

Vol. 18: Computer Architectures for Spatially Distributed Data. Edited by H. Freeman and G. G. Pieroni. VIII, 391 pages. 1985.

Vol. 19: Pictorial Information Systems in Medicine. Edited by K. H. Höhne. XII, 525 pages. 1986.

Vol. 20: Disordered Systems and Biological Organization. Edited by E. Bienenstock, F. Fogelman Soulié, G. Weisbuch. XXI, 405 pages.1986.

Vol. 21: Intelligent Decision Support in Process Environments. Edited by E. Hollnagel, G. Mancini, D.D. Woods. XV, 524 pages. 1986.

NATO ASI Series F

Vol. 22: Software System Design Methods. The Challenge of Advanced Computing Technology. Edited by J.K. Skwirzynski. XIII, 747 pages. 1986.

Vol. 23: Designing Computer-Based Learning Materials. Edited by H. Weinstock and A. Bork. IX, 285 pages. 1986.

Vol. 24: Database Machines. Modern Trends and Applications. Edited by A.K. Sood and A.H. Qureshi. VIII, 570 pages. 1986.

Vol. 25: Pyramidal Systems for Computer Vision. Edited by V. Cantoni and S. Levialdi. VIII, 392 pages. 1986. *(ROB)*

Vol. 26: Modelling and Analysis in Arms Control. Edited by R. Avenhaus, R.K. Huber and J.D. Kettelle. VIII, 488 pages. 1986.

Vol. 27: Computer Aided Optimal Design: Structural and Mechanical Systems. Edited by C.A. Mota Soares. XIII, 1029 pages. 1987.

Vol. 28: Distributed Operating Systems. Theory und Practice. Edited by Y. Paker, J.-P. Banatre and M. Bozyiğit. X, 379 pages. 1987.

Vol. 29: Languages for Sensor-Based Control in Robotics. Edited by U. Rembold and K. Hörmann. IX, 625 pages. 1987. *(ROB)*

Vol. 30: Pattern Recognition Theory and Applications. Edited by P.A. Devijver and J. Kittler. XI, 543 pages. 1987.

Vol. 31: Decision Support Systems: Theory and Application. Edited by C.W. Holsapple and A.B. Whinston. X, 500 pages. 1987.

Vol. 32: Information Systems: Failure Analysis. Edited by J.A. Wise and A. Debons. XV, 338 pages. 1987.

Vol. 33: Machine Intelligence and Knowledge Engineering for Robotic Applications. Edited by A.K.C. Wong and A. Pugh. XIV, 486 pages. 1987. *(ROB)*

Vol. 34: Modelling, Robustness and Sensitivity Reduction in Control Systems. Edited by R.F. Curtain. IX, 492 pages. 1987.

Vol. 35: Expert Judgment and Expert Systems. Edited by J.L. Mumpower, L.D. Phillips, O. Renn and V.R.R. Uppuluri. VIII, 361 pages. 1987.

Vol. 36: Logic of Programming and Calculi of Discrete Design. Edited by M. Broy. VII, 415 pages. 1987.

Vol. 37: Dynamics of Infinite Dimensional Systems. Edited by S.-N. Chow and J.K. Hale. IX, 514 pages. 1987.

Vol. 38: Flow Control of Congested Networks. Edited by A.R. Odoni, L. Bianco and G. Szegö. XII, 355 pages. 1987.

Vol. 39: Mathematics and Computer Science in Medical Imaging. Edited by M.A. Viergever and A. Todd-Pokropek. VIII, 546 pages. 1988.

Vol. 40: Theoretical Foundations of Computer Graphics and CAD. Edited by R.A. Earnshaw. XX, 1246 pages. 1988.

Vol. 41: Neural Computers. Edited by R. Eckmiller and Ch. v. d. Malsburg. XIII, 566 pages. 1988.

NATO ASI Series F

Vol. 42: Real-Time Object Measurement and Classification. Edited by A. K. Jain. VIII, 407 pages. 1988. *(ROB)*

Vol. 43: Sensors and Sensory Systems for Advanced Robots. Edited by P. Dario. XI, 597 pages. 1988. *(ROB)*

Vol. 44: Signal Processing and Pattern Recognition in Nondestructive Evaluation of Materials. Edited by C. H. Chen. VIII, 344 pages. 1988. *(ROB)*

Vol. 45: Syntactic and Structural Pattern Recognition. Edited by G. Ferraté, T. Pavlidis, A. Sanfeliu and H. Bunke. XVI, 467 pages. 1988. *(ROB)*

Vol. 46: Recent Advances in Speech Understanding and Dialog Systems. Edited by H. Niemann, M. Lang and G. Sagerer. X, 521 pages. 1988.

Vol. 47: Advanced Compuling Concepts and Techniques in Control Engineering. Edited by M. J. Denham and A. J. Laub. XI, 518 pages. 1988.

Vol. 48: Mathematical Models for Decision Support. Edited by G. Mitra. IX, 762 pages. 1988.

Vol. 49: Computer Integrated Manufacturing. Edited by I. B. Turksen. VIII, 568 pages. 1988.

Vol. 50: CAD Based Programming for Sensory Robots. Edited by B. Ravani. IX, 565 pages. 1988. *(ROB)*

Vol. 51: Algorithms and Model Formulations in Mathematical Programming. Edited by S. W. Wallace. IX, 190 pages. 1989.

Vol. 52: Sensor Devices and Systems for Robotics. Edited by A. Casals. IX, 362 pages. 1989. *(ROB)*

Vol. 53: Advanced Information Technologies for Industrial Material Flow Systems. Edited by S. Y. Nof and C. L. Moodie. IX, 710 pages. 1989.

Vol. 54: A Reappraisal of the Efficiency of Financial Markets. Edited by R. M. C. Guimarães, B. G. Kingsman and S. J. Taylor. X, 804 pages. 1989.

Vol. 55: Constructive Methods in Computing Science. Edited by M. Broy. VII, 478 pages. 1989.

Vol. 56: Multiple Criteria Decision Making and Risk Analysis Using Microcomputers. Edited by B. Karpak and S. Zionts. VII, 399 pages. 1989.

Vol. 57: Kinematics and Dynamic Issues in Sensor Based Control. Edited by G. E. Taylor. XI, 456 pages. 1990. *(ROB)*

Vol. 58: Highly Redundant Sensing in Robotic Systems. Edited by J. T. Tou and J. G. Balchen. X, 322 pages. 1990. *(ROB)*

Vol. 59: Superconducting Electronics. Edited by H. Weinstock and M. Nisenoff. X, 441 pages. 1989.

Vol. 60: 3D Imaging in Medicine. Algorithms, Systems, Applications. Edited by K. H. Höhne, H. Fuchs and S. M. Pizer. IX, 460 pages. 1990.

Vol. 61: Knowledge, Data and Computer-Assisted Decisions. Edited by M. Schader and W. Gaul. VIII, 421 pages. 1990.

Vol. 62: Supercomputing. Edited by J. S. Kowalik. X, 425 pages. 1990.

Vol. 63: Traditional and Non-Traditional Robotic Sensors. Edited by T. C. Henderson. VIII, 468 pages. 1990. *(ROB)*

Vol. 64: Sensory Robotics for the Handling of Limp Materials. Edited by P. M. Taylor. IX, 343 pages. 1990. *(ROB)*

Vol. 65: Mapping and Spatial Modelling for Navigation. Edited by L. F. Pau. VIII, 357 pages. 1990. *(ROB)*

NATO ASI Series F

Vol. 66: Sensor-Based Robots: Algorithms and Architectures. Edited by C. S. G. Lee. X, 285 pages. 1991. *(ROB)*

Vol. 67: Designing Hypermedia for Learning. Edited by D. H. Jonassen and H. Mandl. XXV, 457 pages. 1990. *(AET)*

Vol. 68: Neuro-Computing. Algorithms, Architectures and Applications. Edited by F. Fogelman Soulié and J. Hérault. XI, 455 pages. 1990.

Vol. 69: Real-Time Integration Methods for Mechanical System Simulation. Edited by E. J. Haug and R. C. Deyo. VIII, 352 pages. 1991.

MIX
Papier aus verantwortungsvollen Quellen
Paper from responsible sources
FSC® C105338

If you have any concerns about our products,
you can contact us on
ProductSafety@springernature.com

In case Publisher is established outside the EU,
the EU authorized representative is:
Springer Nature Customer Service Center GmbH
Europaplatz 3, 69115 Heidelberg, Germany

Printed by Libri Plureos GmbH
in Hamburg, Germany